AF779954

Renewable-Based Microgrids: Design, Control and Optimization

Renewable-Based Microgrids: Design, Control and Optimization

Editors

Marcos Tostado-Véliz
Paul Arévalo
Salah Kamel
Ragab A. El-Sehiemy
Tomonobu Senjyu

MDPI

Basel • Beijing • Wuhan • Barcelona • Belgrade • Novi Sad • Cluj • Manchester

Editors

Marcos Tostado-Véliz
University of Jaén
Linares, Spain

Paul Arévalo
University of Jaén
Linares, Spain

Salah Kamel
Aswan University
Aswan, Egypt

Ragab A. El-Sehiemy
Kafrelsheikh University
Kafrelsheikh, Egypt

Tomonobu Senjyu
University of the Ryukyus
Okinawa, Japan

Editorial Office
MDPI
St. Alban-Anlage 66
4052 Basel, Switzerland

This is a reprint of articles from the Special Issue published online in the open access journal *Applied Sciences* (ISSN 2076-3417) (available at: https://www.mdpi.com/journal/applsci/special_issues/renewable_based_microgrids).

For citation purposes, cite each article independently as indicated on the article page online and as indicated below:

Lastname, A.A.; Lastname, B.B. Article Title. *Journal Name* **Year**, *Volume Number*, Page Range.

ISBN 978-3-0365-8576-5 (Hbk)
ISBN 978-3-0365-8577-2 (PDF)
doi.org/10.3390/books978-3-0365-8577-2

Cover image courtesy of Paul Arévalo

© 2023 by the authors. Articles in this book are Open Access and distributed under the Creative Commons Attribution (CC BY) license. The book as a whole is distributed by MDPI under the terms and conditions of the Creative Commons Attribution-NonCommercial-NoDerivs (CC BY-NC-ND) license.

Contents

About the Editors

Marcos Tostado-Véliz

Tostado-Véliz works at the Department of Electrical Engineering, the University of Jaén, 23700 Linares, Spain. He received B.Sc. (Hons.) and M.Sc. degrees from the University of Seville in 2016 and 2017, respectively, as well as a Ph.D. degree from the University of Jaén in 2020. He is currently an assistant professor at the Electrical Engineering Department, the University of Jaén. He actively partakes in various research projects funded by the Spanish Government. His research interests include the application of numerical algorithms for power system analysis and optimal planning and control of energy systems.

Dr. Tostado-Véliz also works as a reviewer for different editorials and journals. In recognition of his constant effort in this field, he was awarded the 'Outstanding Reviewer Award' by the journal *Energies* (MDPI) in 2023. Moreover, he works as an assistant editor for the *Ain Shams Engineering Journal*, which is a multidisciplinary publication published by Elsevier that is currently ranked in the first quartile of the *Journal Citation Report*.

Paul Arévalo

Arévalo works at the Department of Electrical Engineering, the University of Jaén, 23700 Linares, Spain. Paul Arévalo, who was born in Ecuador in 1990, is an accomplished electrical engineer. He earned his bachelor's degree in electrical engineering from the University of Cuenca, Ecuador, in 2015. Subsequently, he pursued further academic study, obtaining M.Sc. and Ph.D. degrees in 2017 and 2021, respectively, from Rey Juan Carlos University and the University of Jaen, which are both located in Spain. Currently, Paul Arévalo is participating in a postdoctoral research program at the University of Jaen, Spain. His research focuses on various aspects of power systems, with a particular emphasis on modeling and renewable energy. Through his work, he strives to contribute to advance sustainable energy solutions and the optimization of power systems.

Salah Kamel

Kamel works at the Department of Electrical Engineering, the Faculty of Engineering, Aswan University, Aswan 81542, Egypt. He received an international Ph.D. degree from the University of Jaén, Spain (Main), and Aalborg University, Denmark (Host), in 2014. He is currently works as an associate professor at the Department of Electrical Engineering, Aswan University. He is also a leader at the Power Systems Research Group, the Advanced Power Systems Research Laboratory (APSR Laboratory), Aswan, Egypt. His research interests include the analysis and optimization of power systems, smart grids, and renewable energy systems.

Ragab A. El-Sehiemy

El-Sehiemy works at the Electrical Engineering Department, the Faculty of Engineering, Kafrelsheikh University, Kafrelsheikh 33516, Egypt. (Senior Member, IEEE) He was born in Menoufia, Egypt, in 1973. He received a B.Sc. degree in electrical engineering, as well as M.Sc. and Ph.D. degrees, from Menoufia University, Egypt, in 1996, 2005, and 2008, respectively. He is currently a Full Professor at the Electrical Engineering Department, the Faculty of Engineering, Kafrelsheikh University, Egypt. He was awarded the Prof. Mahmoud Khalifia Award for Power System Engineering by the Academy of Research and Technology, Egypt, in 2016. He also awarded the El-Shorouk Academy Award in the Computer and Information Applications in Industry by the Academy of Research and Technology in 2020. He was rated as one of the top 2% most effective researchers in the world in 2021. He has published more than 190 articles in international journals and at industry conferences. His research interests include the operation power systems, planning and control, smart grids, renewable energy, and AI and its application to power systems.

Tomonobu Senjyu

Senjyu works at the Department of Electrical and Electronics Engineering, the University of the Ryukyus, Nishihara, Okinawa 903-0213, Japan. (Fellow, IEEE) He received B.Sc. and M.Sc. degrees in electrical engineering from the University of the Ryukyus in 1986 and 1988, respectively, as well as a Ph.D. degree in electrical engineering from Nagoya University in 1994. Since 1988, he has worked at the Department of Electrical and Electronics Engineering, the Faculty of Engineering, the University of the Ryukyus, where he currently works as a professor. His research interests include power system optimization and operation, advanced control, renewable energy, the IoT for energy management, ZEH/ZEB, smart cities, and power electronics. He is a member of the Institute of Electrical Engineers of Japan.

applied sciences

Editorial

Renewable-Based Microgrids: Design, Control and Optimization

Marcos Tostado-Véliz [1,*], Paul Arévalo [1], Salah Kamel [2], Ragab A. El-Sehiemy [3] and Tomonobu Senjyu [4]

1 Department of Electrical Engineering, University of Jaén, EPS Linares, 23700 Jaén, Spain; warevalo@ujaen.es
2 Department of Electrical Engineering, Faculty of Engineering, Aswan University, Aswan 81542, Egypt; skamel@aswu.edu.eg
3 Electrical Engineering Department, Faculty of Engineering, Kafrelsheikh University, Kafrelsheikh 33516, Egypt; elsehiemy@eng.kfs.edu.eg
4 Department of Electrical and Electronics Engineering, University of the Ryukyus, Nishihara 903-0213, Okinawa, Japan; b985542@tec.u-ryukyu.ac.jp
* Correspondence: mtostado@ujaen.es

1. Introduction

To achieve carbon neutrality by 2050, additional measures must be taken, including the extensive incorporation of renewable energy sources (RESs). Additionally, we must pave the way for new market and grid structures that facilitate the integration of these clean technologies, along with initiatives for demand response (DR), electric vehicles (EVs), and energy storage.

In this context, microgrids (MGs) have emerged as a highly valuable framework. Specifically, a microgrid can be described as a small-scale network equipped with local power generation capabilities, typically connected to a larger power grid but being capable of operating independently [1]. As a result, microgrids offer numerous opportunities for various entities to optimize the utilization of local resources while ensuring a reliable electricity supply in remote and isolated areas [2].

2. Design Control and Optimization of Renewable-Based MGs

For the effective advancement, integration, and implementation of renewable-based MGs in existing power systems, the development of appropriate tools for their optimal design and control is crucial. This Special Issue aims to gather the latest advancements in this field, particularly focusing on optimization tools with various objectives. A total of 16 papers were submitted to this Special Issue, out of which, 10 were accepted for publication, indicating a 62.5% acceptance rate. This demonstrates the significant interest of the research community in this particular topic.

The first paper, authored by S. Panda et al. [3], presents a comprehensive overview of the recent advancements in modern DR initiatives and their interaction with renewable sources and energy storage systems. The second paper, written by A. Malki et al. [4], proposes a procedure based on metaheuristic algorithms to identify the key parameters of photovoltaic (PV) modules. It should be emphasized that accurately determining these parameters is crucial for the development of advanced optimization tools in PV-based systems. Another paper [5], authored by S. Mikkili et al., investigates electronic converters, specifically focusing on the performance analysis of a modular multilevel converter with NPC sub-modules. Such converters have numerous applications in PV-based systems.

The main focus of this Special Issue is the development and study of various optimization tools. These tools have been specifically designed for potential application in MGs. The majority of the papers in this issue concentrate on different strategies for the management of energy in systems that rely on renewable sources. For example, H. Kraiem et al. [6] compare two metaheuristic methodologies aimed at enhancing the autonomy of PV-battery

Citation: Tostado-Véliz, M.; Arévalo, P.; Kamel, S.; El-Sehiemy, R.A.; Senjyu, T. Renewable-Based Microgrids: Design, Control and Optimization. *Appl. Sci.* **2023**, *13*, 8235. https://doi.org/10.3390/app13148235

Received: 14 July 2023
Accepted: 14 July 2023
Published: 15 July 2023

Copyright: © 2023 by the authors. Licensee MDPI, Basel, Switzerland. This article is an open access article distributed under the terms and conditions of the Creative Commons Attribution (CC BY) license (https://creativecommons.org/licenses/by/4.0/).

systems. Z. Xu et al. [7] present a cooperative multi-objective optimization approach for DC multi-MG systems, where multiple MGs work together to achieve common goals. Furthermore, authors in [8] introduce a novel optimization methodology applied to a real case study conducted in Alrashda Village in Egypt. Lastly, M. Espitia-Ibarra et al. [9] introduce a linear-programming methodology for optimal energy management in MGs, notable for its efficient computational performance, making it suitable for online applications.

The previous references primarily focus on electricity systems and do not take into account other energy forms like hydrogen or heat. However, it is essential to consider the interaction between various energy carriers for the future development of renewable energy systems. It has been extensively demonstrated that different energy vectors such as gas, electricity, or hydrogen can effectively interact and offer numerous advantages through established technologies like power-to-gas systems or microturbines [10]. In this Special Issue, a paper is dedicated to multi-energy MGs [11]. It presents an energy management strategy for a thermal–gas–electricity MG, employing Mixed-Integer Linear Programming as the underlying methodology.

While the aforementioned references primarily concentrate on energy management strategies applicable during the operational phase of the system, the successful development of MGs necessitates thorough planning, highlighting the significance of optimal design tools. This Special Issue includes two papers centered around the optimal planning of renewable-based MGs. The first paper authored by H. Abdel-Mawgoud et al. [12] proposes an optimal strategy for integrating PV and storage systems into distribution networks. Notably, this reference takes into account uncertainties in generation and demand, which is crucial in MGs in which the intermittent behavior of renewable generation increases environmental risks and uncertainties. Conversely, reference [13] by A. Almaleh et al. focuses on resilient aspects, specifically aimed at enhancing the autonomy of MGs, particularly in the face of severe weather events or natural disasters.

3. Future Challenges and Perspectives

Despite the closure of this Special Issue, we still anticipate further comprehensive research will be carried out in this field. Specifically, there is a continuing need for investigations into energy management techniques that address uncertainty in renewable-based MGs. Additionally, the integration of EVs into these systems is considered a promising area for future research. Lastly, cooperative game-based methodologies are expected to gain prominence in addressing emerging topics such as MG clusters in the coming years [14].

Conflicts of Interest: The authors declare no conflict of interest.

References

1. Tostado-Véliz, M.; Kamel, S.; Hasanien, H.M.; Turky, R.A.; Jurado, F. Uncertainty-aware day-ahead scheduling of microgrids considering response fatigue: An IGDT approach. *Appl. Energy* **2022**, *310*, 118611. [CrossRef]
2. Tostado-Véliz, M.; Jordehi, A.R.; Fernández-Lobato, L.; Jurado, F. Robust energy management in isolated microgrids with hydrogen storage and demand response. *Appl. Energy* **2023**, *345*, 121319. [CrossRef]
3. Panda, S.; Mohanty, S.; Rout, P.K.; Sahu, B.K.; Parida, S.M.; Kotb, H.; Flah, A.; Tostado-Véliz, M.; Abdul Samad, B.; Shouran, M. An Insight into the Integration of Distributed Energy Resources and Energy Storage Systems with Smart Distribution Networks Using Demand-Side Management. *Appl. Sci.* **2022**, *12*, 8914. [CrossRef]
4. Malki, A.; Mohamed, A.A.; Rashwan, Y.I.; El-Sehiemy, R.A.; Elhosseini, M.A. Parameter Identification of Photovoltaic Cell Model Using Modified Elephant Herding Optimization-Based Algorithms. *Appl. Sci.* **2021**, *11*, 11929. [CrossRef]
5. Mikkili, S.; Krishna, R.V.; Bonthagorla, P.K.; Senjyu, T. Performance Analysis of Modular Multilevel Converter with NPC Sub-Modules in Photovoltaic Grid-Integration. *Appl. Sci.* **2022**, *12*, 1219. [CrossRef]
6. Kraiem, H.; Aymen, F.; Yahya, L.; Triviño, A.; Alharthi, M.; Ghoneim, S.S.M. A Comparison between Particle Swarm and Grey Wolf Optimization Algorithms for Improving the Battery Autonomy in a Photovoltaic System. *Appl. Sci.* **2021**, *11*, 7732. [CrossRef]
7. Xu, Z.; Chen, C.; Dong, M.; Zhang, J.; Han, D.; Chen, H. Cooperative Multi-Objective Optimization of DC Multi-Microgrid Systems in Distribution Networks. *Appl. Sci.* **2021**, *11*, 8916. [CrossRef]
8. El-Sattar, H.A.; Kamel, S.; Sultan, H.; Tostado-Véliz, M.; Eltamaly, A.M.; Jurado, F. Performance Analysis of a Stand-Alone PV/WT/Biomass/Bat System in Alrashda Village in Egypt. *Appl. Sci.* **2021**, *11*, 10191. [CrossRef]

9. Espitia-Ibarra, M.; Maya-Duque, P.; Jaramillo-Duque, Á. Design and Development of a Management System for Energy Microgrids Using Linear Programming. *Appl. Sci.* **2022**, *12*, 3980. [CrossRef]
10. Zhou, L.; Zhou, Y. Study on thermo-electric-hydrogen conversion mechanisms and synergistic operation on hydrogen fuel cell and electrochemical battery in energy flexible buildings. *Energy Convers. Manag.* **2023**, *277*, 116610. [CrossRef]
11. Faghiri, M.; Samizadeh, S.; Nikoofard, A.; Khosravy, M.; Senjyu, T. Mixed-Integer Linear Programming for Decentralized Multi-Carrier Optimal Energy Management of a Micro-Grid. *Appl. Sci.* **2022**, *12*, 3262. [CrossRef]
12. Abdel-Mawgoud, H.; Kamel, S.; Tostado-Véliz, M.; Elattar, E.E.; Hussein, M.M. Optimal Incorporation of Photovoltaic Energy and Battery Energy Storage Systems in Distribution Networks Considering Uncertainties of Demand and Generation. *Appl. Sci.* **2021**, *11*, 8231. [CrossRef]
13. Almaleh, A.; Tipper, D.; Al-Gahtani, S.F.; El-Sehiemy, R. A Novel Model for Enhancing the Resilience of Smart MicroGrids' Critical Infrastructures with Multi-Criteria Decision Techniques. *Appl. Sci.* **2022**, *12*, 9756. [CrossRef]
14. Tostado-Véliz, M.; Hasanien, H.M.; Jordehi, A.R.; Turky, R.A.; Jurado, F. Risk-averse optimal participation of a DR-intensive microgrid in competitive clusters considering response fatigue. *Appl. Energy* **2023**, *339*, 120960. [CrossRef]

Disclaimer/Publisher's Note: The statements, opinions and data contained in all publications are solely those of the individual author(s) and contributor(s) and not of MDPI and/or the editor(s). MDPI and/or the editor(s) disclaim responsibility for any injury to people or property resulting from any ideas, methods, instructions or products referred to in the content.

applied sciences

MDPI

Review

An Insight into the Integration of Distributed Energy Resources and Energy Storage Systems with Smart Distribution Networks Using Demand-Side Management

Subhasis Panda [1], Sarthak Mohanty [1], Pravat Kumar Rout [2], Binod Kumar Sahu [1], Shubhranshu Mohan Parida [1], Hossam Kotb [3], Aymen Flah [4], Marcos Tostado-Véliz [5], Bdereddin Abdul Samad [6,*] and Mokhtar Shouran [6]

1 Department of Electrical Engineering, Siksha 'O' Anusandhan University, Bhubaneswar 751030, Odisha, India
2 Department of Electrical and Electronics Engineering, Siksha 'O' Anusandhan University, Bhubaneswar 751030, Odisha, India
3 Department of Electrical Power and Machines, Faculty of Engineering, Alexandria University, Alexandria 21544, Egypt
4 Energy Processes Environment and Electrical Systems Unit, National Engineering School of Gabes, University of Gabes, Gabes 6072, Tunisia
5 Department of Electrical Engineering, University of Jaén, 23700 Linares, Spain
6 School of Engineering, Cardiff University, Cardiff CF24 3AA, UK
* Correspondence: abdulsamadbf@cardiff.ac.uk

Abstract: Demand-side management (DSM) is a significant component of the smart grid. DSM without sufficient generation capabilities cannot be realized; taking that concern into account, the integration of distributed energy resources (solar, wind, waste-to-energy, EV, or storage systems) has brought effective transformation and challenges to the smart grid. In this review article, it is noted that to overcome these issues, it is crucial to analyze demand-side management from the generation point of view in considering various operational constraints and objectives and identifying multiple factors that affect better planning, scheduling, and management. In this paper, gaps in the research and possible prospects are discussed briefly to provide a proper insight into the current implementation of DSM using distributed energy resources and storage. With the expectation of an increase in the adoption of various types of distributed generation, it is estimated that DSM operations can offer a valuable opportunity for customers and utility aggregators to become active participants in the scheduling, dispatch, and market-oriented trading of energy. This review of DSM will help develop better energy management strategies and reduce system uncertainties, variations, and constraints.

Keywords: demand-side management (DSM); distributed generations (DGs); energy management systems (EMS); renewable energy sources (RES); optimization; waste to energy (W2E)

Citation: Panda, S.; Mohanty, S.; Rout, P.K.; Sahu, B.K.; Parida, S.M.; Kotb, H.; Flah, A.; Tostado-Véliz, M.; Abdul Samad, B.; Shouran, M. An Insight into the Integration of Distributed Energy Resources and Energy Storage Systems with Smart Distribution Networks Using Demand-Side Management. *Appl. Sci.* **2022**, *12*, 8914. https://doi.org/10.3390/app12178914

Academic Editor: Amalia Miliou

Received: 11 August 2022
Accepted: 31 August 2022
Published: 5 September 2022

Publisher's Note: MDPI stays neutral with regard to jurisdictional claims in published maps and institutional affiliations.

Copyright: © 2022 by the authors. Licensee MDPI, Basel, Switzerland. This article is an open access article distributed under the terms and conditions of the Creative Commons Attribution (CC BY) license (https://creativecommons.org/licenses/by/4.0/).

1. Introduction

The management of energy consumption is a critical challenge pertaining to the current load consumption schedule of the electrical power system. With the introduction of several efficient and intelligent devices for use by diverse customers and prosumers participating in a power flow network at the residential and industrial usage load levels, there is a necessity for standard and robust energy management architecture and implementation at the prosumer and the generation levels. The main focus is on load consumption management on the demand side, which can be accomplished by integrating various programs focused on efficiency and minimizing loss at both the appliance and the intelligent grid system level. The consumers and the energy-generating organizations participating at the energy market levels will gain significantly from such an adjustment in the load profile. Introducing standardization protocols for efficiency and consumption management approaches can help resolve severe concerns such as fossil fuel use, carbon emissions, energy costs, and other sustainability elements, to some extent. Integrating multiple communication and

Internet of Things (IoT) protocols in renovating conventional grid systems into intelligent grids has enabled a bidirectional information exchange [1]. This data can be utilized for a variety of energy management strategies. On the demand side, by incorporating various digital sensing and communication protocols, smart device control, and connectivity between utilities and geographically distant grid organizations, appliances can leverage this information to strategically provide an optimal strategy for better operation and efficiency characteristics. Understanding the problems related to integrating different sources and technology can provide ideas to establish synchrony between generation and load.

The notion of demand-side management (DSM) is a solution to these significant challenges related to grid sustainability, security, reliability, and load profile management from the perspective of consumption and for providing strategies for load reduction. DSM is a collection of load management solutions that plan, integrate, and monitor preassigned routine operations on the basis of a consumer's consumption behavior [2]. The DSM architecture can conservatively dispatch available generation capacity, lowering emissions and peak load usage while allowing users to use their preferred energy type [3]. DSM was launched in 1970 [4] when the electrical sector offered the DSM model and architecture to manage time-of-use (ToU) and peak electricity consumption and to analyze consumer load usage profiles. DSM can establish synchrony between generation and load, taking on maximum cases of obstacles.

There are substantial incentives to employ distributed generation (DG) to reduce greenhouse gas emissions, improve power system efficiency and reliability, implement competitive energy policies, and delay transmission and distribution system upgrades. DGs are made up of renewable units such as wind turbines (WTs), photovoltaics (PV), fuel cells (FCs), and biomass, as well as non-renewable units such as micro-turbines (MTs), gas engines (GEs), diesel generators (DiGs), etc. By being near the clients, DGs avoid needing a transmission system. The integration and control of DGs, storage devices, and flexible loads can form a microgrid, a low voltage distribution network that can operate in isolated or grid-connected modes [5]. Due to a lack of sufficient energy generation sources, microgrids frequently struggle to meet demand. The intermittent nature of loads and renewable energy sources create this barrier [6]. As a result, to address this issue, an energy management system (EMS) is required. Using an EMS for a microgrid is a relatively new and trendy issue that has recently received much attention.

1.1. Motivation behind the Adoption of DSM

The necessities of the load–grid from the perspectives of synchronization, stability of operation, security and data protection from external attacks, reliability issues, and profit maximization requirements have prompted attention in various areas of DSM research. The following are motivations for the rising interest in the application of DSM techniques:

- To reduce consumer annoyance during the adoption of DSM by incorporating demand reduction bidding during peak hours, incentive DSM, and demand response (DR) programs.
- To create an interactive load management market, which is a prosumer-based market in which each customer plays a part in achieving low-cost energy usage.
- To match energy supplies and dispatch additional available sources within the current system and regulate the required demand.
- To enable proper demand and supply balance by either reducing or shifting energy use from critical loading periods to fewer off-peak times, factoring in economic standards and active control methods.
- To consider electricity generation and trading tariffs, environmental considerations, demand-based usage patterns, and prosumer convenience levels when creating optimal load dispatch and usage scheduling.
- To adapt to changes brought about by erratic consumption and a lack of understanding of the operational state of daily-use devices and machines [6].

- To conduct forecasting based on weather data assessing client comfort levels and convenience.
- To achieve the lowest possible electricity cost from an economic standpoint, maximizing energy consumption from geographically nearby renewable energy sources (RES) from an environmental perspective and preventing power quality issues.
- To raise consumer awareness of DSM's benefits, which can stimulate adoption or improve electricity usage patterns.
- To combine operational flexibility for an individual home with the flexibility of other residential customers in the neighborhood to achieve operational flexibility for a unique family.
- To improve grid efficiency and reliability by minimizing the peak-to-average ratio (PAR) by offloading optional loads during peak periods [7].

1.2. Benefits of DSM

DSM comes into play to solve such difficulties concerning the current situation on the load end and to enable greater flexibility and the robust scheduling of specific devices and gadgets at an autonomous stage through intelligent control mechanisms. DSM can provide several advantages, such as:

- To help minimize voltage fluctuations on a poor distribution feeder by providing grid support [6].
- To resist environmental concerns by lowering peak demand, which decreases the need for new traditional generating plants.
- To allow the principles of DSM to be successfully implemented, where it can benefit both customers and the utilities economically.
- To guarantee steady and sustainable power delivery within the system, thereby avoiding shortfalls.
- To provide cost savings in energy usage while also assisting in achieving positive environmental goals.
- To decrease load profiles by intelligently managing loads [7].

1.3. Issues and Challenges in Implementing DSM

The path to DSM integration is littered with several challenges and issues that must be solved for the program to be executed effectively and efficiently among the participating institutions. Some of the concerns and difficulties that will be discussed are as follows:

- Residential loads frequently contribute a major portion of load demand owing to seasonal and daily peak load consumption, causing the available grid system to be under-sized in handling peak energy usage.
- Pricing blocks that can be adapted according to consumption at multiple levels can be implemented smoothly.
- To use the best load scheduling approaches possible.
- Centralized controllers for both control choices and control actions are required to implement direct load controls (DLCs), interruptible tariffs, demand-bidding programs, and emergency programs. Because the client wants to save money on energy, and the utility wants to maximize profit from the available energy, the goal is to balance energy and save money.
- Consumer response to the price signals supplied by the utility and market tariffs, which modifies consumer behavior, fluctuates unexpectedly depending on their ability and willingness to adapt quickly, indifference to minor tariff adjustments, and pricing system awareness.
- To address the opposing objectives of consumer convenience and reduced-cost consumption, decrease load consumption for customers and increase revenues for utility companies with accessible energy generation sources, etc., while formulating energy regulations.

- Inadequate system-wide scalability measures to address the multi-vendor dilemma, upgrade, and expansion.
- Usage of robust system privacy measures to secure the vital information of participating customers.
- To address the neighbor effect, some consumers over-estimate other consumers' price rates, where any change affecting a consumer influences the choice and preference of nearby present customers.
- A generalized operational framework of DSM is necessary owing to the characteristics and objectives of DSM participants and loads operating in an independent system in order to provide the customers with more control over their energy consumption.
- The reduction in peak load requirements and the minimization of overall load usage tariffs for residential occupants while maintaining an acceptable degree of comfort and choice for the user.
- Integrated volatile power sources such as wind and solar impact grid stability and create issues.
- The difficulty of balancing supply and demand for electricity in the face of uncertain demand and uncontrollable sources.
- DR faces four significant operational issues: scalability, distribution of control, unpredictability, and aggregation.
- Supply and demand may become imbalanced at different locations along a changeable demand curve [6].
- The need to build a model of energy generation is essential due to the effects of traditional power generation and global climate change.
- Lower peak demand and overall load consumption costs while maintaining appropriate comfort and convenience for residents. Integrated unreliable power sources such as wind and solar impact system stability and create issues.

1.4. Suggested Solutions in DSM Implementation

The following suggested solutions are viable for implementation and for driving grid integration programs in a more effective and coordinated way to deal with the above concerns and obstacles faced during the implementation of various policies concerning DSM using DGs systems:

- An adequately designed pricing structure will result in a flexible electricity system, allowing residential customers and utilities to achieve their goals.
- Time-of-day (ToD) pricing can incentivize large-scale residential and commercial users to conserve energy.
- The load profile forecast mechanism can serve as a transitive feedback signal, and the tariff associated with it can serve as a transitive incentive signal.
- A stochastic and multi-objective optimization technique for the optimal scheduling of various domestic appliances utilizing model predictive control (MPC) optimization.
- The transitive energy concept is a viable coordination paradigm for maximizing the importance placed on prosumers and operators at the utility level and their overall participation in the market structure.
- From the perspective of trading entities present in the market and their involvement and market-based signaling, extensive changes in government laws consider both energy providers and customers.
- A variety of sophisticated methodologies can factor in individual residential prosumer overhead and comfort levels, optimize individual consumption schedules, and offer positive DSM impacts [7].
- To allow the electricity markets to generate higher revenues, an incentive-based program can change conventional consumers into new era prosumers by modifying their behavior and habits of use [8].
- For the improved functioning of DR ideas in residential utilities, measurement and verification protocols and an automated procedure are required [9].

- Complete service-oriented topology and structure is a necessity to allow for provisions of appropriate infrastructure oriented toward dynamic integration techniques and for a more flexible operation to bring out the best in the power system scenario [10].
- DSM contributors can consume or generate energy in a coordinated operational state as cooperative agents or virtual power plant models, which can simulate the performance of an aggregated virtual single power source indirectly incorporated into the power system [11].
- The introduction of generation systems such as solar photovoltaics (SPV) and energy storage system combinations for usage during peak hours.

1.5. Outline of This Paper

This paper is presented to address the issues and solutions of DSM using a methodological and critical survey-based exploration of the implementation of DSM using DERs. Efforts are made to put forth the following points concisely: Firstly, to assess and study the suggested optimization techniques and implementations of DSM in the present literature. This will allow the researchers with the necessary exposure to arrive at more practical and better optimization techniques to establish a proper energy management system (EMS). In addition, energy management modeling studies are examined in terms of uncertainty modeling techniques, objective functions, constraints, and optimization techniques. Lastly, EMS-related papers are reviewed and analyzed correctly to help the researcher find out the problems and the solutions.

The remaining part of this review article is designed as follows: Section 2 represents the detailed review methodology used to formulate this paper, Section 3 states the brief introduction to DSM and DR, Section 4 briefly states the various DGs possible in an intelligent grid network, Section 5 represents the DSM with different types of cleaner energy, Section 6 briefly illustrates the energy management system and some standards related to DG integration, Section 7 explains some issues related to different types of DGs being integrated with DSM techniques, Section 8 represents the various optimization techniques ascribed to DSM with objective and objective functions, Section 9 deals with the different research gaps and critical analyses, and the future scope and conclusion are analyzed in Sections 10 and 11, respectively.

2. Review Methodology

Any research project's primary focus is on three key elements: the purpose, study technique, and outcome, as well as future implementation prospects. An approach based on an analytic-based search technique was undertaken on numerous scientific and interpretive sources such as Google Scholar, ResearchGate, IEEE Explorer, and Scopus to gain a detailed and complete overview of existing research publications. Combinations of thematic words, such as "Demand-side management distributed energy resources", "Demand response", "Energy management using distributed energy sources", "Optimization", "Scheduling", "Distributed energy sources integration in microgrids", and so on, were used to filter out the critical articles using search engines. Specific search engine parameters were employed to find relevant, on-point, particular research papers for the review study. Exact keywords, peer-reviewed publications published in English mainly in the last ten years, and open-access articles were the deciding factors.

Based on the research articles, an eight-point prospect was developed:

- DSM techniques in general, with sub-strategies investigated from a modification standpoint.
- Incentivized and price-based programs, as well as demand response strategies.
- The customer rationale for employing distributed generation to implement DSM.
- Researching the architecture and topology of the EMS system, as well as comparing it to alternative DSM methodologies.
- The scope of limitations and constraints associated with implementing DSM using DER architecture with present issues.

- Published research methods and optimization approaches.
- Analysis and conclusions from the study of the approaches employed in the optimization challenges stated.
- Action plan for the future.

As shown in Figure 1, 31 review papers, 40 case studies, 15 news articles, 107 technical papers, and 10 research reports were reviewed and placed in this paper.

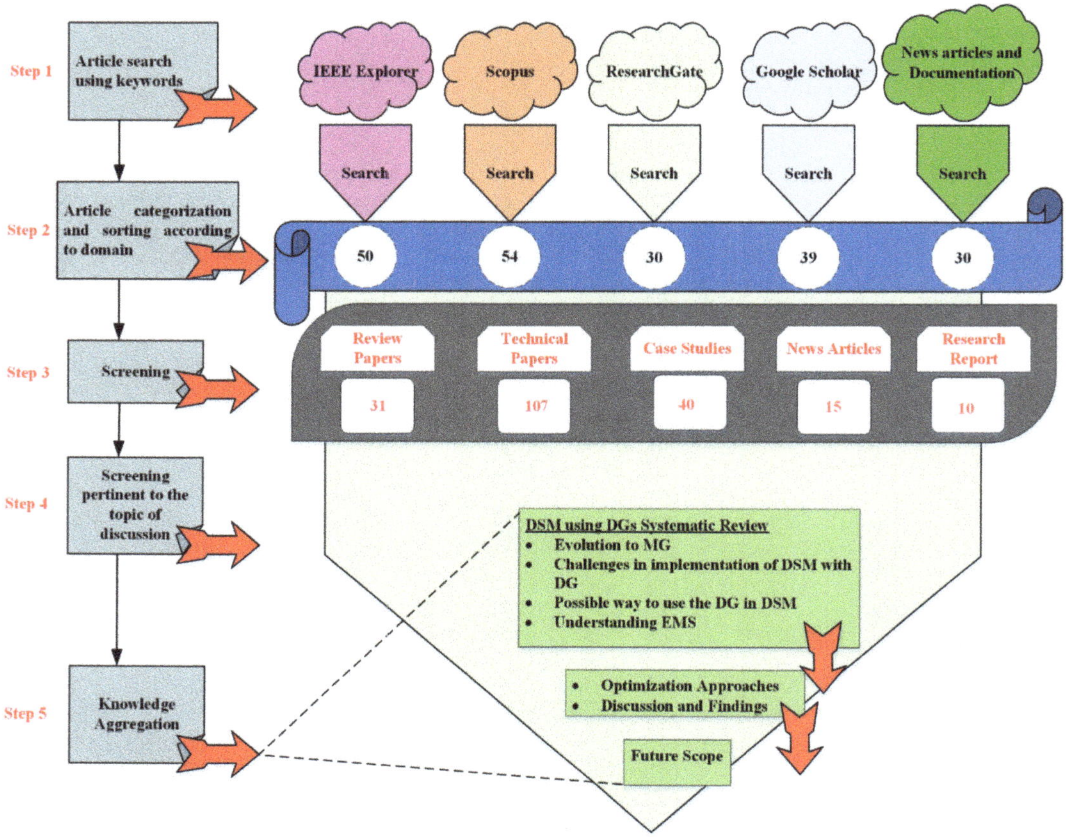

Figure 1. Review methodology for this paper.

3. Demand-Side Management

Demand-side management is an essential part of an intelligent grid architecture because it allows consumers to adjust their load consumption patterns, making it a critical feature of an energy management system in power delivery networks [11,12]. "The planning, implementation, and monitoring of those daily activities designed to influence customer use of electricity in ways that will produce desired changes in the utility's load shape, i.e., time pattern and magnitude of a utility's load," according to the Electric Power Research Institute (EPRI) [13]. Instead of relying on additional generation to meet demand, DSM prioritizes the integration of power-saving techniques, the implementation of variable or dynamic unit pricing, and the adoption of DR-based programs to minimize peak load, managing the DGs to establish a proper power balance, as shown in Figure 2.

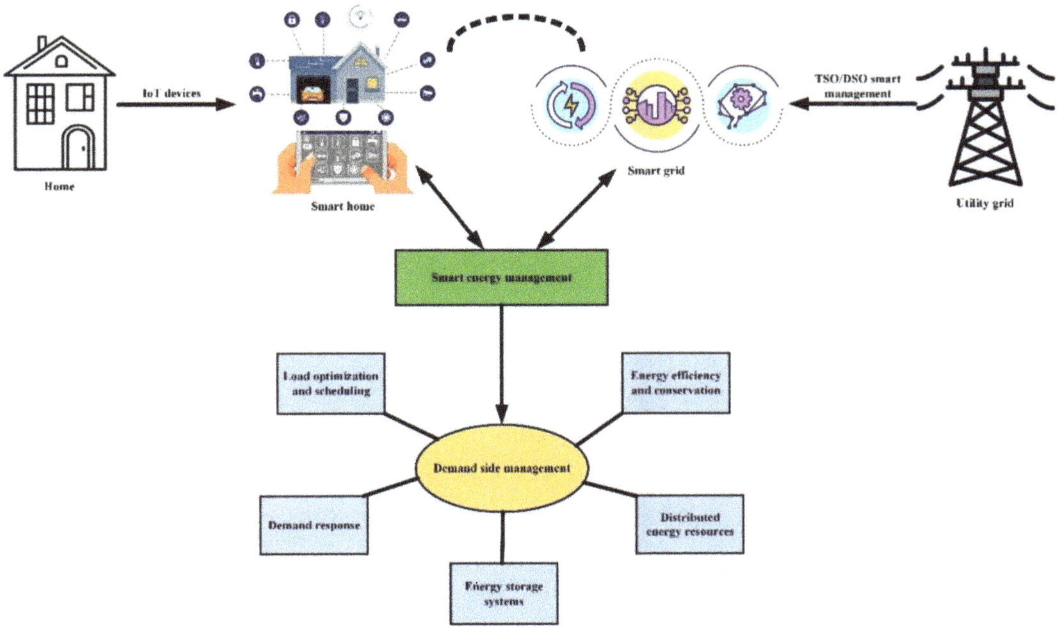

Figure 2. Principle of DSM in the smart grid environment.

The four methodologies outlined below and illustrated in Figure 3 can be used to classify various alterations that can be used to shape and define the electricity load profiles:

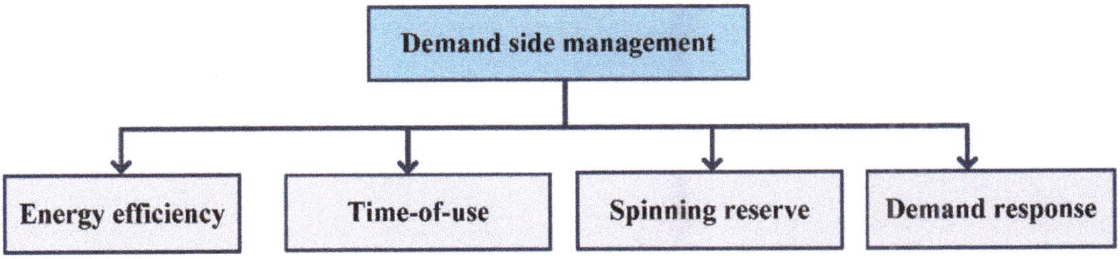

Figure 3. Basic principle of DSM.

Energy efficiency (EE): These are end-user, appliance-specific controls intended to reduce load utilization over time by employing energy-saving methods on the device level. Rather than relying on an event-triggered strategy for consumption profile minimization, energy efficiency refers to the reduction in overall load consumption achieved by providing more efficient power delivery for each unit with respect to the supplied input power to the appliance, decreasing consumption over time. An in-depth look at the energy efficiency improvement profiles, measurements, and roadblocks can be found in [14,15].

Time of use (ToU): The ToU pricing method divides the utility's fixed tariff into 24 h time blocks and then assigns a variable pricing profile for each period [16,17]. This method can help keep peak load rates and seasonal fluctuations in pricing tariffs under control based on the hourly block-based signal tariff of electricity units.

Spinning reserve: In the case of a drastic shortfall in generating levels, the spinning reserve is synonymously recognized with the electric power system's backup power, which may be utilized by the distribution network operator (DNO) to balance the difference or gaps between demand and supply in generation [18]. Power outages can be caused by various factors, including damage to producing units, inadequate load prediction, and dispatch scheduling [19]. In general, there are two types of spinning reserves: primary and secondary [16], with the central spinning reserve employing frequency regulation to limit active power output and the secondary spinning reserve injecting extra active power.

Demand response: Energy users depart from their usual use patterns in response to unit rate variations over time or incentive programs. The primary focus is to reduce load profiles during critical tariff periods in the energy wholesale market or when grid reliability is uncertain [20]. Short-term variations throughout the day's critical peak pricing/usage times, when demand is low and spinning reserve capacity is scarce, are of primary interest to DR. DSM is more concerned with long-term load profiles, which may be accomplished on the demand side by improving energy efficiency or adopting consumer-centric usage behavior.

4. Distributed Generations in Smart Grid

A distributed energy resource (DER) is an aggregation of distributed generators, as shown in Figure 4, or controllable loads (conventional or smart) connected to the network in a smart grid. A DER unit, or distributed generation (DG), often blends a variety of energy sources. They are classified into, essentially, two sorts of sources based on their dispatch capacity and source of generation type:

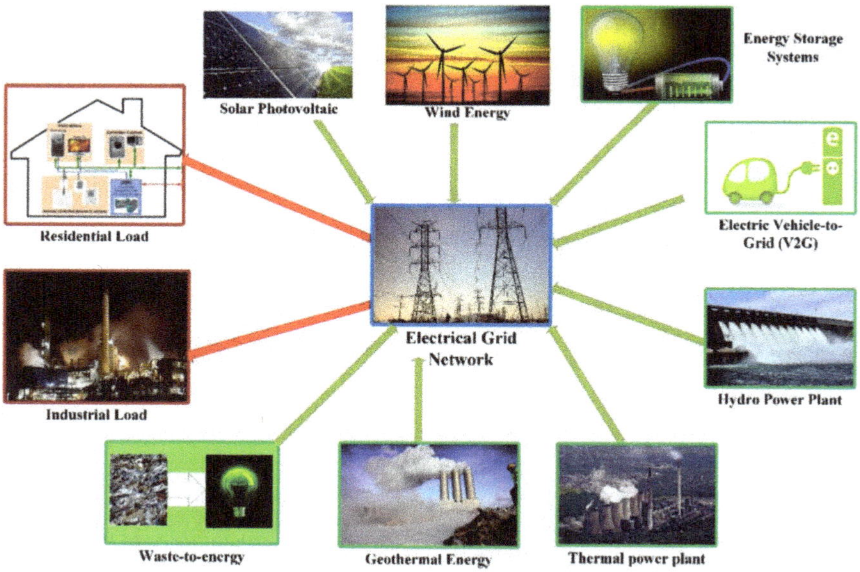

Figure 4. Various types of energy sources in the smart grid.

4.1. Renewable Energy Sources (RES)

Solar Photovoltaics: Converting solar energy to electrical energy via mounted semiconductor panels is the primary renewable generating source across the globe. It can produce energy in any mode, stand-alone or grid-integrated, and small-scale (such as rooftop PV in residential areas) or large-scale (centralized power plants). Scheduling, as per weather condition forecasting, boosts production capabilities.

Solar Thermal: Solar thermal power plants use solar energy to heat a fluid to a high temperature to generate electricity. The heat from this fluid is transferred to water, which subsequently forms superheated steam. In a power plant, steam is utilized to run turbines, and mechanical energy is transformed into electricity by a generator. This sort of generating is similar to electricity generation that uses fossil fuels, except that instead of burning fossil fuels, sunlight is used to heat steam.

Hydropower Plants: The flowing capacity of water is capable of rotating a turbine to generate electricity, which can be the centralized or decentralized mode of operations according to the availability of the water and the water head. Generally, small-scale hydropower stations are used for DSM operations, which are described in Section 5.

Wind Turbines: Wind energy conversion systems (WECs) are also a significant component of DGs where the appropriate wind reach is available. This generation unit is limited to smaller, low-capacity generation units. This allows for small-scale WT unit deployment on the customer side to be possible without affecting the operation of the entire power system as a whole.

Geothermal Energy: Geothermal energy captures the energy from the core of the earth. DGs can be localized around nearby natural geothermal energy sources, such as lava flows, hot springs, geysers, or places that experience direct contact between water and high thermal capacity surfaces. As part of the natural cycle of evaporation and replenishment, geothermal sources can be considered a viable source of renewable energy generation.

4.2. *Traditional Energy Sources*

Combined Heat and Power: Fossil fuels are the primary sources of CHP that are set to run as centralized power stations, mainly to fulfill the baseload requirement. Fossil fuels are burnt to produce steam, which rotates the turbine to produce electricity. Due to substantial carbon emissions, limited availability of sources, and environmental concerns, the focus has shifted toward renewables.

Fuel-based DERs: To supply supplemental power to the grid, diesel generators and fuel cell (FC) generators often employ readily available fossil fuels, waste-derived fuel, and hydrogen-based production, and they are typically run on-demand rather than always-on. Due to their simple dispatch mechanism and controllability, they are suitable DER units to link to a smart grid design. For the purposes of providing power to emergency loads, they are suitable as a DSM option.

4.3. *Energy Storage Systems*

ESS is now viewed as a novel technique for adjusting generating capacity to load demand changes, particularly as energy buffers in the situation of the high availability of non-dispatchable generation sources. These ESS capture and store surplus energy generated during off-peak hours, then dispatch it during peak periods when the extra load is needed. They also allow for the optimal redistribution of PV array and WT unit output power throughout the daily scheduling period. In terms of ESS concerned with energy supply, they are categorized as compressed air energy storage (CAES) and hydraulic pumped energy storage (HPES), depending on the method of application. Similarly, ESS focused on power supply include supercapacitor energy storage (SCES), superconductor magnetic energy storage (SMES), pumped storage, and flywheel energy storage (FWES) [21,22].

4.4. *Waste-to-Energy (Bio-Energy)*

An increase in urbanization is the cause of the generation of a large amount of waste, particularly MSW (municipal solid waste). The thermal treatment of municipal or industrial waste and sludge, as well as medical or industrial hazardous waste, decreases trash disposal in landfill sites dramatically. The produced energy yields a new revenue stream that helps both the local population and the environment through cleaner air, water, and soil. The process starts with collection, followed by segregation, then processing through various stages, such as pyrolysis, and then incineration to produce electricity. Taking into consideration the United Nations' Sustainability Goal of cleaner energy, waste is counted as one of many effective sources of conversion to electricity [23,24]. The different processes, which can be followed to convert the waste, are presented in Figure 5.

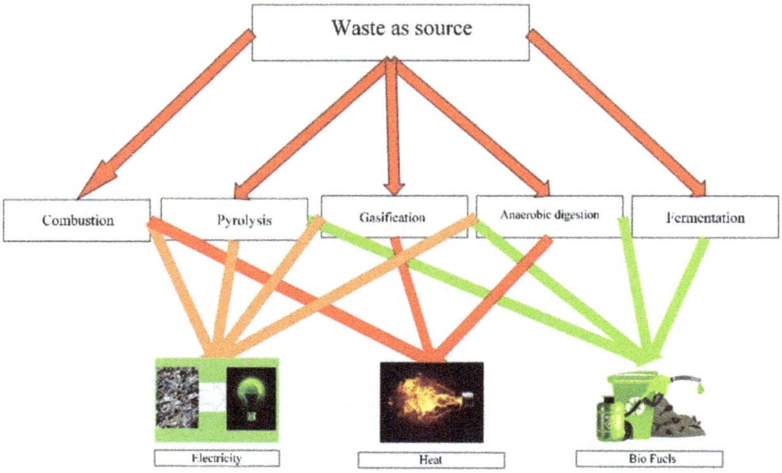

Figure 5. Different processes of waste-to-energy.

4.5. Electric Vehicle (V2G)

Vehicle-to-grid (V2G) is a system in which plug-in electric vehicles (PEVs), such as battery electric vehicles (BEVs), plug-in hybrids (PHEVs), and hydrogen fuel cell electric vehicles (FCEVs), communicate with the power grid to sell demand response services by returning electricity to the grid or throttling their charging rate. Electric vehicles with V2G storage capability can store and discharge power generated from renewable energy sources such as solar and wind, with output that varies based on weather and time of day [25,26]. The process of V2G is the same as in the case of the ESS shown in Figure 6.

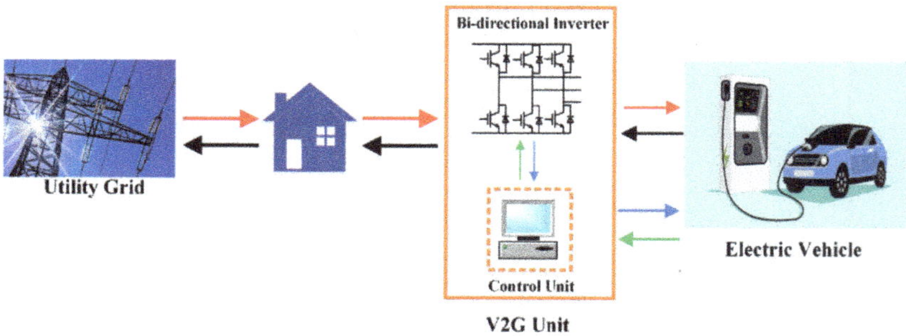

Figure 6. Electric vehicle as a source of energy.

5. DSM Using DGs and ESS

DSM is the systematic energy management in the case of using DGs and ESS. Using DSM can have a lot of benefits to industry, residents, nations, and the globe, which is shown in Figure 7. DSM can be implemented by using distributed energy resources such as solar, wind, waste-to-energy, etc. DSM generally involves load shape modification by applying different optimization techniques [27–29]. This modification is carried out by the significant DSM component, which is the load duration curve (LDC). LDC offers a general and analytical idea of off-peak hours and peak hours. Six techniques are used in load shaping, which are discussed below and in Figure 8.

a. *Peak Clipping*: This technique is used to reduce the peak demand at peak hours. Effective use of this method can reduce the chances of establishing new generating stations. Generation from DERs also helps in balancing load and can reduce the peak demand.

b. *Valley Filling*: This technique is set to rebuild the load during off-peak hours, which helps reduce tariffs. Charging electric vehicles at off-peak hours to work as V2G at the time of need is a possible example of valley filling.
c. *Load Shifting*: This is based on shifting load from peak hours to off-peak hours.
d. *Load Reduction*: This strategy is based on using energy-efficient equipment to reduce load demand. Rooftop solar installation in residential areas can reduce the load overall, which is an example of this technique.
e. *Load Growth*: Building up the load at the time of reduced load conditions or in off-peak hours. This technique is an example of charging ESS or EVs at non-peak times or during non-peak days.
f. *Flexible Load Shaping*: The rearrangement of LDC according to the conditions. WEC system generation is an example of this method.

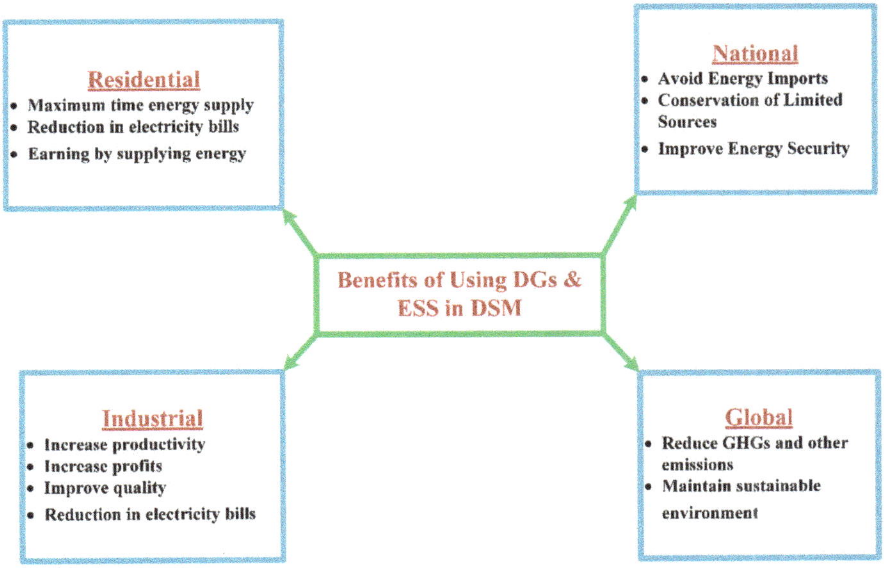

Figure 7. Benefits of using DGs and ESS in DSM.

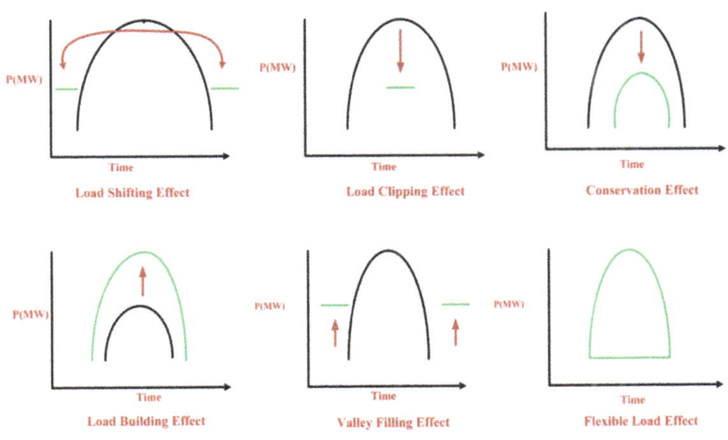

Figure 8. DSM techniques.

The discussed DSM objectives can be achieved by integrating the DERs described in earlier sections in a forecasted manner [29]. The reason for using all these renewables is as simple as the UN's Sustainability Goal of cleaner energy. Different types of available DERs and possible DSM techniques are discussed in Table 1, which is a new and innovative way of expressing the information in this paper.

Table 1. Different types of DERs with DSM applications.

DERs	Available for the Time (24 h)	Possible DSM Techniques	Types of Operations	Benefits
Solar [30,31]	Morning to afternoon (7–9 h)	Peak clipping Load reduction Load shifting Valley filling	Thermal: Converting solar heat energy to electrical energy.	Cleaner energy, reduction in the use of carbon.
			Photovoltaic: Converting solar radiations to electrical energy with solar cells.	Cleaner energy, tariff reduction, decentralized generation, residential mode generation.
Wind Energy [32]	24 h	Load reduction Load shifting Valley filling	Converting wind energy to electrical energy with induction generators	Cleaner energy, decentralized generation.
Hydro Energy [33]	24 h	Peak clipping Load reduction Load shifting Valley filling Flexible load growth	Pumped hydro: Water pumped during off-peak hours generates electricity during peak hours.	Emergency power, cleaner energy, small centralized power generation.
			Small hydro: Decentralized runaway water used for electricity generation.	Emergency power, cleaner energy, small centralized power generation, low-cost generation.
Waste-to-Energy [34,35]	24 h	Peak clipping Load reduction Load shifting Valley filling	Biogas: Anaerobic digestion of biodegradable waste into methane produces energy. Thermal: Combustion of waste to produce energy.	Cleaner energy, small centralized power generation, less carbon production. Cleaner energy, small centralized power generation, less carbon production.
ESS [36]	24 h	Peak clipping Load reduction Load shifting Valley filling Flexible load growth	Energy is stored at off-peak hours in various systems such as electric springs, pumped hydro, fuel cells, hydrogen cells, supercapacitors, etc.	Emergency power, cleaner energy, small centralized power generation, less carbon production, charging stations.
Vehicle-to-Grid [37]	24 h	Peak clipping Load reduction Load shifting Valley filling Flexible load growth	EV charging in off-peak hours can provide power to grid-like ESS at the time of need.	Emergency power, cleaner energy.
Geothermal Energy [38]	Available when water is in contact with lava	Peak clipping Load reduction Valley filling	Providing intermittent boosts to power levels.	Small centralized power generation, spinning reserve.

6. Energy Management System

An energy management system is an operational system used to plan, manage, mitigate, forecast, and continuously improve energy performance to establish a balance in the power flow network, including various DERs, as shown in Figure 9. An EMS optimizes the energy supplied by generating stations to the grid, taking into account various parameters, which are listed below:

- Energy consumption in the power flow network;
- Load behavior pattern on the demand side;
- Consumer energy consumption patterns;
- Seasonal forecasting of consumer data;
- Weather forecasting data;
- Time of pricing when it is highest.

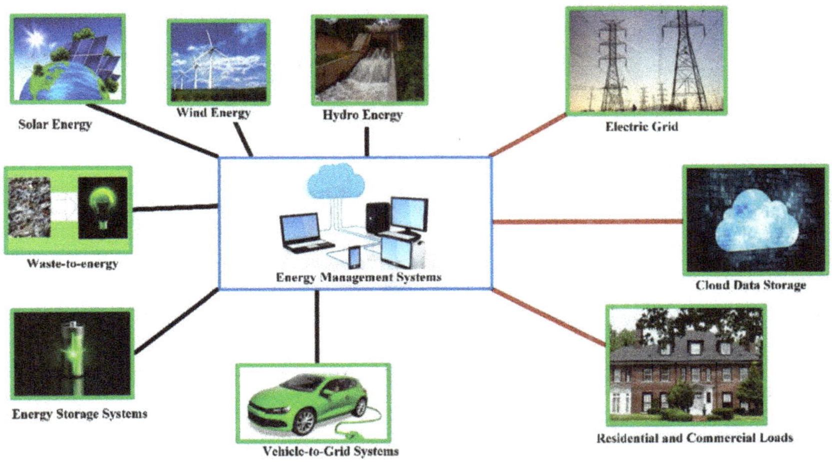

Figure 9. Energy management system.

Major components of EMS are measuring units, IoT-based tools to forecast data from collected data, various types of sources of generation, and generation scheduling. An EMS is operated by various optimization models with specific objectives, taking into account the constraints related to them, which are discussed in the later section.

6.1. Energy Monitoring, Measurement, and Analysis

An EMS includes monitoring, measuring, and analysis as major components to carry forward its operations, which will determine the energy flow performance and help it perform DSM effectively. Its key characteristics include [9,39]:

- Significant energy use in the SG network;
- Variables related to energy use;
- Energy performance indicators;
- Effective energy-efficient plans to achieve objectives and targets.

6.2. Standards Used for Communications in DSM Using DGs and ESS

There are specific standards used for communicating various DGs and ESS to the SG's power flow network, which are provided in Table 2 [40].

Table 2. Standards used for communication in DSM using DGs and ESS.

Code	Year of Implementation	Objective
IEEE 1547	2003	To find a bridge between distributed generation and the electric network.
IEEE 1547.1	2005	Specifies the test procedure for the interconnection.
IEEE 1547.4	2011	Deals with the planning and operation of integration.
IEEE 1547.7	2013	To standardize the DG integration system.
IEEE 1547.8	2014	It identifies and expands the innovative design, process, and operational procedure to achieve flexibility.
IEEE 2030	2011	Integration of information technology into the grid, the establishment of a framework of operation because of the prospects of a smart grid.
IEEE P2030.2	2015	Integration of hybrid energy storage systems into the power flow in the network.
IEEE 2030.3	2015	The test procedure for a single storage device in the power network.
IEEE2030.7	2017	Standards for microgrid energy management.
IEEE 802.1/ 802.3/802.15.4	2003	Interfaces the identifiers, which operate as the interconnecting modes and power control. Information exchange between the components.
EEC 61850-7-2	2003	Sets standards for abstract communication service interface (ACSI) as a paradigm used for vertical and horizontal communication for MC61850.
EEC 61970/ 61968/62325	2013	Sets standards based on information integration and the software framework of EMS for DGs
IEEE 2030.8	2018	Sets standards for microgrid energy management and control in a grid-tied or off-grid system.
IEC 61850	2019	Automation architecture requirement for utility subsystems, enabling communication and semantic interoperability among multi-vendor equipment, communication networking, and the communication front-end for the network.

7. Issues and Challenges

During the review process, several issues related to integrating DGs into the smart grid for DSM purposes were found, which are listed below, categorically segregating different types of DGs [41–48].

7.1. DSM with SPV

SPV energy conversion systems have been used for a long time since their discovery as a significantly cleaner energy source. Much of the maturity in semiconductor technology has allowed for vast improvements in the scope of SPV generation. The issues discussed below are the major difficulties in addressing stable grid operations with DSM.

- These sources of generation, albeit easy to install, are not flexible to operate. This is because the reactive power necessary to complement the generated active power from SPV is not readily sourced and is difficult to integrate when upgrading the primary SPV generation sources.
- SPV generation necessitates the use of ESS to avoid drops in power delivery and the energy buffer as and when power is not readily available for generation.
- The installation safety measures need to be stepped up, as they are prone to damage from meteorological and physical factors such as hurricane winds and rust in the installation equipment. This presents a potential hazard, hindering the safety aspect of SPV installation.
- Maturity in SPV panel technology has vastly improved since its inception, but the technology has still not reached its peak maturity for the maximum extraction of available solar energy using existing power conversion and extraction techniques, viz., MPPT.
- The manufacturing and disposal of SPV equipment leave behind a very high carbon footprint, presenting a deterrent toward adoption owing to environmental concerns.

7.2. DSM with Wind Energy Conversion System

Wind energy conversion systems integrated into DSM have been wholly realized for a long time; this includes wind turbines, wind monitoring systems, and related environmental protection systems, such as environmental protection stations and power generation systems. Still, it has proved challenging to synchronize wind and hydropower effectively in actual operations, which does achieve high efficiency at the same time each year [3]. The complexity can be seen in the following aspects.

There are three significant issues in the current use of wind turbines in DSM that challenge its high efficiency and stable operation [3,24]:

1. The prediction and forecasting are not accurate, as various sources influence the generation capacity. The meteorological uncertainty, coupled with the continuous available wind flow available at the tip level, influences the generation capacity of the wind turbine.
2. The transmission and distribution of the existing power grid are too complicated, making it more challenging to integrate with wind turbines effectively, as they have issues concerning intermittency and frequency deviations from the existing grid requirements.
3. The operation of wind turbines with the existing power grid is not sustainable during the nighttime, eventually leading to an increased load on the grid during the daytime.

7.3. DSM with Hydro Energy Sources

Generation sources pertaining to hydro-based sources are a major part of the power supply to grids. Complex hydropower-generating units installed in dams and tidal-based generation can be leveraged to a high extent, owing to their pollution-free generation and the replenishable source of waterbodies, which are naturally replenished via the water cycle. The major issues that are present in the existing power generation scenario that can affect DSM operation, to a certain extent, are highlighted as follows:

- The availability of water sources for potential generation is not feasible in every possible geographical location. Large-scale generation is only possible if the geographical arrangement allows for dams to be constructed or the waves to be harnessed suitably without causing ecological imbalance to nearby flora and fauna.
- The ratios of cost-to-establishment and revenue generation-to-cost are generally low, owing to high recurring and installation expenditures. However, these can be leveraged by using an environmental outlook to justify the cost.
- Maintaining the frequency of the power generated is complex due to the intermittent nature of water flow at the available head level. This prevents the energy generated from the wind turbines to be directly integrated into the transmission system, as pre-conditioning the power supplied is necessary for reliable grid operation. This adds significantly to the power generation costs as additional power conditioning units are required to bring the frequency and other parameters up to an acceptable generation level.

7.4. DSM with Waste-to-Energy Sources

In recent times, waste-to-energy has gained significant potential in the renewable and biologically eco-friendly energy market owing to its nature-oriented disposal and pollution-free generation. DSM can be implemented at the prosumer level, with the prosumers being active power generation sources putting greater emphasis on waste-to-energy potential as their preferred source of energy to defer them from using conventional grid-supplied electricity. However, the present scenario is plagued with many issues, some of which affect DSM integration with waste-to-energy potential as a potent source of power generation. Some of these issues are:

- Waste segregation and management on the ground level are the primary tasks that need to be focused on. In developed countries, this is not a major issue, as the general population is aware at a high level in comparison to countries with developing economies. Better awareness among the general masses can be a solution to the segregation and management of waste.

- Complex techniques are involved in waste-to-energy-based generation, such as pyrolysis in controlled environments and the use of a specific mixture of substances to keep the entire process pollution-free.
- High investment costs are required for setting up the incineration and biogas plants.

7.5. DSM with EV and ESS

The automobile sector is presently witnessing a surge in sales of EVs with the transition to battery-based power delivery from conventional gasoline and diesel as sources of transportation fuel being the prime focus. This has seen a rise in the adoption of more efficient and high power density battery systems to be implemented at the EV end-user side. Higher capacity batteries can be configured from a backup or buffer storage system standpoint. A bidirectional implementation of these battery-enabled mobile EVs can allow for dispatch strategies to be collectively aggregated and disbursed by DNOs as a virtual power plant system through DSM strategies. Similarly, DSM can allow intelligent control of EVs, charging and discharging according to adaptive schedules, which will benefit the power system. Nonetheless, issues pertaining to EV adoption prevent the large-scale integration of DSM in EV-based programs due to:

- High investment costs and costs of ownership from the perspective of the manufacturers, grid utility operators, and consumers. The careful economic and technical planning of charging stations, the grid's capacity to accept increased loading, and the minimization of losses along the transmission and distribution systems to allow for efficient use of available energy are the primary concerns from the maintenance and setting-up perspective.
- Consumer acceptance is still low in developing countries due to a range of anxieties about reliability and initial expenditure during purchase as compared to fossil fuel-based automotives [46].
- Battery degradation and better health are also major concerns; the periodic maintenance of fossil fuel vehicles is a fuss-free ownership experience in the long run, as battery degradation does not hinder the performance of the vehicle to a great extent. In contrast, in the case of EVs, the batteries eventually require replacement at their end-of-life stage unless they find use as second-life batteries.

These issues and challenges must be addressed to make EMS architecture more robust and reliable. After resolving these challenges, the optimization techniques can be integrated seamlessly into the EMS architecture.

8. Optimization Methods

The earlier literature presents various ideas about different types of mathematical approaches to solve the DSM problem. Techniques such as linear programming, dynamic programming, non-linear programming, game theory approach, and particle swarm optimization set a mark for solving the DSM objectives. Recently, hybrid techniques, such as gray wolf optimization (GWO), harmony search (HS) algorithm, enhanced differential evolution (EDE), etc., have drawn the interest of researchers in this field. Many of these optimization approaches and real-world implementations of DSM (mostly on residential premises) were discussed in [26], with distinct classifications between each approach and classification. The various objectives and constraints are discussed in Table 3, and different types of optimization techniques are broadly listed in Table 4, below.

Table 3. Classical technique-based single objective optimization.

Refs.	Objective	Objective Function	Concepts Employed					
			LS	PC	VF	LS	LG	LR
[47]	To facilitate EMS to reduce the total cost of energy consumption and generation.	$F(x) = \min\left(\sum_{i \in I} E_{con,i}.P_i - \sum_{j \in J} E_{gen,j}.P_j\right)$	✔	✔				
[48]	To assign a thermal management system for peak load shifting.	$F(x) = J(s,p) = \sum_{t=1}^{T} \Delta t(p_{load}(t) - \overline{p_{load}}(t))^T \times K\Delta t(p_{load}(t) - \overline{p_{load}}(t)) + (s(t) - \overline{s}(t))^T M(s(t) - \overline{s}(t))$	✔					
[49]	To reduce power consumption in classroom-based smart buildings.	$F(x) = E^{obj} = \min\left(\sum_{r=1}^{c}\sum_{t=1}^{h} E_r\alpha_{rt} + \sum_{r=1}^{c} E_r^{\alpha}\right)$		✔				✔
[50]	To reduce a building's peak electrical demand through customer-side load control.	$F(x) = \min\left(\sum_{\text{h}} E_{DG}^h.C^h\right) \forall h \in H$ $E_{DG}^h = \left(P_{NINSLs}^h + P_{INSLs}^h + P_{SLs}^h + P_B^h - P_R^h\right) \cdot \left(\frac{D_S}{60}\right)$		✔				
[51]	To propose reduction values for home energy management.	$F(x) = \min\left\{\begin{array}{c}\sum_{Load=1}^{nLoad} \lambda_{Load} \times P_{Load} + \lambda_{Grid} \times P_{Grid} + \lambda_{Down} \times \text{Reg}_{Down} - \\ \sum_{DG=1}^{nDG} \lambda_{DG} \times P_{DG} + \lambda_{Up} \times \text{Reg}_{Up}\end{array}\right\}$		✔		✔	✔	
[52]	To minimize the electricity cost and lower the delay of equipment running.	$F\ (x) = W_1 \frac{(\sum_{u=1}^{120} pr\, Cu\, P_{scd}{}^{(u)})P_{scd}{}^{(u)}}{((\sum_{u=1}^{120} pr\, Cu\, P_{scd}{}^{(u)})P_{scd}{}^{(u)})\max} + W_2 \frac{\sum_{a \in A} \rho^{DT\, R_a}}{(\sum_{a \in A} \rho^{DT\, R_a})\max}$		✔				
[53]	To minimize the cost of use on the generation side.	$F\ (x) = (E_{\max n} e_n \ - \ 1/2\alpha \frac{E_{res\, n}}{E_{\max n}} {e_n}^2 - \beta \frac{p_m}{p_{\min}} S_w e_n$			✔			✔
[54]	To minimize the cost, including overall energy costs, scheduling costs, and climate comfort.	$F\ (x) = \alpha_{EC}\overline{C}_{EC} + \alpha_{PR}\overline{C}_{PR} + \alpha_{CC}\overline{C}_{CC}$						✔
[55]	To minimize the cost of use on the consumer side.	$F\ (x) = \sum_{i,j} F_{i,j} x_{i,j} + \sum_{i,j} G_{i,j} d_{i,j}$		✔				
[56]	To minimize the generation costs, including all possible types of DGs.	$F = \sum_{t=1}^{T}\left\{\sum_{i=1}^{N_g}\left[u_i(t)P_{gi}(t)\left(B_{gi}(t) + K_{OMi}\right) + S_{gi}\lvert u_i(t-1)\rvert\right] + \sum_{j=1}^{N_{ES}}\left[u_i(t)P_{Sj}\ (t)B_{Sj}\ (t)\right] + P_{Grid}\ (t)B_{Grid}\ (t)\right\} + \sum_{t=1}^{T}\left\{\left(\sum_{i=1}^{T_E}\sum_{j=1}^{N} EF_{ij}P_{gi}\ (t)\right) + P_{Grid}\ (t)EF_{grid}\right\}$		✔				
[57]	To reduce the charge and discharge costs.	$F = \sum_{t=1}^{m}\left(C_t^g + Ct_t^g + C_t^{ES-} - C_t^l - C_t^{ES+} + \Omega_t\right) \times \Delta t$		✔				

Table 3. *Cont.*

Refs.	Objective	Objective Function	Concepts Employed					
			LS	PC	VF	LS	LG	LR
[58]	To reduce NPC, taking into account all types of sources.	$F = NPC + \sum_{t=1}^{8760} P_b(t) + \sum_{t=1}^{8760} P_{H_2}(t) + \sum_{t=1}^{8760} P_w(t) + P_{wt} + P_{H_2T}$	✔					
[59]	To reduce operation costs, emissions, and the reliability of SG.	$F = CF_t^{OPR} + CF_t^{EMI} + CF_t^{RLB}$ $F = C_{in}^{MG} + C_{op}^{MG}$	✔					
[60]	To reduce the investment and operating costs.	$C_{op}^{MG} = \sum_{i=1}^{L} (C_{Fi} + C_{OMi} + C_{Si} + C_{Ei}) + \sum_{j=1}^{M} C_{OMj}^{ESS} - C_G^{MG}$				✔		
[61]	To minimize the operating and emission costs, including startup and shutdown costs, reverse costs, and exchange of power costs.	$F = \mathrm{Cost}^{Operating} + \mathrm{Cost}^{Emission}$ $\mathrm{Cost}^{Operating} = \sum_{t=1}^{T} (\cos t_{DG}(t) + ST_{DG}(t) + \cos t_s(t) + \cos t_{Grid}(t) + \cos t_{DR}(t))$ $\mathrm{Cost}^{Emission} = \sum_{t=1}^{T} \{emission_{DG}(t) + emissions_S(t) + emission_{Grid}(t)\}$ $F = F_{\mathrm{Cost}}^{start-up} + F_{\mathrm{Cost}}^{reserve} + F_{\mathrm{Cost}}^{generation} + F_{\mathrm{Cost}}^{DR} + F_{Emission}$		✔				
[62]	To minimize the overall costs of generation.	$F = \sum_{t=1}^{ND} \left\{ \sum_{a=1}^{A} [(AT_{at}.ut_{at} + (MTC_a + BT_{at}).pt_{at}).H/ND \right.$ $+DT_a.yt_{at} + FT_a.zt_{at}] + \sum_{b=1}^{B} [((MFC_b + CF_b).pf_{bt} + \zeta_b.dpf_{bt})$ $..H/ND + EF_b.yf_{bf} + GF_b.zf_{bt}\big] + \sum_{c=1}^{C} [(CC_c.pdc_{ct})..H/ND]$ $\left. +[BP_t.pgb_t - SP_t.pgs_t + CD.pde_t + CE.pex_t]..H/ND \right\}$		✔				
[63]	To reduce the operating costs.	$F = \sum_{s\in S} \lambda_s \left[\sum_{k\in K} \sum_{j\in J} (C_j(P_{j,k,s}) + SU_{j,k}) + \sum_{k\in K} C_{ES}.\left(V_{k,s}^{CH} + V_{k,s}^{DCH}\right) \right.$ $\left. + \sum_{k\in K} P_{k,s}^{Int,R-C-I}.C_k^{Int,R-C-I} + \sum_{k\in K} \Delta P_{k,s}^{do,R-C-I}.C_k^{DR,R-C-I} \right]$	✔					
[64]	To minimize short-term variable generation costs.	$Minimize \quad f = C_{HV}(P_{HV}, Q_{HV})$		✔				
[65]	To maximize economic benefit by integrating small CPPs, ESSs, RES, and interruptible demand loads.	$\max \sum_{t=1}^{T} \sum_{i=1}^{I} \rho_{s,t} P_{dl,i,t} - C_{dg,i,t}(P_{dg,i,t}) - C_{es,i,t}(P_{es,i,t}) - \rho_{b,t} P_{b,t}$		✔				
[66]	To provide a self-scheduling program for an SG.	$Maximize_{t=1}^{T} (\sum_{w=1}^{n_w} \pi(w). \sum_{s=1}^{N_s} \pi(s). \sum_{p=1}^{N_p} \pi(p).(\lambda^p(t).G^{wsp}(t)$ $-C_{conv}^{wsp}(t) - y_{conv}^{wsp}(t).S_{conv}))$	✔					
[67]	To maximize the SG's short-term profit.	$Max \sum_{t=1}^{T} \sum_{w=1}^{n_w} \pi_w \sum_{p=1}^{n_p} \pi_p \sum_{r^{down}=1}^{n_r^{down}} \pi_r^{down} \sum_{r^{up}=1}^{n_r^{up}} \pi_r^{up} [\lambda_p(t)(G_{wp}(t) + bm_{wp}^{down}(t).\psi_r^{down}(t)$ $-bm_{wp}^{up}(t).\psi_r^{up}(t)) - C_{wp}^c(t) - SUC^c.v_{wp}^c(t)]$	✔					
[68]	To minimize the cost, as well as the carbon emission percentage.	$Profit =$ $\max \sum_{s=1}^{N_s} \pi_s \times \sum_{t=1}^{24} \left\{ \begin{array}{c} -\sum_{p=1}^{N_p} \left\{ C_{t,p}^{chp} + C_{t,p}^{ho} \right\} \\ \left\{ \rho_{s,t}^{em} \times P_{s,t}^{grid} + \sum_{p=1}^{N_p} \left\{ \rho_t^{ret\_e} \times P_{s,t,p}^{sel} + \rho_t^h \times H_{t,p}^{tl} - C_{s,t,p}^{ens} \right\} \right. \end{array} \right\}$ $Emission = \min \sum_{s=1}^{N_s} \pi_s \times \sum_{t=1}^{24} \left\{ \begin{array}{c} \sum_{p=1}^{N_p} \left\{ E_{t,p}^{chp} + E_{t,p}^{ho} \right. \\ E_{s,t}^{grid} \end{array} \right\}$		✔		✔		

Table 3. *Cont.*

Refs.	Objective	Objective Function	Concepts Employed					
			LS	PC	VF	LS	LG	LR
[69]	To minimize of average generation cost of DG units.	$MinS_{VPP} = \frac{\sum_{i=1}^{n} P_{DG\_C,i} * v_{DG\_C,i}(P_{DG\_C,i})}{\sum_{i=1}^{n} P_{DG\_C,i}}$ $MinC_{VPP} = \sum_{i=1}^{n} P_{DG\_C,i} * v_{DG\_C,i}(P_{DG\_C,i})$		✔				
[70]	To maximize the worst-condition expected profit of SG.	$\max_{\psi M} \sum_{\omega \in \Omega} \pi_\omega \left[\sum_{t \in \tau} [\lambda^E_{t\omega} p^E_t \Delta t + \hat{\lambda}^{R}_{t\omega} + p^R_t + \hat{\lambda}^R_{t\omega} - p^R_t - \right.$ $-(C^{C,F} u^C_t + SUC^C v^{C,SU}_t + SDC^C v^{C,SD}_t)] + v$		✔				✔
[71]	To maximize the SG profit.	$Max\ Z_{\text{Profit}} = \sum_{t=1}^{T} (P^D_t.\lambda^{DSO,charge}_t + \sum_{k \in GSP} P^{Upstream}_{kt}.\lambda^{LMP}_{kt}$ $- \sum_{i \in DG} (P^{DG}_{it}.\lambda^{DG,\cos t}_i + y^{DG,start}_{it}.\lambda^{DG,start\cos t}_i + z^{DG,shut}_{it}.\lambda^{DG,shut\cos t}_i$ $- \sum_{j \in SG} P^{SG}_{jt}.\lambda^{SG,\cos t}_j - P^{FL}_t.\lambda^{FL,\cos t}_t)$		✔		✔		
[72]	To integrate EV, ESS, and wind generation for participation in the day-ahead and reserve electricity market.	$\max Y = \sum_{t=1}^{T} \left(\lambda^{DA}_t e^{DA}_t + P_{res} \lambda^{\text{Res}}_t e^{\text{Res}}_t + call_t \lambda^{\text{Res}}_t e^{\text{Res}}_t - \sum_{n=1}^{N} \cos t^{\deg}_{t,n} \right) + \sum_{n=1}^{N} (\lambda^S E^{dem}_n)$	✔					
[73]	To minimize the SG cost and emissions using day-ahead scheduling.	$Minimize\ C^{DN} = \Delta T.\sum_{t=1}^{T} \begin{pmatrix} k^P_t.P^{DN}_t.Q^{DN}_t + \sum_{i=1}^{NB} \left[c^{RES}_i.P^{RES}_{i,t} + F^{DG}_i \left(P^{DG}_{i,t} \right) \right] \\ - \sum_{k=1}^{N_{VPP}} \left(\pi^P_{k,t}.P^{VPP}_{k,t} + \pi^Q_{k,t}.Q^{VPP}_{k,t} \right) \end{pmatrix}$ $Maximize\ B^{VPP,k}_i = \Delta T.\sum_{t=1}^{T} \begin{pmatrix} U^{DR}_k \left(P^{DR}_{k,t} \right) - F^{DG}_k \left(P^{DG}_{k,t} \right) \\ -c^{RES}_k.P^{RES}_{k,t} - \left(\pi^P_{k,t}.P^{VPP}_{k,t} + \pi^Q_{k,t}.Q^{VPP}_{k,t} \right) \end{pmatrix}$		✔		✔		
[74]	To minimize the total operating cost of SG.	$Minimize \quad f = \sum_{t=1}^{T} Cost = \sum_{t=1}^{T} \begin{pmatrix} P_{Grid}(t) \times C_{Grid}(t) \\ +U_{WT}(t) \times P_{WT}(t) \times C_{WT}(t) \\ +U_{PV}(t) \times P_{PV}(t) \times C_{PV}(t) \\ +U_{FC}(t) \times P_{FC}(t) \times C_{FC}(t) \\ +U_{MT}(t) \times P_{MT}(t) \times C_{MT}(t) \\ \sum_{j=1}^{N_s} U_j(t) \times P_{Sj}(t) \times C_{Sj}(t) \\ +\sum_{i=1}^{N_g} S_{Gi} \lvert U_i(t) - U_i(t-1) \rvert \\ +\sum_{i=1}^{N_s} S_{Sj} \lvert U_j(t) - U_j(t-1) \rvert \\ -\Delta P(t) \times C_{\Delta P}(t) \end{pmatrix}$	✔					
[75]	To maximize the profit.	$Profit_{increase} = \left[\sum_{j=1}^{n} profit_j (1+i)^{-j} \right] - C_{cap}$ $= \left[\sum_{j=1}^{n} (price_{VPP} \times power_{VPP} - Income_{baseline}) \times (1+i)^{-j} \right] - C_{cap}$		✔		✔		
[76]	To minimize the generation costs.	$F = \sum_{t=1}^{T} \left(\sum_{i=1}^{N^{cpp}} F^{cpp}_{i,t} \left(P^{cpp}_{Gi,t} \right) + \sum_{j=1}^{N^{vpp}} F^{vpp}_{j,t} \left(P^{vpp}_{Gj,t} \right) \right)$		✔				
[77]	To minimize congestion based on the day-ahead scheduling of DERs.	$\min \sum_{s=1}^{N_s} p_s (\sum_{t_h=1\tau_{da}}^{H_{da}} (\sum_{i=1}^{N_{chp}} C_{f,chp,i,s}(t_h)$ $+ \sum_{i=1}^{N_{e,sto}} C_{op,e,sto,i,s}(t_h) - P_{m,da}(t_h) c_{m,da}(t_h) \tau_{da}))$ $\min \sum_{t_h=1\tau_{id}}^{H_{id}} (\sum_{i=1}^{N_{chp}} C_{f,chp,i}(t_h) + \sum_{i=1}^{N_{e,sto}} C_{op,e,sto,i}(t_h) + C_{pen,imb}(t_h))$		✔	✔			

Table 3. *Cont.*

Refs.	Objective	Objective Function	Concepts Employed					
			LS	PC	VF	LS	LG	LR
[78]	To schedule optimally using EMS, taking into account all possible types of DGs aimed toward profit and the minimization of carbon emissions.	$Maximize \quad Income = \sum_{t=1}^{T}\left[\begin{pmatrix} \sum_{L=1}^{N_L} P_{Load(L,t)}.MP_{Load(L,t)} + \sum_{M=1}^{N_M} P_{Sell(M,t)}.MP_{Sell(M,t)} + \\ \sum_{E=1}^{N_E} P_{Discharge(E,t)}.MP_{Discharge(E,t)} + \sum_{V=1}^{N_V} P_{Discharge(V,t)}.MP_{Discharge(V,t)} \end{pmatrix}.\Delta t\right]$ $Minimize \quad OperatingCost = \sum_{t=1}^{T}\left[\begin{pmatrix} \sum_{I=1}^{N_{DG}} P_{DG(I,t)}.c_{DG(I,t)} + \sum_{S=1}^{N_S} P_{Supplier(S,t)}.c_{Supplier(S,t)} + \\ \sum_{L=1}^{N_L} P_{LoadDR(L,t)}.c_{LoadDR(L,t)} + \sum_{I=1}^{N_{DG}} P_{Discharge(E,t)}.c_{Discharge(E,t)} \\ \sum_{V=1}^{N_V} P_{Discharge(V,t)}.c_{Discharge(V,t)} + \sum_{L=1}^{N_L} P_{NSD(L,t)}.c_{NSD(L,t)} \\ \sum_{I=1}^{N_{DG}} P_{GCP(I,t)}.c_{GCP(I,t)} \end{pmatrix}.\Delta t\right]$ $Minimize \quad E = \sum_{t=1}^{T}\left[\begin{pmatrix} \sum_{I=1}^{\Omega_{DG}^{t}} P_{DG(I,t)} \times E_{DG(I,t)} + \\ \sum_{S=1}^{\Omega_{SP}^{t}} P_{Supplier(S,t)} \times E_{Supplier(S,t)} \end{pmatrix}.\Delta t\right]$		✔		✔		
[79]	To minimize the operating cost of SG over 24 h.	$Minimize \;\; F = \left[\sum_{t=1}^{24}\left[\begin{pmatrix} P_{WT}(t).K_{WT}(t) + P_{PV}(t).K_{PV}(t) \\ +P_{FC}(t).K_{FC}(t) - P_{Ch}(t).K_{Ch}(t) \\ +P_{Dch}(t).K_{Dch}(t) - P_{Sp}(t).K_{Sp}(t) \\ +P_{Ns}(t).K_{Ns}(t) \end{pmatrix}\Delta t\right]\right]$			✔		✔	

Table 4. The optimization papers surveyed across DGs DSM optimization problems.

Refs.	Optimization Algorithm	DR Programs Used	Objective Function	Constraints	Decision Variables
[80]	ANN	• RTP	• Minimization of the total cost to consumers	• Battery SoC • Battery charge/discharge power	• RTP pricing values • PEV charging/discharging rate
[81]		• Peak shaving (PS)	• Minimization of transformer loading	• Transformer limits • Line current carrying capacity	• Transformer parameters • Number of EVs • EV charging power
[82]		• PS • IBR	• Minimization of energy cost • Minimization of network losses • Minimization of voltage magnitude deviation	• EV battery SoC • EV charging power • DG power balance limits	• Participating active loads • Power injected into the grid
[83]		• PS	• Maximization of revenue	• Load charge/discharge limits • EV SoC • Maximum charge/discharge power • Charging time constraints	• Charging tariff • Day-ahead forecasted prices • EV drive cycle
[84]		• ToU	• Minimization of the total cost • Maximization of revenue	• RES generation limits • DG unit operating costs	• Available tradeable power • RES generation-dependent parameters

Table 4. *Cont.*

Refs.	Optimization Algorithm	DR Programs Used	Objective Function	Constraints	Decision Variables
[85]	DP	• Load shifting (LS)	• Minimization of energy costs without sacrificing user preferences and satisfaction	• EV charge/discharge power • EV battery SoC	• RES generation parameters • Utility tariff rates
[86]	Fuzzy Logic (FL)	• LS • PS	• Minimization of total operation cost	• Power balance constraints • Spinning reserve constraints • Generator limits • Wind power penetration rate	• Fuel cost • Startup cost
[87]		• ToU • CPP • Valley filling (VF)	• Minimization of peak load demand	• EV SoC • Bus voltage limits	• EV charge/discharge time • Market pricing signals
[88]		• ToU • VF • LS	• Maximization of profit of consumers through maximum EV integration	• EV SoC • Charging preference limits consumers	• Electricity tariff • EV availability
[89]		• LS	• Minimization of generation costs, emissions, and energy losses	• Active power output limits • Generator limits • Total flexible load limits	• Flexible load operation time
[90]		• VF	• Minimization of high ramp rates in G2V mode	• EV SoC • Ramp rate limits • Wind power output limits	• EV charging current
[91]	Game Theory	• LS	• Minimization of cost for residential users	• The discharge rate of PEV	• Hourly electricity tariff • PEV energy consumption
[92]		• PS	• Minimization of energy cost	• Transmission limits • EV charge/discharge limits	• Total load demand • Cost function • Welfare function
[93]		• LS • PS	• Minimization of peak demand using distributed EV integration	• Charging outlet limits • Energy trading limits	• EV charging time • Number of participating EVs under the same aggregator

Table 4. *Cont.*

Refs.	Optimization Algorithm	DR Programs Used	Objective Function	Constraints	Decision Variables
[94]	Game Theory	• PS • ToU	• Minimization of electricity costs • Minimization of deviation between predicted and actual load demand	• EV storage limits • ESS storage limits • EV SoC limits	• EV availability • Load demand
[95]		• ToU • PS	• Minimization of the peak-to-average ratio (PAR) of the total energy demand	• Energy balance limits • PEV discharge limits • Charging/ discharging time limits	• Cost function • Load demand
[96]		• ToU	• Maximization of profits in the market environment	• EV charging limits	• Number of participating EVs • Bidding tariff
[97]		• PS • ToU • VF	• Minimization of charging the cost of EV	• Grid power limits • EV SoC limits	• Satisfaction income of EVs • Battery loss of EV • Charging cost
[98]		• PS • VF	• Minimization of energy cost • Minimization of battery degradation	• Client usage parameters	• Cost function • Residential load demand • PHEV driving behavior
[99]		• ToU • RTP	• Minimization of electricity tariff for the customers	• Hourly power demand limits • Total energy consumption limits	• Availability of EVs in the parking lot • SoC of EVs • Battery power rate • Load demand
[100]		• RTP • ToU	• Maximization of system stability • Maximization of profits	• Average power generation limits • Daily energy usage limits	• EV availability • Load demand
[101]		• RTP	• Maximization of profits of utility companies	• Charging rate limits	• Price function of utility
[102]		• RTP	• Maximization of retailer profits • Minimization of generation cost	• Charging rate limits	• Charging period of EVs • Battery charging efficiency
[103]	LP	• PS • VF	• Minimization of energy expenses of individual customer	• Charging rate limits • Battery SoC for driving cycle	• Appliance operating time • Appliance power demand

Table 4. *Cont.*

Refs.	Optimization Algorithm	DR Programs Used	Objective Function	Constraints	Decision Variables
[104]	LP	• LS • RTP • ToU	• Minimization of peak load in the distribution network • Minimization of consumer tariff	• Power limit of EV	• Availability of appliances • Power generation
[105]		• PS • VF • ToU	• Minimization of difference between peak and off-peak tariff • Minimization of EV charging cost	• Base tariff limits • Price deviation limits • EV SoC limits • EV charging power limits • Feeder baseload limits	• Electricity tariff • Operation time slot
[106]		• VF	• Maximization of EVs availability in charging • Minimization of monetary expenses	• Charging load limits • EV SoC limits	• Charging decision value/vector
[107]		• PS • ToU	• Minimization of home electricity expenses	• EV availability period	• EV demand • Electricity tariff
[108]		• RTP • ToU	• Minimization of the operation cost of EVCS and energy management system (EMS)	• Power supply constraints • ESS constraints • Heating system constraints • EV power balance limits	• Load demand • EV and ESS reserve tariff • Heating compensation prices
[109]		• PS	• Minimization of variation of the load curve	• EV SoC	• EV charging load
[110]		• PS • LS	• Maximization of revenues	• EV charging level limits • Grid power limits	• Hourly tariff • DG power generation capacity • Hourly critical load demand
[111]		• RTP • ToU	• Minimization of PAR and system costs	• PV power trade limit • EV SoC limits	• PV generation capacity • EV charging load • EV availability
[112]		• RTP	• Minimization of costs, peak charging load • Maximization of PV integration	• EV charging demand limit	• EV availability • EV charging load

Table 4. *Cont.*

Refs.	Optimization Algorithm	DR Programs Used	Objective Function	Constraints	Decision Variables
[113]	LP	• RTP	• Minimization of cost of the system	• EV charging limits • EV SoC limits	• Grid power consumption • Appliance schedule • Hourly tariff
[114]		• PS	• Minimization of operational costs and emissions	• Thermal unit limits • Power flow and grid constraints • PEV constraints • Power balance limits	• EV SoC • Thermal generation requirement
[115]		• ToU	• Minimization of the total cost for the consumer	• Power balance limits • EV SoC limits • Power transaction limits	• EV charging/discharging time • The usable capacity of EV ESS
[116]		• ToU	• Minimization of the total cost for the consumer	• EV charging limits • EV operation time limits • EV battery capacity limits	• Real-time tariff
[117]		• RTP	• Minimization of the total energy cost of a smart home	• Power balance limits • Power trading limits • EV SoC limits • PV generation limits	• PV generated power • EV availability
[118]		• PS • LS	• Minimization of individual consumer costs at lower participation levels	• EV SoC limits • ESS storage limits • DER generation limits	• Price indicators • Customer fairness index
[119]		• ToU	• Maximization of EVCS operating profits	• EV SoC limits • ESS charge/discharge power limits • Efficiency limits	• Short-term forecasted loads • Load reduction signal
[120]		• ToU • PS	• Minimization of energy cost	• EV charging limits	• Cost function • Total charging demand
[121]		• RTP • LS	• Minimization of generation costs for the customer and utility	• Shiftable load power limits • EV SoC limits	• EV availability
[122]		• PS	• Minimization of PAR of the system	• Grid power injection limits • EV SoC limits	• EV charging efficiency

Table 4. *Cont.*

Refs.	Optimization Algorithm	DR Programs Used	Objective Function	Constraints	Decision Variables
[123]	LP	• ToU	• Minimization of overall system cost	• ESS power limits • EV charge/discharge power limits	• Cost function
[124]	PSO	• ToU	• Maximization of revenues • Minimization of load fluctuation	• EV aggregator power limits • Grid power limits • EV charge/discharge power limits	• Charging tariffs from the grid • Service revenues of EV aggregator
[125]		• LS • PS	• Minimization of operating costs for the network operator	• Grid power balance limits • Bus voltage limits • Line thermal limits • EV charge/discharge limits	• EV SoC • Network power injection • DG power injection
[126]		• PS	• Minimization of fuel and startup costs	• Power balance constraints • Generation limits • Up/downtime constraints • Spinning reserve limits • EV charge/discharge power limits	• Fuel economics cost • Startup/shutdown time
[127]		• RTP	• Minimization of the load curve • Maximization of customer profit	• Power capacity and balance constraints • EV charge/discharge limits • EV charging time limits	• EV availability • The power exchanged from the grid
[128]	Evolutionary PSO	• ToU	• Minimization of system cost	• Active and reactive power generation limits • Grid voltage limits	• Power flow from the grid • EV availability
[129]	ACO	• PS	• Minimization of overall system cost	• DG generation limits • Grid power balance limits	• Cost function
[130]	GA	• PS	• Minimization of cost variance • Maximization of user satisfaction	• EV SoC limits • EV charge/discharge power limits	• EV availability • Load demand from the grid
[131]		• VF • PS	• Minimization of PAR	• EV SoC limits	• EV availability

Table 4. *Cont.*

Refs.	Optimization Algorithm	DR Programs Used	Objective Function	Constraints	Decision Variables
[132]	GA	• PS	• Minimization of PAR	• EV SoC limits	• Power demand • EV availability
[133]		• ToU	• Maximization of profit • Minimization of PAR • Minimization of variance	• EV SoC limits • EV charge/discharge power limits	• EV availability • EV charging power
[134]	Improved partheno-genetic algorithm (IPGA)	• LS	• Minimization of annual construction maintenance cost	• Grid power limits • System reliability constraints • DG and ESS penetration limits • EVCS charging power limits	• EV availability at EVCS • DG power generation capacity
[135]	Hyper-heuristic optimization	• LS	• Minimization of total cost and emission	• EV SoC limits • Electricity tariff limits	• Emissions from CPP • DG is active in the grid
[136]	DE	• PS	• Maximization of energy consumption using EV-ESS • Minimization of PAR	• EV SoC limits	• EV availability
[137]	Virus colony search (VCS) optimization	• PS	• Minimization of smart parking costs	• Upstream grid power limits • EV SoC limits • Power equilibrium limits	• Cost function
[138]	Hybrid GA and PSO	• LS • ToU	• Minimization of total tariff for customers in 24 h	• Energy balance limits	• EV availability
[139]	Model predictive control (MPC)	• PS • RTP	• Minimization of total operational cost for energy management	• Heat pump capacity limits • Heat pump thermal capacity limits • SoC of EV limit	• Heat pump generated power • EV availability • Fuel price • Natural gas price
[140]		• RTP	• Minimization of ramping requirements from power plant	• Power balance constraints • Service quality constraints of EVs	• EV charging load request vector
[141]		• RTP	• Minimization of cost of energy consumption considering EV owner preferences	• EV SoC limits	• EV SoC level • Price signal • Volume signal

Table 4. *Cont.*

Refs.	Optimization Algorithm	DR Programs Used	Objective Function	Constraints	Decision Variables
[142]	Model predictive control (MPC)	• PS	• Minimization of electricity bills and peak load	• EV SoC limits • EV charge/discharge power limits • Grid power balance limits	• Energy tariff • Capacity tariff
[143]	Nonlinear programming (NLP)	• RTP	• Maximization of total profit considering social welfare	• EVCS EV loading limits • EV SoC limits • EV BESS temperature limits	• EVCS operation time
[144]	Robust programming	• PS • LS	• Maximization of EV-V2G power integration	• Grid power balance limits • EV power trajectory limits	• EV availability
[145]	Robust mixed-integer linear programming (RMILP)	• LS	• Minimization of total operational costs and emissions	• CAES operational limits • BESS charge/discharge limits • EV SoC limits • RES generation limits	• EV availability • Grid power injection
[146]	Robust mixed-integer quadratic programming (RMIQP)	• PS • LS	• Minimization of PAR and energy cost for the users	• RES generation limits • Appliance loading limits • EV SoC limits • Power demand-supply balance limits	• Appliance operation time • Grid power exchange tariff
[147]	Stochastic programming	• RTP • PS	• Minimization of operational cost	• DG power limits • Fuel cell power limits • EV SoC limits • Grid power balance limits	• Cost of power at DG units
[148]		• ToU	• Maximization of expected profits of EV aggregator	• Bidding amount capacity limits • EV charger capacity limits	• EV charge/discharge power • Grid electricity tariff
[149]		• ToU • CPP • RTP • Incentive-based pricing	• Maximization of a parking lot profit	• EV SoC limits • Parking lot stored energy limits	• EV arrival and departure SoC

Table 4. *Cont.*

Refs.	Optimization Algorithm	DR Programs Used	Objective Function	Constraints	Decision Variables
[150]	Stochastic programming	• ToU • DLC	• Minimization of maximum transformer loading during the charging operation	• EV SoC limits • EV charge/discharge limits • Grid power balance limits	• Load demand curve • EV availability • Transformer loading capacity
[151]		• ToU • Incentive-based pricing	• Maximization of a parking lot profit	• EV SOC limits • EV battery efficiency	• EV battery capacity • Cost of degradation • Availability of EVs • EV charge/discharge tariff
[152]		• ToU	• Maximization of EV aggregation profit	• EV SoC limits	• Market electricity tariff • Spinning reserve capacity • EV availability
[153]		• ToU	• Maximization of expected profit • Minimization of risks and costs associated with DR	• Available DR limits • EV charging/discharging power limits • EV SoC limits	• Intraday price • RES generation capacity
[154]	Conditional value at risk (CVaR) function optimization	• RTP	• Minimization of EV charge/discharge cost	• EV charge/discharge rate limits • EV SoC limits • EV charging time limits	• EV charge/discharge power
[155]	CVaR-based stochastic programming	• LS	• Minimization of operation cost, emissions, and renewable power curtailment	• Active and reactive power limits • Power flow and balance limits • EV SoC limits	• Shiftable appliance schedule • EV availability
[156]	Multi-period security constraint optimal power flow (MPSOPF)	• ToU	• Minimization of generation costs, contingency costs, load-following costs, and load shedding costs	• EV SoC limits • Distributed energy resource (DER) generation limits • Load shedding and load following reserve limits	• Electricity ToU tariff • Electricity load curve
[157]	Techno-economic optimization	• ToU • CPP • RTP	• Maximization of income of distribution operator • Minimization of operational costs	• RES generation limits • Bus and line voltage limits • Available DR limits • EV SoC, efficiency, and power exchange limits	• EV energy trading tariff • Bidirectional power flow tariff • Battery depreciation cost

Table 4. *Cont.*

Refs.	Optimization Algorithm	DR Programs Used	Objective Function	Constraints	Decision Variables
[158]	Stochastic dynamic programming (SDP)	• ToU	• Minimization of customer's energy charges considering residential power demand and EV charging	• EV SoC limits • EV charger power limits • Grid power injection limits	• Time index • Residential load demand
[159]	Deep learning (DL)	• ToU • PS	• Minimization of overall vehicle energy cost	• EV SoC limits • EV charger efficiency limits	• Cost function • Real-time electricity tariff • EV availability
[160]		• ToU	• Minimization of energy costs in the real-time market	• Voltage and current limits • EV SoC limits	• Real-time electricity tariff • EV load demand
[161]	Robust adversarial reinforcement learning (RARL)	• ToU	• Minimization of customer's electricity bill considering privacy concerns	• RES generation limits • EV SoC limits	• Dynamic electricity tariff • Appliance schedule
[162]	Reinforcement learning (RL)	• PS	• Minimization of monetary and non-monetary costs in DSM	• EV battery SoC limits	• Energy prices • Load demand curve • Total cost function
[163]		• ToU	• Minimization of the load demand curve of the system	• EV SoC limits • EV charge/discharge power limits	• Charging reward function
[164]		• ToU	• Minimization of charging cost over the day-ahead time frame	• EV BESS charge/discharge time limits • EV charge/discharge rate limits	• EV availability • Real-time electricity tariff
[165]	Hierarchical reinforcement learning (HRL)	• PS	• Minimization of hydrogen consumption	• EV SoC limits • Fuel cell operation limits	• Fuel consumption • Fuel cell operation status
[166]	RL-based pursuit algorithm (PA)	• RTP • ToU	• Minimization of total energy cost	• EV SoC limits • EV charge/discharge time limits	• Reward function
[167]	Correlation optimization algorithm (COA)	• ToU	• Minimization of electricity cost of the consumers considering PV generation and ToU pricing	• PV generation limits • EV operation time limit	• Grid supply of power • Electricity price

Table 4. *Cont.*

Refs.	Optimization Algorithm	DR Programs Used	Objective Function	Constraints	Decision Variables
[168]	Market-based multi-agent system optimization	• PS	• Minimization of total operation costs	• Aggregated energy constraints • Power limit of EV fleet • EV battery capacity limits	• Cost function • Demand function
[169]	Alternating direction method of multipliers (ADMM)-based decentralized optimization algorithm	• PS	• Minimization of the load curve	• EV SoC limits • EV charging efficiency limits • EV charge rate limits • Network constraints	• EV charging load
[170]		• VF • LS • PS	• Minimization of total generation cost	• EV charging/discharging efficiency limits • EV ESS capacity limits • EV SoC limits	• Load demand curve • EV availability
[171]	Multi-EV reference and single-EV real-time response (MRS2R) online algorithm	• PS • VF	• Minimization of payment by EV customers	• EV SoC limits • EV BESS capacity limits	• EV availability
[172]	Interior point optimization	• VF	• Minimization of peak valley difference and improvement of stability	• EVCS charging/discharging time limits • Grid power limits	• Active power load • Grid bus voltage magnitude
[173]	Constrained nonlinear optimization problem with Karush–Kuhn–Tucker (KKT) conditions	• PS	• Minimization of charging the cost for EV owners	• Charging power limits • Grid power limits	• Cost function
[174]	Decision-table-based control optimization	• PS	• Maximization of economic benefits • Minimization of grid power consumption	• EV BESS SoC limits • Balancing current limits	• PV generation during daytime • SoH of BESS
[175]	Monte Carlo simulation using mixed-integer linear programming (MILP)	• ToU	• Minimization of building energy consumption	• EV charging time limitations • EV SoC limits • Energy balance limits	• Load demand curve

Table 4. *Cont.*

Refs.	Optimization Algorithm	DR Programs Used	Objective Function	Constraints	Decision Variables
[176]	Convex optimization	• VF	• Minimization of EV charging costs	• ESS charging rate constraints • ESS SoC limits	• ESS availability
[177]		• RTP • VF	• Minimization of electricity costs	• Consumer comfort limits • EV charging time constraints	• ESS availability
[178]		• PS	• Minimization of total electric energy costs	• Power balance limits • SoC limits • Home ESS SoC limits	• Number of available EVs
[179]		• RTP	• Minimization of electricity cost for the consumer	• Charge/discharge power limits • Load threshold • SoC limits	• Number of available EVs • Real-time energy tariff
[180]	Quadratic programming	• PS	• Maximization of vehicle's fuel economy	• Power flow limits • SoC limits	• Cost function
[181]	Non-intrusive load extracting (NILE) algorithm	• LS • PS	• Minimization of the daily cluster charging costs of EVs	• Power balance limits • Ramping rate limits • User comfort constraints	• EV charging power • Availability of EVs
[182]	Monte Carlo-based risk-averse charge scheduling optimization	• ToU • RTP	• Maximization of profits	• SoC limits • Charging period limits	• Electricity tariff • EV drive cycle

9. Discussion and Findings

During the systematic review of the papers as a part of the literature survey, several research gaps were identified in the present research scenario, as well as implementations in various projects across the research domain. Some of the key findings identified during the survey include:

- Most of the research papers addressed DSM formulation in the EV scenario by incorporating bidirectional power flow, but the uncertainty in demand and supply forecasting leads to inefficient control over power flow.
- The limited participation of DGs, mainly on the distribution level, restrains the individual customers, and they cannot directly participate in ancillary services and energy markets [183,184]. Clustered DGs must be able to collectively participate in the formation and maintenance of such groups in the proper sizing and architecture, which should be scalable in future implementations.
- The clustering of uncoordinated DGs, which generally operate in a decentralized setup among different utility operators, seems challenging. It is necessary to implement a proper service-oriented architecture to group together the operation and participa-

tion of different DG aggregating companies to make DG-DSM integration into the commercial markets more profitable and easier to implement on a technical front.

- The drive cycle of the EV owners, on an individual basis, has not been taken into consideration on an end-user level. The optimization of charging and discharging can be improved, to a great extent, with the personalized scheduling of EV and ESS charge/discharge operations based on the user's comfort and usage cycle.
- ICT technologies are currently implemented mainly on the transmission system operator (TSO) and DSO levels. They need to be integrated directly into the end-user location with a two-way communication channel to ensure more engaging and detailed EV and ESS charge scheduling operations. The EV and ESS can provide personalized data collected during diagnostic and data collection schedules to supply the EV aggregator with proper charge schedule data. This will allow the EV aggregator to optimally dispatch loads based on detailed SoC, SoH, BESS capacity, and drive cycle condition data.
- The customer's security and privacy are prioritized in the public domain. Consumers need to be made aware that their privacy is assured when they avail themselves of services in public locations, such as when sharing the consumer's charging location history and charging and discharging profile. The public charge scheduling setup presents the issue of DGs sending private information or erroneous data to affect grid operation and load dispatch scheduling. Even though there is research on DGs, communication strategies concerning privacy issues, their effect on DG DSM scheduling in coordination with secure communication protocols, and procedures to mitigate them have not been explored in detail.
- Meta-heuristic optimization techniques have been studied in a few research formulations, and their efficiency in forecasting the load and charge schedule of DGs in DSM operation can be exploited to a greater extent with the discovery of newer and more efficient meta-heuristic techniques. This would ensure better computation with less complexity in arriving at a proper solution.
- Consumer comfort needs to be given a higher priority in DSM operations regarding their drive cycle usage and charge/discharge patterns.
- The maximum penetration of EVs in the grid system can facilitate the better usage of RES generation, and the high capacity of EV BESS can provide ample reserves for power relaying, which are necessary in cases of intermittent generation sources. The DSM operation, in the case of DGs, ensures the maximum utilization of the BESS capacity in conjunction with RES generation.
- The centralized control architecture of DGs is necessary for setting up standards of DSM operation and charge scheduling.
- The higher penetration of DGs into the distribution grid and the DSM operation associated with them can cause problems during peak usage periods, when other factors such as voltage drops and thermal overloading of transformer equipment and cables might occur.
- Robust control and device monitoring and remote upgrade capabilities in DG DSM architecture are important, as they may facilitate further upgradation and provide better and more reliable operation and communication.
- Most DGs can be connected to the Internet through the global system for mobiles (GSM), Wi-Fi, ZigBee, and other communication networks, which aggregators can exploit and coordinate the operation thereof among constituent EVs as dispatchable loads to the distribution grid [185].
- In the DSM environment, DGs lack methodologies to maximize revenues and grid utilization. The primary reason can be attributed to the lack of policies for participating entities in wholesale electricity markets, and low priority being given to commercializing DSM due to environmental, economic, and social barriers [186,187].

10. Future Research Direction

This literature survey carefully examined the current research and the advances in the domain of EV-based DSM, and after thoughtful discussion, based on the identified research gaps, some valuable suggestions regarding future research directions and prospective areas of research are suggested:

- DG integration on a system-wide scale can be a beneficial front for the maximum utilization of intelligent loads and appliances to participate in DSM, with EVs being smart energy hubs concerning energy dispatch and storage [188,189].
- Data collection and data handling for relevant information extraction and calculation should be prioritized in the future since information gathering and processing have a significant influence on performance.
- Hybrid incentive-based and tariff-based financial models can be formulated for the optimization of load control features, such as the DSM response speed, the duration of the program, advanced alert and notification systems, geolocation sensitivity-based analysis, and real-time load monitoring rates [190–192].
- Meta-heuristic-based optimization can be hybridized, or newer, more efficient heuristic algorithms can be used for better computation in the scheduling of DSM operations. PSO, GA, wavelet transform-modified ANN, adaptive FL, support vector machine computation, and autoregressive moving average value integration with models can be implemented to obtain higher load forecast accuracy considering the regulation of loads, dispatch, scheduling, and the unit commitment problems of smart grids [193,194].
- K-map algorithms, fuzzy constrained algorithms, self-reorganizing maps, multilevel hierarchy-based clustering techniques, artificial bee colony (ABC) optimization, and an ACO can be implemented for the extraction of crucial information from aggregated load consumption profiles and in the classification of various load types in intelligent distribution systems [195].
- EV DSM models need to be more comprehensive in their operation for better practical implementation, i.e., varying charging rates, standards implemented on EVCS premises, standardized BESS swapping station methodologies, and the active participation of EVs in overall market trading and ancillary service support scenarios. More research needs to be focused on obtaining an optimized tradeoff between the performance of the system and computational complexity.
- Through data-mining and decision-making processes, diverse and hybrid optimization techniques, such as game theory and Bayesian probability theory, among others, should be explored further for internal energy dispatch, external market participation, risk evaluation, information and strategy coordination, and bidding strategy.
- The practical and easy implementation of the management of charging demand during peak/off-peak usage periods, with price-sensitive scheduling, is an excellent prospect for DSM aggregators. With large-scale EV integration into smart grids, it is a very feasible research direction to be focused upon, with an emphasis on EV charging strategies based on price response and price elasticity dynamics [196].
- Climate-based EV-DSM scheduling should be researched further, as it would affect RES generation to a large extent, and forecasting-based scheduling could help the RES to be dispatched more efficiently based on meteorological data [197].
- There is a severe lack of datasets necessary for training machine learning and deep learning models. Only five well-known EV charge scheduling datasets are available in the open research domain for researchers [198–202]. Other datasets that have been developed are available to commercial companies. More machine learning models and bio-inspired optimization techniques need to be developed to represent varying architectures and geographical locations [203].
- Big-data analysis should be emphasized to establish appropriate information to improve the perception of the energy market to bring compatibility, universality, and competitiveness.

11. Conclusions

In this review paper, existing research on DSM operations, including the various DGs and an area that has witnessed significant interest in the energy management domain in the last few years, was reviewed extensively. The general structure, operation, and optimization models of DSM and DG-DSM integration into the present smart grid scenario were discussed and represented. New concepts such as waste-to-energy were explored through a brief study, as were their implementations in test case scenarios. The optimization aspect of DG-DSM scheduling was tabulated and represented, with emphasis placed on the objective function formulation, constraints or limitations, and the selection and parameterization of decision variables. With the expectation of an increase in the adoption of various types of DG, it is estimated that DSM operations can play a valuable opportunity for the customers and utility aggregators to be active participants in the scheduling, dispatch, and market-oriented trading of energy. The research directions that this review article provides can help researchers identify potential gaps, which have been discussed previously, and they can be given due importance in finding solutions to the existing issues and challenges.

Author Contributions: S.P.: Conceptualization, Writing—Review & Editing, Writing—Original Draft. S.M.: Writing—Review & Editing. P.K.R.: Investigation, Formal Analysis, Supervision. B.K.S.: Writing, Review & Editing, Supervision. S.M.P.: Review & Editing. H.K.: Writing, Review & Editing. A.F.: Supervision. M.T.-V.: Supervision. B.A.S.: Supervision, Review & Editing. M.S.: Supervision, Review & Editing. All authors have read and agreed to the published version of the manuscript.

Funding: This research received no external funding.

Institutional Review Board Statement: Not applicable.

Informed Consent Statement: Not applicable.

Data Availability Statement: The study did not report any data.

Conflicts of Interest: The authors declare no conflict of interest.

Nomenclature

DG	Distributed Generation
DR	Demand Response
RES	Renewable Energy Sources
DERs	Distributed Energy Resources
EMS	Energy Management System
CPP	Critical Peak Pricing
V2G	Vehicle to Grid
EVCS	Electric Vehicle Charging Station
BESS	Battery Energy Storage System
PHEV	Plug-in Hybrid Electric Vehicle
PV	Photovoltaic
PEV	Plug-In Electric Vehicle
DG	Distributed Generation
DP	Dynamic Programming
PSO	Particle Swarm Optimization
GA	Genetic Algorithm
FL	Fuzzy Logic
PAR	Peak-to-Average Ratio
VCS	Virus Colony Search
NLP	Nonlinear Programming
RMIQP	Robust Mixed-Integer Quadratic Programming
DER	Distributed Energy Resource
SBP	Stochastic Dynamic Programming
RARL	Robust Adversarial Reinforcement Learning

HRL	Hierarchical Reinforcement Learning
ADMM	Alternating Direction Method of Multipliers
KKT	Karush–Kuhn–Tucker
PC	Peak Clipping
VF	Valley Filling
LG	Load Growth
$P_{grid}(h)$	Transfer of Power from the Grid to Load (kW)
$D_e(h)$	Electrical Energy Demand at Hour h (kWh)
$SoC_{min}(h)$	Minimum SoC at Hour h
$E^h{}_{batt}$	The Battery Energy at Hour h
$\mathrm{d_r}$	Load Duration
$P^{max}{}_{grid}(h)$	The Maximum Power Draw by Load from the Grid at Hour h
$B_{sj}(t)$	The Energy of jth Storage Device
P_{sj}	Power Emission from jth Storage
$Cr_t{}^g$	Cost of Renewable Energy Production
P_b	Penalty of Battery
P_H	Penalty of Hydrogen
P_{HT}	Penalty Hydride Tank
CFt^{RLB}	Cost of Reliability Operations
DSM	Demand-Side Management
EV	Electric Vehicle
SG	Smart Grid
EE	Energy Efficiency
SoC	State of Charge
SoH	State of Health
RTP	Real-time Pricing
DoD	Depth of Discharge
ISO	Independent System Operator
ADR	Automated Demand Response
UC	Unit Commitment
ANN	Artificial Neural Network
LP	Linear Programming
ACO	Ant Colony Optimization
DE	Differential Evolution
EMS	Energy Management System
IPGA	Improved Parthenogenetic Algorithm
MPC	Model Predictive Control
RMILP	Robust Mixed-Integer Linear Programming
CVaR	Conditional Value at Risk
MPSOPF	Multi-Period Security Constraint Optimal Power Flow
DL	Deep Learning
RL	Reinforcement Learning
PA	Pursuit Algorithm
MRS2R	Multi-EV Reference and Single-EV Real-time Response
MILP	Mixed Integer Linear Programming
TSO	Transmission System Operator
LS	Load Shifting
LS	Flexible Load Shifting
LR	Load Reduction
$P_{batt}(h)$	The Net Output Power of the Battery in (kW)
$SoC_{max}(h)$	Maximum SoC at Hour h
$SoC(h)$	SoC at Hour h
$P_{ch}(h)$	Power for Charging at Hour h (kW)
$P_{max}(h)$	Maximum Power at Hour h (kW)
$B_{gi}(t)$	Energy Bids of ith DG
P_{gi}	Power Generations of ith DG

$C_t{}^{g}$	Cost of Energy Production
$C_t{}^{ES-}$, $C_t{}^{ES+}$	Cost of Energy Storage Charge (+) and Discharge (−)
CFt^{OPR}	Cost of Operations
P_w	Penalty for Water Tank
2 CFt^{EMI}	Cost of Microgrid Installation

References

1. Sarker, E.; Halder, P.; Seyedmahmoudian, M.; Jamei, E.; Horan, B.; Mekhilef, S.; Stojcevski, A. Progress on the demand side management in smart grid and optimization approaches. *Int. J. Energy Res.* **2021**, *45*, 36–64. [CrossRef]
2. Ding, K.; Zhi, J. Wind power peak-valley regulation and frequency control technology. In *Large-Scale Wind Power Grid Integration;* Academic Press: Cambridge, MA, USA, 2016; pp. 211–232.
3. Maharjan, S.; Zhang, Y.; Gjessing, S.; Tsang, D.H.K. User-Centric Demand Response Management in the Smart Grid With Multiple Providers. *IEEE Trans. Emerg. Top. Comput.* **2014**, *5*, 494–505. [CrossRef]
4. Gellings, C.W. Evolving practice of demand-side management. *J. Mod. Power Syst. Clean Energy* **2017**, *5*, 1–9. [CrossRef]
5. Shayeghi, H.; Shahryari, E.; Moradzadeh, M.; Siano, P. A Survey on Microgrid Energy Management Considering Flexible Energy Sources. *Energies* **2019**, *12*, 2156. [CrossRef]
6. Deconinck, G.; Thoelen, K. Lessons From 10 Years of Demand Response Research: Smart Energy for Customers? *IEEE Syst. Man, Cybern. Mag.* **2019**, *5*, 21–30. [CrossRef]
7. Akasiadis, C.; Chalkiadakis, G. Mechanism Design for Demand-Side Management. *IEEE Intell. Syst.* **2017**, *32*, 24–31. [CrossRef]
8. Siano, P. Demand response and smart grids—A survey. *Renew. Sustain. Energy Rev.* **2014**, *30*, 461–478. [CrossRef]
9. Vojdani, A. Smart integration. *IEEE Power Energy Mag.* **2008**, *6*, 71–79. [CrossRef]
10. Mohagheghi, S.; Raji, N. Managing Industrial Energy Intelligently: Demand Response Scheme. *IEEE Ind. Appl. Mag.* **2013**, *20*, 53–62. [CrossRef]
11. Samad, T.; Koch, E.; Stluka, P. Automated Demand Response for Smart Buildings and Microgrids: The State of the Practice and Research Challenges. *Proc. IEEE* **2016**, *104*, 726–744. [CrossRef]
12. Rahman, S. Rinaldy An efficient load model for analyzing demand side management impacts. *IEEE Trans. Power Syst.* **1993**, *8*, 1219–1226. [CrossRef]
13. Cohen, A.; Wang, C. An optimization method for load management scheduling. *IEEE Trans. Power Syst.* **1988**, *3*, 612–618. [CrossRef]
14. Gellings, C. The concept of demand-side management for electric utilities. *Proc. IEEE* **1985**, *73*, 1468–1470. [CrossRef]
15. Chowdhury, J.I.; Hu, Y.; Haltas, I.; Balta-Ozkan, N.; Matthew, G.; Varga, L. Reducing industrial energy demand in the UK: A review of energy efficiency technologies and energy saving potential in selected sectors. *Renew. Sustain. Energy Rev.* **2018**, *94*, 1153–1178. [CrossRef]
16. Tronchin, L.; Manfren, M.; Nastasi, B. Energy efficiency, demand side management and energy storage technologies–A critical analysis of possible paths of integration in the built environment. *Renew. Sustain. Energy Rev.* **2018**, *95*, 341–353. [CrossRef]
17. Logenthiran, T.; Srinivasan, D.; Vanessa, K.W.M. Demand side management of smart grid: Load shifting and incentives. *J. Renew. Sustain. Energy* **2014**, *6*, 033136. [CrossRef]
18. Yang, P.; Chavali, P.; Gilboa, E.; Nehorai, A. Parallel Load Schedule Optimization with Renewable Distributed Generators in Smart Grids. *IEEE Trans. Smart Grid* **2013**, *4*, 1431–1441. [CrossRef]
19. Rebours, Y.; Kirschen, D. *What Is Spinning Reserve;* The University of Manchester: Singapore, 2005; 174p.
20. Ortega-Vazquez, M.A.; Kirschen, D.S. Estimating the Spinning Reserve Requirements in Systems With Significant Wind Power Generation Penetration. *IEEE Trans. Power Syst.* **2008**, *24*, 114–124. [CrossRef]
21. Exarchakos, L.; Leach, M.; Exarchakos, G. Modelling electricity storage systems management under the influence of demand-side management programmes. *Int. J. Energy Res.* **2009**, *33*, 62–76. [CrossRef]
22. Panda, S.; Mohanty, S.; Rout, P.K.; Sahu, B.K. A conceptual review on transformation of micro-grid to virtual power plant: Issues, modeling, solutions, and future prospects. *Int. J. Energy Res.* **2022**, *46*, 7021–7054. [CrossRef]
23. Al-Sakkari, E.G.; Habashy, M.M.; Abdelmigeed, M.O.; Mohammed, M.G. An Overview of Municipal Wastes. In *Waste-to-Energy;* Springer: Cham, Switzerland, 2022; pp. 1–20. [CrossRef]
24. Das, D.; Panda, S.; Kisku, T. A Survey on Waste to Energy Potential in Odisha. *Int. J. Innov. Res. Sci. Eng. Technol.* **2017**, *6*, 16110–16115.
25. Singh, K.; Singh, A. Behavioural modelling for personal and societal benefits of V2G/V2H integration on EV adoption. *Appl. Energy* **2022**, *319*, 119265. [CrossRef]
26. Revathi, R.; Karthikeyan, M.; Senthilnathan, N.; Chinnaiyan, V.K.; Rajesh, J.S.; Lokesh, K. Machine Learning Based Smart Energy Management for Residential Application in Grid Connected System. In Proceedings of the 2022 First International Conference on Electrical, Electronics, Information and Communication Technologies (ICEEICT), Trichy, India, 16–18 February 2022; pp. 1–6. [CrossRef]

27. Kamrani, F.; Fattaheian-Dehkordi, S.; Gholami, M.; Abbaspour, A.; Fotuhi-Firuzabad, M.; Lehtonen, M. A Two-Stage Flexibility-Oriented Stochastic Energy Management Strategy for Multi-Microgrids Considering Interaction With Gas Grid. *IEEE Trans. Eng. Manag.* **2021**, 1–14. [CrossRef]
28. Jani, A.; Karimi, H.; Jadid, S. Multi-time scale energy management of multi-microgrid systems considering energy storage systems: A multi-objective two-stage optimization framework. *J. Energy Storage* **2022**, *51*, 104554. [CrossRef]
29. Panda, S.; Mohanty, S.; Rout, P.K.; Sahu, B.K.; Bajaj, M.; Zawbaa, H.M.; Kamel, S. Residential Demand Side Management model, optimization and future perspective: A review. *Energy Rep.* **2022**, *8*, 3727–3766. [CrossRef]
30. Panda, S.; Rout, P.K.; Sahu, B.K. Demand Side Management by PV Integration to Micro-grid Power Distribution System: A Review and Case Study Analysis. *Green Technol. Smart City Soc.* **2020**, *151*, 417–432. [CrossRef]
31. Genis Mendoza, F.; Konstantopoulos, G.; Bauso, D. Online pricing for demand-side management in a low-voltage resistive micro-grid via a Stackelberg game with incentive strategies. *IET Smart Grid* **2022**, *5*, 76–89. [CrossRef]
32. Kumar, K.K.P.; Soren, N.; Latif, A.; Das, D.C.; Hussain, S.M.S.; Al-Durra, A.; Ustun, T.S. Day-Ahead DSM-Integrated Hybrid-Power-Management-Incorporated CEED of Solar Thermal/Wind/Wave/BESS System Using HFPSO. *Sustainability* **2022**, *14*, 1169. [CrossRef]
33. Jena, C.; Guerrero, J.M.; Abusorrah, A.; Al-Turki, Y.; Khan, B. Multi-Objective Generation Scheduling of Hydro-Thermal System Incorporating Energy Storage With Demand Side Management Considering Renewable Energy Uncertainties. *IEEE Access* **2022**, *10*, 52343–52357. [CrossRef]
34. Bolurian, A.; Akbari, H.; Mousavi, S. Day-ahead optimal scheduling of microgrid with considering demand side management under uncertainty. *Electr. Power Syst. Res.* **2022**, *209*, 107965. [CrossRef]
35. Teixeira, N.; Barreto, R.; Gomes, L.; Faria, P.; Vale, Z. A Trustworthy Building Energy Management System to Enable Direct IoT Devices' Participation in Demand Response Programs. *Electronics* **2022**, *11*, 897. [CrossRef]
36. Xydis, G.; Nanaki, E.A.; Koroneos, C. Low-enthalpy geothermal resources for electricity production: A demand-side management study for intelligent communities. *Energy Policy* **2013**, *62*, 118–123. [CrossRef]
37. Chandak, S.; Rout, P.K. The implementation framework of a microgrid: A review. *Int. J. Energy Res.* **2021**, *45*, 3523–3547. [CrossRef]
38. Hashmi, M.H.S.M.K.; Hänninen, S.; Mäki, K. Survey of smart grid concepts, architectures, and technological demonstrations worldwide. In Proceedings of the 2011 IEEE PES Conference on Innovative Smart Grid Technologies Latin America (ISGT LA), Medellin, Colombia, 19–21 October 2011; pp. 1–7.
39. Dranka, G.G.; Ferreira, P. Towards a smart grid power system in Brazil: Challenges and opportunities. *Energy Policy* **2020**, *136*, 111033. [CrossRef]
40. Shaukat, N.; Ali, S.; Mehmood, C.; Khan, B.; Jawad, M.; Farid, U.; Ullah, Z.; Anwar, S.; Majid, M. A survey on consumers empowerment, communication technologies, and renewable generation penetration within Smart Grid. *Renew. Sustain. Energy Rev.* **2018**, *81*, 1453–1475. [CrossRef]
41. Kanakadhurga, D.; Prabaharan, N. Demand side management in microgrid: A critical review of key issues and recent trends. *Renew. Sustain. Energy Rev.* **2022**, *156*, 111915. [CrossRef]
42. Judge, M.A.; Khan, A.; Manzoor, A.; Khattak, H.A. Overview of smart grid implementation: Frameworks, impact, performance and challenges. *J. Energy Storage* **2022**, *49*, 104056. [CrossRef]
43. Stanelyte, D.; Radziukyniene, N.; Radziukynas, V. Overview of Demand-Response Services: A Review. *Energies* **2022**, *15*, 1659. [CrossRef]
44. Mishra, S.; Saini, G.; Saha, S.; Chauhan, A.; Kumar, A.; Maity, S. A survey on multi-criterion decision parameters, integration layout, storage technologies, sizing methodologies and control strategies for integrated renewable energy system. *Sustain. Energy Technol. Assess.* **2022**, *52*, 102246. [CrossRef]
45. Chiri, G.; Bergey, M.; Mackie, T.I. Deserving but not entitled: The social construction of autism spectrum disorder in federal policy. *Soc. Sci. Med.* **2022**, *301*, 114974. [CrossRef]
46. Panda, S.; Rout, P.K.; Sahu, B.K. Residential sector demand side management: A review. In Proceedings of the 2021 1st Odisha International Conference on Electrical Power Engineering, Communication and Computing Technology (ODICON), Bhubaneswar, India, 8–9 January 2021; pp. 1–6.
47. Qiu, J.; Meng, K.; Zheng, Y.; Dong, Z.Y. Optimal scheduling of distributed energy resources as a virtual power plant in a transactive energy framework. *IET Gener. Transm. Distrib.* **2017**, *11*, 3417–3427. [CrossRef]
48. Thavlov, A.; Bindner, H.W. Utilization of Flexible Demand in a Virtual Power Plant Set-Up. *IEEE Trans. Smart Grid* **2014**, *6*, 640–647. [CrossRef]
49. Moghaddam, I.G.; Nick, M.; Fallahi, F.; Sanei, M.; Mortazavi, S. Risk-averse profit-based optimal operation strategy of a combined wind farm–cascade hydro system in an electricity market. *Renew. Energy* **2013**, *55*, 252–259. [CrossRef]
50. Pandžić, H.; Morales, J.M.; Conejo, A.J.; Kuzle, I. Offering model for a virtual power plant based on stochastic programming. *Appl. Energy* **2013**, *105*, 282–292. [CrossRef]
51. Mohammadi, J.; Rahimi-Kian, A.; Ghazizadeh, M.-S. Aggregated wind power and flexible load offering strategy. *IET Renew. Power Gener.* **2011**, *5*, 439–447. [CrossRef]
52. Pousinho, H.M.I.; Mendes, V.M.F.; Catalão, J.P.D.S. A risk-averse optimization model for trading wind energy in a market environment under uncertainty. *Energy* **2011**, *36*, 4935–4942. [CrossRef]

53. Gong, J.; Xie, D.; Jiang, C.; Zhang, Y. Multiple Objective Compromised Method for Power Management in Virtual Power Plants. *Energies* **2011**, *4*, 700. [CrossRef]
54. Dimeas, A.L.; Hatziargyriou, N.D. Agent based control of virtual power plants. In Proceedings of the 2007 International Conference on Intelligent Systems Applications to Power Systems, Kaohsiung, Taiwan, 5–8 November 2007; pp. 1–6.
55. Yousaf, W.; Asghar, E.; Meng, H.; Songyuan, Y.; Fang, F. Intelligent control method of distributed generation for power sharing in virtual power plant. In Proceedings of the 2017 IEEE International Conference on Unmanned Systems (ICUS), Beijing, China, 27–29 October 2017; pp. 576–581. [CrossRef]
56. Peik-Herfeh, M.; Seifi, H.; Sheikh-El-Eslami, M.K. Two-stage approach for optimal dispatch of distributed energy resources in distribution networks considering virtual power plant concept. *Int. Trans. Electr. Energy Syst.* **2012**, *24*, 43–63. [CrossRef]
57. Peik-Herfeh, M.; Seifi, H.; Sheikh-El-Eslami, M. Decision making of a virtual power plant under uncertainties for bidding in a day-ahead market using point estimate method. *Int. J. Electr. Power Energy Syst.* **2013**, *44*, 88–98. [CrossRef]
58. Kardakos, E.G.; Simoglou, C.K.; Bakirtzis, A.G. Optimal Offering Strategy of a Virtual Power Plant: A Stochastic Bi-Level Approach. *IEEE Trans. Smart Grid* **2015**, *7*, 794–806. [CrossRef]
59. Zdrilić, M.; Pandžić, H.; Kuzle, I. The mixed-integer linear optimization model of virtual power plant operation. In Proceedings of the 2011 8th International Conference on the European Energy Market (EEM), Zagreb, Croatia, 25–27 May 2011; pp. 467–471.
60. Rahimiyan, M.; Baringo, L. Strategic Bidding for a Virtual Power Plant in the Day-Ahead and Real-Time Markets: A Price-Taker Robust Optimization Approach. *IEEE Trans. Power Syst.* **2015**, *31*, 2676–2687. [CrossRef]
61. Shafiekhani, M.; Badri, A.; Shafie-Khah, M.; Catalão, J.P. Strategic bidding of virtual power plant in energy markets: A bi-level multi-objective approach. *Int. J. Electr. Power Energy Syst.* **2019**, *113*, 208–219. [CrossRef]
62. Pandžić, H.; Kuzle, I.; Capuder, T. Virtual power plant mid-term dispatch optimization. *Appl. Energy* **2013**, *101*, 134–141. [CrossRef]
63. Tajeddini, M.A.; Rahimi-Kian, A.; Soroudi, A. Risk averse optimal operation of a virtual power plant using two stage stochastic programming. *Energy* **2014**, *73*, 958–967. [CrossRef]
64. Liu, Y.; Xin, H.; Wang, Z.; Gan, D. Control of virtual power plant in microgrids: A coordinated approach based on photovoltaic systems and controllable loads. *IET Gener. Transm. Distrib.* **2015**, *9*, 921–928. [CrossRef]
65. Shabanzadeh, M.; Sheikh-El-Eslami, M.; Haghifam, M. Risk-based medium-term trading strategy for a virtual power plant with first-order stochastic dominance constraints. *IET Gener. Transm. Distrib.* **2017**, *11*, 520–529. [CrossRef]
66. Sowa, T.; Krengel, S.; Koopmann, S.; Nowak, J. Multi-criteria Operation Strategies of Power-to-Heat-Systems in Virtual Power Plants with a High Penetration of Renewable Energies. *Energy Procedia* **2014**, *46*, 237–245. [CrossRef]
67. Zhou, B.; Liu, X.; Cao, Y.; Li, C.; Chung, C.Y.; Chan, K.W. Optimal scheduling of virtual power plant with battery degradation cost. *IET Gener. Transm. Distrib.* **2016**, *10*, 712–725. [CrossRef]
68. Ruiz, N.; Cobelo, I.; Oyarzabal, J. A Direct Load Control Model for Virtual Power Plant Management. *IEEE Trans. Power Syst.* **2009**, *24*, 959–966. [CrossRef]
69. Pasetti, M.; Rinaldi, S.; Manerba, D. A Virtual Power Plant Architecture for the Demand-Side Management of Smart Prosumers. *Appl. Sci.* **2018**, *8*, 432. [CrossRef]
70. Kuzle, I.; Zdrilić, M.; Pandžić, H. Virtual power plant dispatch optimization using linear programming. In Proceedings of the 2011 10th International Conference on Environment and Electrical Engineering, Rome, Italy, 8–11 May 2011; pp. 1–4.
71. Morais, H.; Kadar, P.; Faria, P.; Vale, Z.; Khodr, H. Optimal scheduling of a renewable micro-grid in an isolated load area using mixed-integer linear programming. *Renew. Energy* **2010**, *35*, 151–156. [CrossRef]
72. Garcia-Gonzalez, J.; De la Muela, R.M.R.; Santos, L.M.; Gonzalez, A.M. Stochastic Joint Optimization of Wind Generation and Pumped-Storage Units in an Electricity Market. *IEEE Trans. Power Syst.* **2008**, *23*, 460–468. [CrossRef]
73. Dabbagh, S.R.; Sheikh-El-Eslami, M.K. Risk-based profit allocation to DERs integrated with a virtual power plant using cooperative Game theory. *Electr. Power Syst. Res.* **2015**, *121*, 368–378. [CrossRef]
74. Ju, L.; Tan, Z.; Yuan, J.; Tan, Q.; Li, H.; Dong, F. A bi-level stochastic scheduling optimization model for a virtual power plant connected to a wind–photovoltaic–energy storage system considering the uncertainty and demand response. *Appl. Energy* **2016**, *171*, 184–199. [CrossRef]
75. Candra, D.I.; Hartmann, K.; Nelles, M. Economic Optimal Implementation of Virtual Power Plants in the German Power Market. *Energies* **2018**, *11*, 2365. [CrossRef]
76. Zamani, A.G.; Zakariazadeh, A.; Jadid, S.; Kazemi, A. Stochastic operational scheduling of distributed energy resources in a large scale virtual power plant. *Int. J. Electr. Power Energy Syst.* **2016**, *82*, 608–620. [CrossRef]
77. Yu, J.; Feng, Q.; Li, Y.; Cao, J. Stochastic Optimal Dispatch of Virtual Power Plant considering Correlation of Distributed Generations. *Math. Probl. Eng.* **2015**, *2015*, 135673. [CrossRef]
78. Caldon, R.; Patria, A.R.; Turri, R. Optimal control of a distribution system with a virtual power plant. In Proceedings of the Bulk Power System Dynamics and Control - VI, Cortina d'Ampezzo, Italy, 22–27 August 2004; pp. 278–284.
79. Chen, G.; Li, J. A fully distributed ADMM-based dispatch approach for virtual power plant problems. *Appl. Math. Model.* **2018**, *58*, 300–312. [CrossRef]
80. Zong, X.; Wang, H.; He, X. A neurodynamic algorithm to optimize residential demand response problem of plug-in electric vehicle. *Neurocomputing* **2020**, *405*, 1–11. [CrossRef]

81. Ferdowsi, M.; Monti, A.; Ponci, F.; Fathi, G. Demand side management verification system for electric vehicles. In Proceedings of the 2014 IEEE International Workshop on Applied Measurements for Power Systems Proceedings (AMPS), Aachen, Germany, 24–26 September 2014; pp. 1–6.
82. Bakhshinejad, A.; Tavakoli, A.; Moghaddam, M.M. Modeling and simultaneous management of electric vehicle penetration and demand response to improve distribution network performance. *Electr. Eng.* **2021**, *103*, 325–340. [CrossRef]
83. Lin, J.; Dong, P.; Liu, M.; Huang, X.; Deng, W. Research on Demand Response of Electric Vehicle Agents Based on Multi-Layer Machine Learning Algorithm. *IEEE Access* **2020**, *8*, 224224–224234. [CrossRef]
84. Zeynali, S.; Rostami, N.; Ahmadian, A.; Elkamel, A. Stochastic energy management of an electricity retailer with a novel plug-in electric vehicle-based demand response program and energy storage system: A linearized battery degradation cost model. *Sustain. Cities Soc.* **2021**, *74*, 103154. [CrossRef]
85. Korkas, C.D.; Terzopoulos, M.; Tsaknakis, C.; Kosmatopoulos, E.B. Nearly optimal demand side management for energy, thermal, EV and storage loads: An Approximate Dynamic Programming approach for smarter buildings. *Energy Build.* **2022**, *255*, 111676. [CrossRef]
86. Zhang, N.; Hu, Z.; Han, X.; Zhang, J.; Zhou, Y. A fuzzy chance-constrained program for unit commitment problem considering demand response, electric vehicle and wind power. *Int. J. Electr. Power Energy Syst.* **2015**, *65*, 201–209. [CrossRef]
87. Faddel, S.G.; Mohammed, O.A. Automated Distributed Electric Vehicle Controller for Residential Demand Side Management. *IEEE Trans. Ind. Appl.* **2018**, *55*, 16–25. [CrossRef]
88. Vujasinovic, J.; Savic, G. Demand Side Management and Integration of a Renewable Sources Powered Station for Electric Vehicle Charging into a Smart Grid. In Proceedings of the 2021 International Conference on Applied and Theoretical Electricity (ICATE), Craiova, Romania, 27–29 May 2021; pp. 1–6. [CrossRef]
89. Narimani, M.R.; Maigha; Joo, J.-Y.; Crow, M. Multi-Objective Dynamic Economic Dispatch with Demand Side Management of Residential Loads and Electric Vehicles. *Energies* **2017**, *10*, 624. [CrossRef]
90. Raoofat, M.; Saad, M.; Lefebvre, S.; Asber, D.; Mehrjedri, H.; Lenoir, L. Wind power smoothing using demand response of electric vehicles. *Int. J. Electr. Power Energy Syst.* **2018**, *99*, 164–174. [CrossRef]
91. Liu, X.; Gao, B.; Wu, C.; Tang, Y. Bayesian game-theoretic energy management for residential users in smart grid. In Proceedings of the 2016 IEEE International Conference on Cyber Technology in Automation, Control, and Intelligent Systems (CYBER), Chengdu, China, 19–22 June 2016; pp. 67–71. [CrossRef]
92. Xu, N.Z.; Chung, C.Y. Challenges in Future Competition of Electric Vehicle Charging Management and Solutions. *IEEE Trans. Smart Grid* **2014**, *6*, 1323–1331. [CrossRef]
93. Rassaei, F.; Soh, W.-S.; Chua, K.-C. Demand Response for Residential Electric Vehicles With Random Usage Patterns in Smart Grids. *IEEE Trans. Sustain. Energy* **2015**, *6*, 1367–1376. [CrossRef]
94. Tushar, M.H.K.; Zeineddine, A.W.; Assi, C.M. Demand-Side Management by Regulating Charging and Discharging of the EV, ESS, and Utilizing Renewable Energy. *IEEE Trans. Ind. Inform.* **2017**, *14*, 117–126. [CrossRef]
95. Liu, X.; Gao, B.; Wu, C.; Tang, Y. Demand-Side Management With Household Plug-In Electric Vehicles: A Bayesian Game-Theoretic Approach. *IEEE Syst. J.* **2017**, *12*, 2894–2904. [CrossRef]
96. Kong, F.; Liu, X. Distributed deadline and renewable aware electric vehicle demand response in the smart grid. In Proceedings of the 2015 IEEE Real-Time Systems Symposium, San Antonio, TX, USA, 1–4 December 2015; pp. 23–32.
97. Zheng, Y.; Luo, J.; Yang, X.; Yang, Y. Intelligent Regulation on Demand Response for Electric Vehicle Charging: A Dynamic Game Method. *IEEE Access* **2020**, *8*, 66105–66115. [CrossRef]
98. Shokri, M.; Kebriaei, H. Mean Field Optimal Energy Management of Plug-In Hybrid Electric Vehicles. *IEEE Trans. Veh. Technol.* **2018**, *68*, 113–120. [CrossRef]
99. Salyani, P.; Abapour, M.; Zare, K. Stackelberg based optimal planning of DGs and electric vehicle parking lot by implementing demand response program. *Sustain. Cities Soc.* **2019**, *51*, 101743. [CrossRef]
100. Azimian, B.; Fijani, R.F.; Ghotbi, E.; Wang, X. Stackelberg game approach on modeling of supply demand behavior considering BEV uncertainty. In Proceedings of the 2018 IEEE International Conference on Probabilistic Methods Applied to Power Systems (PMAPS), Boise, ID, USA, 24–28 June 2018; pp. 1–6.
101. Shinde, P.; Swarup, K.S. Stackelberg game-based demand response in multiple utility environments for electric vehicle charging. *IET Electr. Syst. Transp.* **2018**, *8*, 167–174. [CrossRef]
102. Yoon, S.-G.; Choi, Y.-J.; Park, J.-K.; Bahk, S. Stackelberg-Game-Based Demand Response for At-Home Electric Vehicle Charging. *IEEE Trans. Veh. Technol.* **2015**, *65*, 4172–4184. [CrossRef]
103. Safdarian, A.; Fotuhi-Firuzabad, M.; Lehtonen, M. A Distributed Algorithm for Managing Residential Demand Response in Smart Grids. *IEEE Trans. Ind. Inform.* **2014**, *10*, 2385–2393. [CrossRef]
104. Gang, J.; Lin, X. A novel demand side management strategy on electric vehicle charging behavior. In Proceedings of the 2018 IEEE 15th International Conference on Networking, Sensing and Control (ICNSC), Zhuhai, China, 27–29 March 2018; pp. 1–5. [CrossRef]
105. Astaneh, M.F.; Bhattarai, B.P.; Bak-Jensen, B.; Hu, W.; Pillai, J.R.; Chen, Z. A novel technique to enhance demand responsiveness: An EV based test case. In Proceedings of the 2015 IEEE PES Asia-Pacific Power and Energy Engineering Conference (APPEEC), Brisbane, Australia, 15–18 November 2015; pp. 1–5.

106. Yao, L.; Lim, W.H.; Tsai, T.S. A Real-Time Charging Scheme for Demand Response in Electric Vehicle Parking Station. *IEEE Trans. Smart Grid* **2016**, *8*, 52–62. [CrossRef]
107. Wang, Z.; Paranjape, R. An Evaluation of Electric Vehicle Penetration under Demand Response in a Multi-Agent Based Simulation. In Proceedings of the 2014 IEEE Electrical Power and Energy Conference, Washington, DC, USA, 12–14 November 2014; pp. 220–225. [CrossRef]
108. Li, Y.; Han, M.; Yang, Z.; Li, G. Coordinating Flexible Demand Response and Renewable Uncertainties for Scheduling of Community Integrated Energy Systems with an Electric Vehicle Charging Station: A Bi-level Approach. *IEEE Trans. Sustain. Energy* **2021**, *12*, 2321–2331. [CrossRef]
109. Mou, Y.; Xing, H.; Lin, Z.; Fu, M. Decentralized Optimal Demand-Side Management for PHEV Charging in a Smart Grid. *IEEE Trans. Smart Grid* **2014**, *6*, 726–736. [CrossRef]
110. Battistelli, C. Demand Response program for electric vehicle service with physical aggregators. In Proceedings of the IEEE PES ISGT Europe 2013, Copenhagen, Denmark, 6–9 October 2013; pp. 1–5. [CrossRef]
111. Cheng, P.-H.; Huang, T.-H.; Chien, Y.-W.; Wu, C.-L.; Tai, C.-S.; Fu, L.-C. Demand-side management in residential community realizing sharing economy with bidirectional PEV while additionally considering commercial area. *Int. J. Electr. Power Energy Syst.* **2019**, *116*, 105512. [CrossRef]
112. Chen, Q.; Wang, F.; Hodge, B.M.; Zhang, J.; Li, Z.; Shafie-Khah, M.; Catalão, J.P. Dynamic price vector formation model-based automatic demand response strategy for PV-assisted EV charging stations. *IEEE Trans. Smart Grid* **2017**, *8*, 2903–2915. [CrossRef]
113. Luo, X.; Zhu, X.; Lim, E.G. Dynamic pricing based and electric vehicle assisted demand response strategy. In Proceedings of the 2017 IEEE International Conference on Smart Grid Communications (SmartGridComm), Dresden, Germany, 23–26 October 2017; pp. 357–362.
114. Guo, Q.; Liang, X.; Xie, D.; Jermsittiparsert, K. Efficient integration of demand response and plug-in electrical vehicle in microgrid: Environmental and economic assessment. *J. Clean. Prod.* **2021**, *291*, 125581. [CrossRef]
115. Pal, S.; Kumar, R. Electric Vehicle Scheduling Strategy in Residential Demand Response Programs with Neighbor Connection. *IEEE Trans. Ind. Inform.* **2017**, *14*, 980–988. [CrossRef]
116. Luo, X.; Zhu, X.; Lim, E.; Kellerer, W. Electric Vehicles Assisted Multi-Household Cooperative Demand Response Strategy. In Proceedings of the 2019 IEEE 89th Vehicular Technology Conference (VTC2019-Spring), Kuala Lumpur, Malaysia, 28 April–1 May 2019; pp. 1–5. [CrossRef]
117. Chandra, L.; Chanana, S. Energy management of smart homes with energy storage, rooftop PV and electric vehicle. In Proceedings of the 2018 IEEE International Students' Conference on Electrical, Electronics and Computer Science (SCEECS), Bhopal, India, 24–25 February 2018; pp. 1–6.
118. Chen, Y.-W.; Chang, J. Fair Demand Response With Electric Vehicles for the Cloud Based Energy Management Service. *IEEE Trans. Smart Grid* **2016**, *9*, 458–468. [CrossRef]
119. Upadhaya, B.; Feng, D.; Zhou, Y.; Gui, Q.; Zhao, X.; Wu, D. Optimal decision making model of battery energy storage-assisted electric vehicle charging station considering incentive demand response. In Proceedings of the 8th Renewable Power Generation Conference (RPG 2019), Shanghai, China, 24–25 October 2019. [CrossRef]
120. Zhang, L.; Li, Y. Optimal Management for Parking-Lot Electric Vehicle Charging by Two-Stage Approximate Dynamic Programming. *IEEE Trans. Smart Grid* **2015**, *8*, 1722–1730. [CrossRef]
121. Pal, S.; Kumar, M.; Kumar, R. Price aware residential demand response with renewable sources and electric vehicle. In Proceedings of the 2017 IEEE International WIE Conference on Electrical and Computer Engineering (WIECON-ECE), Dehradun, India, 18–19 December 2017; pp. 211–214.
122. Sengor, I.; Guner, S.; Erdinc, O. Real-Time Algorithm Based Intelligent EV Parking Lot Charging Management Strategy Providing PLL Type Demand Response Program. *IEEE Trans. Sustain. Energy* **2020**, *12*, 1256–1264. [CrossRef]
123. Agrawal, S.; Patne, N.R.; Jadhav, A.M. Study of storage and demand response coordination for electric vehicle charging station by using ESS. *JREAS* **2018**, *3*, 57–62.
124. Hou, H.; Wang, Y.; Xie, C.; Xiong, B.; Zhang, Q.; Huang, L. A dispatching strategy for electric vehicle aggregator combined price and incentive demand response. *IET Energy Syst. Integr.* **2021**, *3*, 508–519. [CrossRef]
125. Soares, J.; Morais, H.; Sousa, T.; Vale, Z.; Faria, P. Day-Ahead Resource Scheduling Including Demand Response for Electric Vehicles. *IEEE Trans. Smart Grid* **2013**, *4*, 596–605. [CrossRef]
126. Wang, Y.; Yang, Z.; Mourshed, M.; Guo, Y.; Niu, Q.; Zhu, X. Demand side management of plug-in electric vehicles and coordinated unit commitment: A novel parallel competitive swarm optimization method. *Energy Convers. Manag.* **2019**, *196*, 935–949. [CrossRef]
127. Xu, K.; Ren, S.; Xie, B.; Dou, X.; Zhang, P. Optimization Strategy of Electric Vehicle Charging Based on Electricity Price Demand Response. In Proceedings of the 2020 12th IEEE PES Asia-Pacific Power and Energy Engineering Conference (APPEEC), Nanjing, China, 20–23 September 2020; pp. 1–5. [CrossRef]
128. Gomes, P.V.; Saraiva, J.T. Transmission system planning considering solar distributed generation penetration. In Proceedings of the 2017 14th International Conference on the European Energy Market (EEM), Dresden, Germany, 6–9 June 2017; pp. 1–6. [CrossRef]
129. Rotering, N.; Kellermann, J.; Moser, A. Algorithm for Simultaneous Medium Voltage Grid Planning and Electric Vehicle Scheduling. *IEEE Trans. Smart Grid* **2021**, *12*, 3305–3313. [CrossRef]

130. Rasheed, M.B.; Awais, M.; Alquthami, T.; Khan, I. An Optimal Scheduling and Distributed Pricing Mechanism for Multi-Region Electric Vehicle Charging in Smart Grid. *IEEE Access* **2020**, *8*, 40298–40312. [CrossRef]
131. Liu, H.; Zeng, P.; Guo, J.; Wu, H.; Ge, S. An optimization strategy of controlled electric vehicle charging considering demand side response and regional wind and photovoltaic. *J. Mod. Power Syst. Clean Energy* **2015**, *3*, 232–239. [CrossRef]
132. Lee, J.; Park, G.L. Genetic algorithm-based demand response scheme for electric vehicle charging. *Int. J. Intell. Inf. Database Syst.* **2013**, *7*, 535–549. [CrossRef]
133. Li, T.; Tao, S.; He, K.; Fan, H.; Zhang, Y.; Sun, Y. Multi-objective Optimal Dispatching of Electric Vehicle Cluster Considering User Demand Response. In Proceedings of the 2021 IEEE 4th International Conference on Electronics Technology (ICET), Chengdu, China, 7–10 May 2021; pp. 1003–1008. [CrossRef]
134. Li, X.; Song, Y.; Shan, W. Joint Planning of Distributed Generation, Electric Vehicle Charging Station, and Active Distribution Network Framework. In Proceedings of the 2019 IEEE 14th International Conference on Intelligent Systems and Knowledge Engineering (ISKE), Dalian, China, 14–16 November 2019; pp. 470–475.
135. Ramachandran, B.; Ramanathan, A. Decentralized demand side management and control of PEVs connected to a smart grid. In Proceedings of the Power Systems Conference (PSC), Clemson, SC, USA, 10–13 March 2015; pp. 1–7. [CrossRef]
136. Galván-López, E.; Curran, T.; McDermott, J.; Carroll, P. Design of an autonomous intelligent Demand-Side Management system using stochastic optimisation evolutionary algorithms. *Neurocomputing* **2015**, *170*, 270–285. [CrossRef]
137. Hao, D.; Ren, X.; Mohammed, A.S. Optimal Design of Electric Vehicle Parking Lot based on Energy Management Considering Hydrogen Storage System and Demand Side Management. *J. Energy Storage* **2021**, *42*, 103045. [CrossRef]
138. Pal, A.; Chatterjee, S.; Bhattacharya, A.; Chakraborty, A.K. Optimal Design of Microgrid with Demand Side Management in Presence of Electric Vehicle. In Proceedings of the 2020 IEEE First International Conference on Smart Technologies for Power, Energy and Control (STPEC), Nagpur, India, 25–26 September 2020; pp. 1–6. [CrossRef]
139. Ji, Z.; Huang, X.; Xu, C.; Sun, H. Accelerated Model Predictive Control for Electric Vehicle Integrated Microgrid Energy Management: A Hybrid Robust and Stochastic Approach. *Energies* **2016**, *9*, 973. [CrossRef]
140. Khatami, R.; Parvania, M.; Bagherinezhad, A. Continuous-time Model Predictive Control for Real-time Flexibility Scheduling of Plugin Electric Vehicles. *IFAC-PapersOnLine* **2018**, *51*, 498–503. [CrossRef]
141. Di Giorgio, A.; Liberati, F.; Canale, S. Electric vehicles charging control in a smart grid: A model predictive control approach. *Control Eng. Pract.* **2014**, *22*, 147–162. [CrossRef]
142. Van Kriekinge, G.; De Cauwer, C.; Sapountzoglou, N.; Coosemans, T.; Messagie, M. Peak shaving and cost minimization using model predictive control for uni- and bi-directional charging of electric vehicles. *Energy Rep.* **2021**, *7*, 8760–8771. [CrossRef]
143. Mehrabi, A.; Nunna, H.S.V.S.K.; Dadlani, A.; Moon, S.; Kim, K. Decentralized Greedy-Based Algorithm for Smart Energy Management in Plug-in Electric Vehicle Energy Distribution Systems. *IEEE Access* **2020**, *8*, 75666–75681. [CrossRef]
144. Bashash, S.; Fathy, H.K. Robust demand-side plug-in electric vehicle load control for renewable energy management. In Proceedings of the 2011 American Control Conference, San Francisco, CA, USA, 29 June–1 July 2011; pp. 929–934. [CrossRef]
145. Zeynali, S.; Rostami, N.; Ahmadian, A.; Elkamel, A. Robust multi-objective thermal and electrical energy hub management integrating hybrid battery-compressed air energy storage systems and plug-in-electric-vehicle-based demand response. *J. Energy Storage* **2021**, *35*, 102265. [CrossRef]
146. Hosseini, S.M.; Carli, R.; Dotoli, M. Robust Optimal Energy Management of a Residential Microgrid Under Uncertainties on Demand and Renewable Power Generation. *IEEE Trans. Autom. Sci. Eng.* **2020**, *18*, 618–637. [CrossRef]
147. Farsangi, A.S.; Hadayeghparast, S.; Mehdinejad, M.; Shayanfar, H. A novel stochastic energy management of a microgrid with various types of distributed energy resources in presence of demand response programs. *Energy* **2018**, *160*, 257–274. [CrossRef]
148. Zheng, Y.; Yu, H.; Shao, Z.; Jian, L. Day-ahead bidding strategy for electric vehicle aggregator enabling multiple agent modes in uncertain electricity markets. *Appl. Energy* **2020**, *280*, 115977. [CrossRef]
149. Shafie-Khah, M.; Gil, F.; Catalao, J.; Aghaei, J.; Barani, M. Impacts of participating in different demand response programs on operational behavior of plug-in electric vehicle parking lots. In Proceedings of the 2015 International Symposium on Smart Electric Distribution Systems and Technologies (EDST), Vienna, Austria, 7–11 September 2015; pp. 99–104. [CrossRef]
150. Cao, C.; Wang, L.; Chen, B. Mitigation of the Impact of High Plug-in Electric Vehicle Penetration on Residential Distribution Grid Using Smart Charging Strategies. *Energies* **2016**, *9*, 1024. [CrossRef]
151. Shafie-Khah, M.; Heydarian-Forushani, E.; Osório, G.J.; Gil, F.A.; Aghaei, J.; Barani, M.; Catalão, J.P. Optimal behavior of electric vehicle parking lots as demand response aggregation agents. *IEEE Trans. Smart Grid* **2015**, *7*, 2654–2665. [CrossRef]
152. Shafie-Khah, M.; Heydarian-Forushani, E.; Golshan ME, H.; Siano, P.; Moghaddam, M.P.; Sheikh-El-Eslami, M.K.; Catalão JP, S. Optimal trading of plug-in electric vehicle aggregation agents in a market environment for sustainability. *Appl. Energy* **2016**, *162*, 601–612. [CrossRef]
153. Afzali, P.; Rashidinejad, M.; Abdollahi, A.; Bakhshai, A. Risk-Constrained Bidding Strategy for Demand Response, Green Energy Resources, and Plug-In Electric Vehicle in a Flexible Smart Grid. *IEEE Syst. J.* **2020**, *15*, 338–345. [CrossRef]
154. Wang, F.; Zheng, T.; Sun, B.; Gao, J.; Yang, Y. Based on regional integrated energy automatic demand response optimization scheduling of electric vehicle. *IOP Conf. Ser. Earth Environ. Sci.* **2021**, *675*, 012116. [CrossRef]
155. Guo, Q.; Nojavan, S.; Lei, S.; Liang, X. Economic-environmental analysis of renewable-based microgrid under a CVaR-based two-stage stochastic model with efficient integration of plug-in electric vehicle and demand response. *Sustain. Cities Soc.* **2021**, *75*, 103276. [CrossRef]

156. Jeon, W.; Cho, S.; Lee, S. Estimating the Impact of Electric Vehicle Demand Response Programs in a Grid with Varying Levels of Renewable Energy Sources: Time-of-Use Tariff versus Smart Charging. *Energies* **2020**, *13*, 4365. [CrossRef]
157. Sadati, S.M.B.; Moshtagh, J.; Shafie-khah, M.; Catalão, J.P. Smart distribution system operational scheduling considering electric vehicle parking lot and demand response programs. *Electr. Power Syst. Res.* **2018**, *160*, 404–418. [CrossRef]
158. Wu, X.; Hu, X.; Moura, S.; Yin, X.; Pickert, V. Stochastic control of smart home energy management with plug-in electric vehicle battery energy storage and photovoltaic array. *J. Power Sources* **2016**, *333*, 203–212. [CrossRef]
159. Lopez, K.L.; Gagne, C.; Gardner, M.-A. Demand-Side Management Using Deep Learning for Smart Charging of Electric Vehicles. *IEEE Trans. Smart Grid* **2018**, *10*, 2683–2691. [CrossRef]
160. Jahangir, H.; Gougheri, S.S.; Vatandoust, B.; Golkar, M.A.; Ahmadian, A.; Hajizadeh, A. Plug-in Electric Vehicle Behavior Modeling in Energy Market: A Novel Deep Learning-Based Approach with Clustering Technique. *IEEE Trans. Smart Grid* **2020**, *11*, 4738–4748. [CrossRef]
161. Reka, S.S.; Venugopal, P.; Alhelou, H.H.; Siano, P.; Golshan, M.E.H. Real Time Demand Response Modeling for Residential Consumers in Smart Grid Considering Renewable Energy with Deep Learning Approach. *IEEE Access* **2021**, *9*, 56551–56562. [CrossRef]
162. Sheikhi, A.; Rayati, M.; Ranjbar, A.M. Demand side management for a residential customer in multi-energy systems. *Sustain. Cities Soc.* **2016**, *22*, 63–77. [CrossRef]
163. Jiang, C.; Jing, Z.; Cui, X.; Ji, T.; Wu, Q. Multiple agents and reinforcement learning for modelling charging loads of electric taxis. *Appl. Energy* **2018**, *222*, 158–168. [CrossRef]
164. Chis, A.; Lunden, J.; Koivunen, V. Reinforcement Learning-Based Plug-in Electric Vehicle Charging with Forecasted Price. *IEEE Trans. Veh. Technol.* **2016**, *66*, 3674–3684. [CrossRef]
165. Yuan, J.; Yang, L.; Chen, Q. Intelligent energy management strategy based on hierarchical approximate global optimization for plug-in fuel cell hybrid electric vehicles. *Int. J. Hydrogen Energy* **2018**, *43*, 8063–8078. [CrossRef]
166. Arif, A.I.; Babar, M.; Ahamed, T.P.I.; Al-Ammar, E.A.; Nguyen, P.H.; Kamphuis, I.G.R.; Malik, N.H. Online scheduling of plug-in vehicles in dynamic pricing schemes. *Sustain. Energy Grids Netw.* **2016**, *7*, 25–36. [CrossRef]
167. Wang, J.; Yang, Y.; He, W. A research of the strategy of electric vehicle ordered charging based on the demand side response. In Proceedings of the 2015 6th International Conference on Power Electronics Systems and Applications (PESA), Hong Kong, China, 15–17 December 2015; pp. 1–5. [CrossRef]
168. Vandael, S.; Claessens, B.; Hommelberg, M.; Holvoet, T.; Deconinck, G. A Scalable Three-Step Approach for Demand Side Management of Plug-in Hybrid Vehicles. *IEEE Trans. Smart Grid* **2012**, *4*, 720–728. [CrossRef]
169. Li, Z.; Guo, Q.; Sun, H.; Su, H. ADMM-based decentralized demand response method in electric vehicle virtual power plant. In Proceedings of the 2016 IEEE Power and Energy Society General Meeting (PESGM), Boston, MA, USA, 17–21 July 2016; pp. 1–5. [CrossRef]
170. Tan, Z.; Yang, P.; Nehorai, A. An Optimal and Distributed Demand Response Strategy With Electric Vehicles in the Smart Grid. *IEEE Trans. Smart Grid* **2014**, *5*, 861–869. [CrossRef]
171. Zhou, B.; Yao, F.; Littler, T.; Zhang, H. An electric vehicle dispatch module for demand-side energy participation. *Appl. Energy* **2016**, *177*, 464–474. [CrossRef]
172. Gan, H.; Zheng, C. An electric vehicle operation optimization method based on demand-side management. *Concurr. Comput. Pract. Exp.* **2019**, *32*, 5532. [CrossRef]
173. Rahbari-Asr, N.; Chow, M.-Y. Cooperative Distributed Demand Management for Community Charging of PHEV/PEVs Based on KKT Conditions and Consensus Networks. *IEEE Trans. Ind. Inform.* **2014**, *10*, 1907–1916. [CrossRef]
174. Tong, S.; Fung, T.; Klein, M.P.; Weisbach, D.A.; Park, J.W. Demonstration of reusing electric vehicle battery for solar energy storage and demand side management. *J. Energy Storage* **2017**, *11*, 200–210. [CrossRef]
175. Thomas, D.; Ioakimidis, C.S.; Klonari, V.; Vallée, F.; Deblecker, O. Effect of electric vehicles' optimal charging-discharging schedule on a building's electricity cost demand considering low voltage network constraints. In Proceedings of the 2016 IEEE PES Innovative Smart Grid Technologies Conference Europe (ISGT-Europe), Ljubljana, Slovenia, 9–12 October 2016; pp. 1–6.
176. Chen, N.; Tan, C.W.; Quek, T.Q.S. Electric Vehicle Charging in Smart Grid: Optimality and Valley-Filling Algorithms. *IEEE J. Sel. Top. Signal Process.* **2014**, *8*, 1073–1083. [CrossRef]
177. Wang, Z.; Paranjape, R. Evaluation of electric vehicle penetration in a residential sector under demand response considering both cost and convenience. In Proceedings of the 2017 IEEE Electrical Power and Energy Conference (EPEC), Saskatoon, SK, Canada, 22–25 October 2017; pp. 1–5. [CrossRef]
178. Wu, X.; Hu, X.; Teng, Y.; Qian, S.; Cheng, R. Optimal integration of a hybrid solar-battery power source into smart home nanogrid with plug-in electric vehicle. *J. Power Sources* **2017**, *363*, 277–283. [CrossRef]
179. Wang, Z.; Paranjape, R. Optimal scheduling algorithm for charging electric vehicle in a residential sector under demand response. In Proceedings of the 2015 IEEE Electrical Power and Energy Conference (EPEC), London, ON, Canada, 24–26 October 2015; pp. 45–49. [CrossRef]
180. Kessels, J.; Koot, M.; De Jager, B.; Bosch, P.V.D.; Aneke, N.; Kok, D. Energy Management for the Electric Powernet in Vehicles With a Conventional Drivetrain. *IEEE Trans. Control Syst. Technol.* **2007**, *15*, 494–505. [CrossRef]
181. Zhao, H.; Yan, X.; Ren, H. Quantifying flexibility of residential electric vehicle charging loads using non-intrusive load extracting algorithm in demand response. *Sustain. Cities Soc.* **2019**, *50*, 101664. [CrossRef]

182. Sharma, S.; Jain, P. Risk-averse integrated demand response and dynamic G2V charge scheduling of an electric vehicle aggregator to support grid stability. *Int. Trans. Electr. Energy Syst.* **2021**, *31*, e12867. [CrossRef]
183. Mukherjee, J.C.; Gupta, A. A Review of Charge Scheduling of Electric Vehicles in Smart Grid. *IEEE Syst. J.* **2014**, *9*, 1541–1553. [CrossRef]
184. Jin, R.; Wang, B.; Zhang, P.; Luh, P.B. Decentralised online charging scheduling for large populations of electric vehicles: A cyber-physical system approach. *Int. J. Parallel Emergent Distrib. Syst.* **2013**, *28*, 29–45. [CrossRef]
185. Falvo, M.C.; Graditi, G.; Siano, P. Electric Vehicles integration in demand response programs. In Proceedings of the 2014 International Symposium on Power Electronics, Electrical Drives, Automation and Motion, Ischia, Italy, 18–20 June 2014; pp. 548–553. [CrossRef]
186. Warren, P. A review of demand-side management policy in the UK. *Renew. Sustain. Energy Rev.* **2014**, *29*, 941–951. [CrossRef]
187. Harish, V.S.K.V.; Kumar, A. Demand side management in India: Action plan, policies and regulations. *Renew. Sustain. Energy Rev.* **2014**, *33*, 613–624. [CrossRef]
188. Mohanty, S.; Panda, S.; Sahu, B.K.; Rout, P.K. A Genetic Algorithm-Based Demand Side Management Program for Implementation of Virtual Power Plant Integrating Distributed Energy Resources. In *Innovation in Electrical Power Engineering, Communication, and Computing Technology*; Springer: Singapore, 2021; pp. 469–481. [CrossRef]
189. Bahloul, M.; Breathnach, L.; Cotter, J.; Daoud, M.; Saif, A.; Khadem, S. Role of Aggregator in Coordinating Residential Virtual Power Plant in "StoreNet": A Pilot Project Case Study. *IEEE Trans. Sustain. Energy* **2022**, 1–10. [CrossRef]
190. Yesilbudak, M.; Colak, A. Integration challenges and solutions for renewable energy sources, electric vehicles and demand-side initiatives in smart grids. In Proceedings of the 2018 7th International Conference on Renewable Energy Research and Applications (ICRERA), Paris, France, 18–21 September 2018; pp. 1407–1412.
191. Behrangrad, M. A review of demand side management business models in the electricity market. *Renew. Sustain. Energy Rev.* **2015**, *47*, 270–283. [CrossRef]
192. Shariatzadeh, F.; Mandal, P.; Srivastava, A.K. Demand response for sustainable energy systems: A review, application and implementation strategy. *Renew. Sustain. Energy Rev.* **2015**, *45*, 343–350. [CrossRef]
193. Khan, A.R.; Mahmood, A.; Safdar, A.; Khan, Z.A.; Khan, N.A. Load forecasting, dynamic pricing and DSM in smart grid: A review. *Renew. Sustain. Energy Rev.* **2016**, *54*, 1311–1322. [CrossRef]
194. Raza, M.Q.; Khosravi, A. A review on artificial intelligence based load demand forecasting techniques for smart grid and buildings. *Renew. Sustain. Energy Rev.* **2015**, *50*, 1352–1372. [CrossRef]
195. Zhou, K.-L.; Yang, S.-L.; Shen, C. A review of electric load classification in smart grid environment. *Renew. Sustain. Energy Rev.* **2013**, *24*, 103–110. [CrossRef]
196. Biviji, M.; Uckun, C.; Bassett, G.; Wang, J.; Ton, D. Patterns of electric vehicle charging with time of use rates: Case studies in California and Portland. In Proceedings of the ISGT 2014, Washington, DC, USA, 19–22 February 2014; pp. 1–5. [CrossRef]
197. Nazaripouya, H.; Wang, B.; Black, D. Electric Vehicles and Climate Change: Additional Contribution and Improved Economic Justification. *IEEE Electrif. Mag.* **2019**, *7*, 33–39. [CrossRef]
198. Fang, Z. Hawaii Public Electric Vehicle Charging Stations—Hawaii Open Data. Available online: https://opendata.hawaii.gov/dataset/hawaii-public-electric-vehicle-charging-stations (accessed on 19 October 2020).
199. Muerdter, N. Alternative Fueling Station Locations—Catalog. Available online: https://catalog.data.gov/dataset/alternative-fueling-station-locations-422f2 (accessed on 30 August 2022).
200. GLA Environment Team. Ultra Low Emissions Zone Expansion—London Datastore. Available online: https://data.london.gov.uk/dataset/ultra_low_emissions_zone_expansion_new (accessed on 30 August 2022).
201. New York State Energy Research and Development Authority (NYSERDA). NYSERDA Electric Vehicle Drive Clean Rebate Data: Beginning 2017 | State of New York.
202. Lee, Z.J.; Li, T.; Low, S.H. ACN-data: Analysis and applications of an open EV charging dataset. In Proceedings of the Tenth ACM International Conference on Future Energy Systems, Phoenix, AZ, USA, 25–28 June 2019; pp. 139–149.
203. Dik, A.; Omer, S.; Boukhanouf, R. Electric Vehicles: V2G for Rapid, Safe, and Green EV Penetration. *Energies* **2022**, *15*, 803. [CrossRef]

applied sciences

MDPI

Article

Parameter Identification of Photovoltaic Cell Model Using Modified Elephant Herding Optimization-Based Algorithms

Amer Malki [1], Abdallah A. Mohamed [1,2], Yasser I. Rashwan [3], Ragab A. El-Sehiemy [4,*] and Mostafa A. Elhosseini [3,*]

[1] College of Computer Science and Engineering, Taibah University, Yanbu 46421, Saudi Arabia; asamalki@taibahu.edu.sa (A.M.); aammohamed@taibahu.edu.sa (A.A.M.)

[2] Mathematics and Computer Science Department, Faculty of Science, Menoufia University, Shebin El Kom 32511, Egypt

[3] Computers and Control Systems Engineering Department, Faculty of Engineering, Mansoura University, Mansoura 35516, Egypt; yasser_e2005@hotmail.com

[4] Electrical Engineering Department, Faculty of Engineering, Kafrelsheikh University, Kafrelsheikh 33516, Egypt

* Correspondence: elsehiemy@eng.kfs.edu.eg (R.A.E.-S.); Melhosseini@ieee.org (M.A.E.)

Abstract: The use of metaheuristics in estimating the exact parameters of solar cell systems contributes greatly to performance improvement. The nonlinear electrical model of the solar cell has some parameters whose values are necessary to design photovoltaic (PV) systems accurately. The metaheuristic algorithms used to determine solar cell parameters have achieved remarkable success; however, most of these algorithms still produce local optimum solutions. In any case, changing to more suitable candidates through elephant herd optimization (EHO) equations is not guaranteed; in addition, instead of making parameter α adaptive throughout the evolution of the EHO, making them adaptive during the evolution of the EHO might be a preferable choice. The EHO technique is used in this work to estimate the optimum values of unknown parameters in single-, double-, and three-diode solar cell models. Models for five, seven, and ten unknown PV cell parameters are presented in these PV cell models. Applications are employed on two types of PV solar cells: the 57 mm diameter RTC Company of France commercial silicon for single- and double-diode models and multi-crystalline PV solar module CS6P-240P for the three-diode model. The total deviations between the actual and estimated result are used in this study as the objective function. The performance measures used in comparisons are the RMSE and relative error. The performance of EHO and the proposed three improved EHO algorithms are evaluated against the well-known optimization algorithms presented in the literature. The experimental results of EHO and the three improved EHO algorithms go as planned and proved to be comparable to recent metaheuristic algorithms. The three EHO-based variants outperform all competitors for the single-diode model, and in particular, the culture-based EHO (CEHO) outperforms others in the double/three-diode model. According the studied cases, the EHO variants have low levels of relative errors and therefore high accuracy compared with other optimization algorithms in the literature.

Keywords: metaheuristics; solar cell systems; elephant herding optimization; alpha tuned EHO; cultural-based; biased initialization; parameter identification; single diode; double diode; three diodes

Citation: Malki, A.; Mohamed, A.A.; Rashwan, Y.I.; El-Sehiemy, R.A.; Elhosseini, M.A. Parameter Identification of Photovoltaic Cell Model Using Modified Elephant Herding Optimization-Based Algorithms. *Appl. Sci.* **2021**, *11*, 11929. https://doi.org/10.3390/app112411929

Academic Editor: Edris Pouresmaeil

Received: 10 November 2021
Accepted: 13 December 2021
Published: 15 December 2021

Publisher's Note: MDPI stays neutral with regard to jurisdictional claims in published maps and institutional affiliations.

Copyright: © 2021 by the authors. Licensee MDPI, Basel, Switzerland. This article is an open access article distributed under the terms and conditions of the Creative Commons Attribution (CC BY) license (https://creativecommons.org/licenses/by/4.0/).

1. Introduction

Energy is an essential component of the universe and is considered one of the forms of existence. Energy is divided into two main types (renewable energy and non-renewable energy); non-renewable energy as fossil fuels has a terrible impact on the environment. Therefore, many nations tend to use renewable energy to produce their electricity. Solar energy is one of the primary and available renewable energy sources on the planet that has no pollution and easy installation as well as being inexpensive and noise-free. The need to add renewable energy sources is increased with the dramatic changes in electricity requirements. Therefore, the effective modeling of renewable energy resources is an important issue for efficient energy management [1].

Solar cells are one of the ways to take advantage of solar energy, so significant attention went to model photovoltaic (PV) cells [2–7]. Several parameters define the nonlinear electrical model of a solar cell, which must be studied in depth to design PV systems. It is vital to understand the current–voltage graph (I-V) before using PV cells. In addition to determining FV's parameters, picking a few points from this curve can also help. Based on the number of diodes, different parameter models are presented. Three different types are available: single diode, double diode, and three diode [8–11].

Parameter identification can be accomplished in two ways, using deterministic methods or using metaheuristics. Examples of traditional approaches are Lambert W-functions [12] and the interior-point method [13]. Although traditional models can solve parameter identification, it has some drawbacks facing nonlinear problems such as sensitivity to the initial solution besides sticking in a local optimum solution with heavy computations and taking a long time to reach this optimum. Therefore, metaheuristics algorithms are used to overcome these drawbacks. Examples of these metaheuristics are the Particle Swarm Optimization (PSO) [6], Genetic Algorithm (GA) [14], Differential Evolution (DE) [15], Harmony Search (HS) [16], Artificial Bee Colony (ABC) [17], and Simulated Annealing (SA) [18].

The continuous development in optimization methods has been notable in recent decades. For example, several optimization methods were developed and applied for different power system problems, as presented in [19,20]. Furthermore, in [21–25], an algorithm that mimics the elephant herding behavior called Elephant Herding Algorithm (EHO) was proposed for different applications. Reference [26] proposes three improved variants of EHO that are developed.

The basic architecture of the PV cell guarantees that two differentially doped semiconductor layers form a PN junction. When irradiation is present, the cell absorbs photons from incoming light and produces carriers (or electron–hole pairs). As a result, there may be a discrepancy at the intersection [27]. In an ideal PV cell model, a photocurrent source and a diode are connected in parallel. Model estimation is made easiest by the fact that there are only three unknown parameters: the ideality factor η, the photocurrent I_{pv}, and the reverse saturation current I_s.

The contact resistance R_s between the silicon and electrode surfaces is described by this resistance. A parallel resistance R_p is attached to the diode to prepare for leakage current in the PN junction. The single-diode model (SDM) model has five parameters that must be estimated: I_{pv}, I_s, R_s, and R_p [28]. The double-diode model (DDM) is a more precise method of modeling PV cells. It takes into account current loss recombination in the depletion area. With the addition of the seventh parallel diode, there are now seven parameters to estimate (I_{pv}, η_1, I_{d1}, η_2, I_{d2}, R_p, and R_s) [8].

These models are of great interest to many researchers. There have been many successful algorithms for adjusting parameters of PV cells in SDM and DDM, but few works in TDM have been published in this area. Reference [29] proposed a solar PV parameter extraction method based on the Flower Pollination Algorithm (FPA). Two diode models are chosen to understand the precision of the computation. The authors experimented with the effectiveness of FPA using RTC France info. Simulated Annealing (SA), Pattern Search (PS), Harmony Search (HS), and Artificial Bee Swarm Optimization (ABSO) techniques are often used to compare the measured root mean square error and relative error for the built model. Researchers [30] proposed a hybridized optimization algorithm (HISA) for accurately estimating the parameters of the PV cells and modules. From the experimental data obtained from five case studies consisting of two cells and three modules for monocrystalline, multi-crystalline, and thin-film PV technologies, single- and double-diode models of PV cells/modules were developed with their respective single I V nonlinear characteristics.

The authors [31] propose two simple metaphor-free algorithms called Rao-2 (R-II) and Rao-3 (R-III) to estimate the parameters of PV cells. Several well-known optimization algorithms are compared to the efficiency of the proposed algorithms. The comparison helps show the merit of the algorithms. Finally, an analysis of statistical data is combined with experimental findings to verify the efficiency of the proposed algorithms. The Grasshopper Optimization Algorithm (GOA) is proposed [32] for parameter extraction of a PV module's three-diode PV model. This GOA-based PV model uses two popular commercial modules: Kyocera KC200GT and Solarex MSX-60.

The single-, double-, and three-diode models have different solar cell parameters. These models have five parameters for the single-diode model and seven parameters for the double- and three-diode models. Each parameter must be obtained accurately based on the objective function to reach the global optimum. In this study, the EHO algorithms have been chosen to solve this problem because they have a few control parameters and smooth implementation. In addition, EHO's simplicity and few parameters made it a suitable choice for achieving such enhancements. Furthermore, by dividing the population into clans, we could avoid becoming trapped in a local optimum and instead converge on reaching a global minimum. Finally, after getting experimental results for this problem, a comparison with other well-known algorithms was presented to prove the result's quality. This comparison is important to ensure that the new variants can solve this problem and compete with other algorithms.

Table 1 reports some of the recent solvers that were applied for PV parameter estimation problems in the recent years

Table 1. Recent optimizers for PV parameter estimation.

Ref #/Year	Algorithm	Ref #/Year	Algorithm	Ref #/Year	Algorithm
[3], 2020	Projectile Search Algorithm	[32], 2020	Grasshopper Optimizer	[33], 2020	Backtracking Search Algorithm
[5], 2020	Cuckoo Search Optimizer	[34], 2020	Flower Pollination	[35], 2021	Marine Predators Optimizer
[6], 2018	Differential Evolution Algorithm	[36], 2021	Newton-Raphson jointed with Heuristic Algorithm	[37], 2020	Improved Wind-Driven Algorithm
[9], 2021	Turbulent Flow of Water Optimizer	[38], 2021	Supply–Demand Optimizer	[39], 2019	Differential Evolution Algorithm
[10], 2021	Forensic Optimizer	[40], 2021	Improved Bonobo Optimizer	[41], 2020	Slime Mold Optimizer
[11], 2021	Gorilla Optimization Algorithm	[42], 2013	Artificial Bee Swarm	[43], 2020	Coyote Optimization Algorithm
[21], 2021	Closed loop PSO and EHO	[44], 2021	Hybrid Whale and PSO Optimizer	[45], 2020	Adaptive Differential Evolution
[31], 2019	Metaphor-Less Algorithms	[46], 2021	Artificial Ecosystem Optimizer	[47], 2019	Gray Wolf Optimization

The RMSE and the relative error are used as the most performance measures developed in the previous methods. The proposed variants of EHO are compared against most of the new well-known algorithms on the parameter identification of different photovoltaics. The performance of these proposed algorithms can be judged according to convergence speed, high estimation of parameters, and low computation time.

The main contributions of this paper can be summarized as follows:

- Proposing three variants of the EHO algorithms for solar cell parameters estimation.
- The EHO and the proposed EHO variants are tested on single-, double-, and three-diode models.
- Verifying the performance of each algorithm by comparing results with those of competitors.

- Proving that the culture-based variant has the most effective performance that improves the EHO.
- Validation of the proposed variants under different environmental conditions for temperature and irradiation. In this regard, the applications are employed on two types of PV solar cells.

The rest of the paper is organized as follows. The second section focuses on solar cells and mathematical models. In Section 3, an elephant-herding algorithm is proposed, and its different versions are discussed. The results, computer simulations, and comparisons are listed and discussed in Section 4. Finally, we conclude in Section 5 with a wrap-up and conclusion.

2. Mathematical Models of Photovoltaic Cell

Solar cell models describing the I-V characteristics typically contain one diode, two diodes, or three diodes. These detailed models are described as follows:

2.1. Single Diode Model (Five-Parameter Model)

A modified Shockley diode equation can describe a single diode model. It is widely used for modeling solar cells because it is simple to implement with five parameters (I_{ph}, I_d, n, R_{sh}, R_s). However, at low illuminations, the single diode model is particularly inaccurate in describing cell behavior [48,49]. Figure 1 shows a single diode model consisting of a current source in parallel with a diode, and the module shunt resistance controls the loss of currents at the junction within the cell.

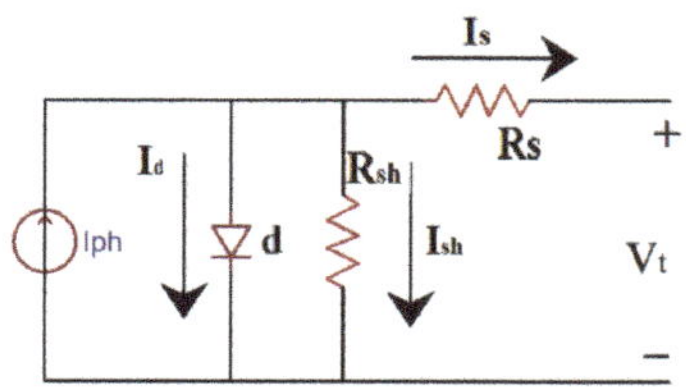

Figure 1. Single diode model.

The mathematical model of the single diode model is given by:

$$I_t = I_{ph} - I_{d1}\left[\exp\left(\frac{q(V_t + R_s \cdot I_t)}{n_1 \cdot k \cdot T}\right) - 1\right] - \frac{V_t + R_s \cdot I_t}{R_{sh}}. \tag{1}$$

2.2. Double-Diode Model (Seven-Parameter Model)

Figure 2 shows the double-diode model as an additional diode is added in parallel with the current source. This additional diode can achieve higher accuracy than a single diode model, but with seven parameters, more computation is needed (I_{ph}, I_{d1}, I_{d2}, n_1, n_2, R_{sh}, R_s).

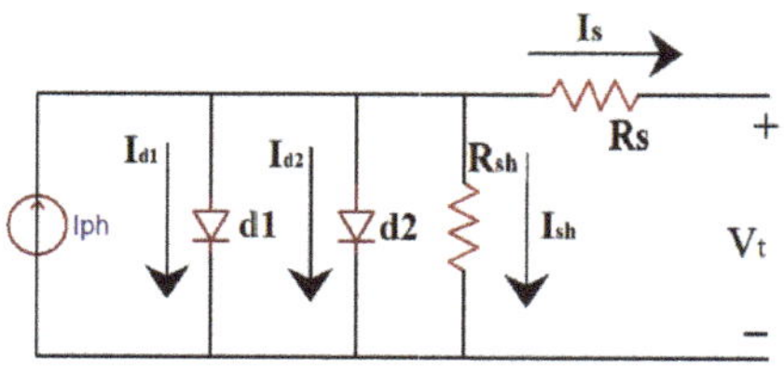

Figure 2. Double-diode model.

The mathematical model of the double-diode model is given below.

$$I_t = I_{ph} - I_{d1}\left[\exp\left(\frac{q(V_t + R_s \cdot I_t)}{n_1 \cdot k \cdot T}\right) - 1\right] - I_{d2}\left[\exp\left(\frac{q(V_t + R_s \cdot I_t)}{n_2 \cdot k \cdot T}\right) - 1\right] - \frac{V_t + R_s \cdot I_t}{R_{sh}} \tag{2}$$

2.3. Three-Diode Model (10-Parameter Model)

The three-diode model shown in Figure 3 extends the double-diode model by adding the third diode in parallel with the two other diodes. The three-diode model has three more parameters than the double-diode model (I_{d3}, n_2, K) [50,51].

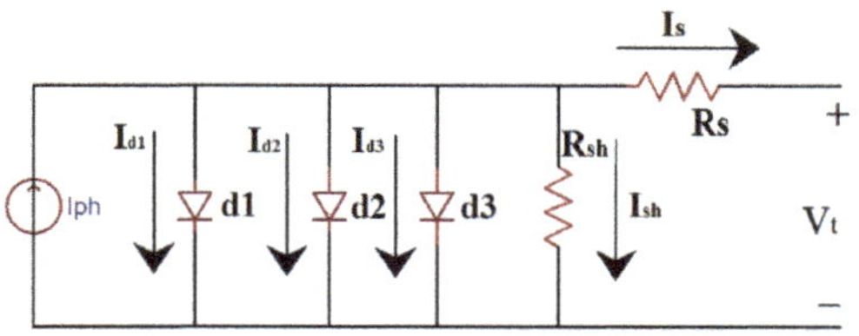

Figure 3. Three-diode model.

The mathematical formulation of the three-diode model is given by Equation (3) as:

$$I_t = I_{ph} - I_{d1}\left[\exp\left(\frac{q(V_t + R_s \cdot I_t)}{n_1 \cdot k \cdot T}\right) - 1\right] - I_{d2}\left[\exp\left(\frac{q(V_t + R_s \cdot I_t)}{n_2 \cdot k \cdot T}\right) - 1\right] - I_{d3}\left[\exp\left(\frac{q(V_t + R_s \cdot I_t)}{n_3 \cdot k \cdot T}\right) - 1\right] - \frac{V_t + R_s \cdot I_t}{R_{sh}}. \quad (3)$$

2.4. Parameter Extraction of the Solar Cell

A set of current–voltage (I–V) experimental data is given to extract the cell parameters. To define an objective function to be used in optimization algorithms, Equations (1)–(3) are reformed as in Equations (4)–(6). Equations (4)–(6) are used to get the error between the experimental and measured currents for the PV models, which are considered as the fitness functions of the three PV models.

$$f_1(V_t, I_t, y) = I_t - I_{ph} + I_{d1}\left[\exp\left(\frac{q(V_t + R_s \cdot I_t)}{n_1 \cdot k \cdot T}\right) - 1\right] + \frac{V_t + R_s \cdot I_t}{R_{sh}} \quad (4)$$

$$f_2(V_t, I_t, y) = I_t - I_{ph} + I_{d1}\left[\exp\left(\frac{q(V_t + R_s \cdot I_t)}{n_1 \cdot k \cdot T}\right) - 1\right] + I_{d2}\left[\exp\left(\frac{q(V_t + R_s \cdot I_t)}{n_2 \cdot k \cdot T}\right) - 1\right] + \frac{V_t + R_s \cdot I_t}{R_{sh}} \quad (5)$$

$$\begin{aligned} f_3(V_t, I_t, y) = I_t - I_{ph} &+ I_{d1}\left[\exp\left(\tfrac{q(V_t+R_s \cdot I_t)}{n_1 \cdot k \cdot T}\right) - 1\right] + I_{d2}\left[\exp\left(\tfrac{q(V_t+R_s \cdot I_t)}{n_2 \cdot k \cdot T}\right) - 1\right] \\ &+ I_{d3}\left[\exp\left(\tfrac{q(V_t+R_s \cdot I_t)}{n_3 \cdot k \cdot T}\right) - 1\right] + \tfrac{V_t+R_s \cdot I_t}{R_{sh}} \end{aligned} \quad (6)$$

The objective function can be implemented as the root mean square error (RMSE) as:

$$\mathrm{F} = \sqrt{\frac{1}{N}\sum_{l=1}^{N} f_l(V_t, I_t, y)^2}. \quad (7)$$

3. EHO-Based Optimization Algorithms

The wild elephant grows in herds. Clans of elephants are organized into groups under the leadership of female leaders. Furthermore, male elephants abandon the herd as they mature. To implement the elephant's behavior to solve nonlinear optimization problems, EHO is summarized into three essential rules:

1. The population has a fixed number of clans; each clan consists of some elephants.
2. The male elephant separates the clan and lives alone away from the group.
3. A leadership of female elephants rules the clan.

There are clans within the elephant population, and within each clan, each elephant is ranked based on its fitness, and then each group is updated separately.

Clan updating operator: For each member in clan ci, the best elephant effect on its next position in clan *c*. We can update elephant *j* in clan *c* by:

$$x_{n,c,j} = x_{c,j} + \alpha \cdot r \cdot \left(x_{best,c} - x_{c,j}\right). \quad (8)$$

The best elephant in each clan can be updated as:

$$x_{n,c,j} = \beta \cdot x_{center,c}. \quad (9)$$

Separating operator: As mentioned, the male elephant will live alone, separately away from the family. This separating process acts as the separating operator, which can be implemented into each generation as the worst fitness. We achieve it as follows:

$$x_{worst,c} = x_{min} + r \cdot (x_{max} - x_{min} + 1). \tag{10}$$

The elephant optimization procedure has been randomly generated based on the pseudocode in Figure 4 and the flowchart in Figure 5. The EHO algorithm has significant merit of a few control parameters. However, the chances of finding a new good elephant vs. a poor one are low; thus, the new candidate solution is unlikely to be as excellent as or better than the old one. The search operator does not consider the knowledge of the best solution or other solutions that may have a beneficial influence on steering EHO toward more promising areas of search space due to the participation of these random variables. However, a closer look at the flowchart and pseudocode of EHO reveals several gaps and shortcomings. These shortcomings may have a bad impact, affecting EHO's performance.

- As depicted in Equation (10), the new generated $x_{worst,ci}$ value may be worse than the original value of F. Thus, in this equation, a better value cannot be guaranteed.
- The constant value alpha (α in Equation (8)) remains consistent during algorithmic steps. Therefore, making the parameter based on the generation number of the elephant makes sense.

This paper aims to improve EHO performance, which is under-reported in the scientific literature. Listed below are three potential enhancements to EHO performance:

- Alpha tuning of αEHO.
- Cultural-based EHO (CEHO).
- Biased initialization EHO (BIEHO).

```
Initialization:
Initialize (Population size, Maximum generation, Boundaries).
Initialize the population.
Calculate the cost for each elephant.
Repeat
  Sort all population according to fitness.
  Clan updating:
  For c=1 to nClan (for each clan) do
    For j=1 to nci (for each elephant in clan c) do
      If x_c,j=x_best,ci then
            Update x_c,j (old member) and calculate x_n,c,j (new member) by Eq. (9).
      Else
            Update x_c,j (old member) and calculate x_n,c,j (new member) by Eq. (8).
      End if
    End for j
  End for c
Separating operator:
For c=1 to nClan (for each clan in population) do
     Replace the worst member in clan c by Eq. (10).
End for c
Evaluate all population by the updated elephants.
Until (Iteration = Maximum generation)
```

Figure 4. Pseudocode for EHO procedure.

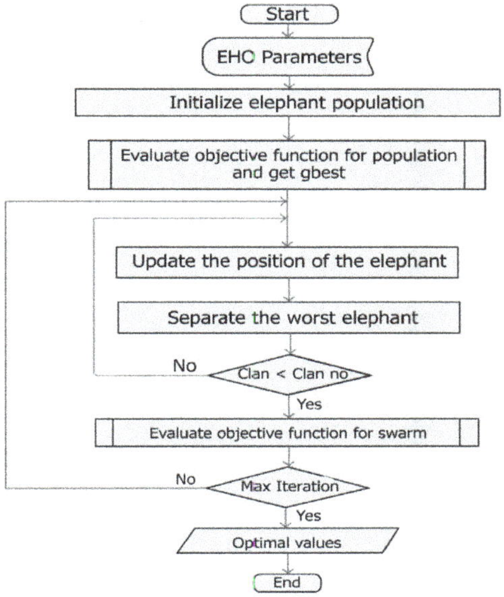

Figure 5. Flowchart of EHO.

3.1. *Alpha Tuning of αEHO*

Careful investigation of EHO parameters recommends setting the scale factor α to be adaptive is more promising than being a constant value in the range [0, 1].

Putting it simply, making alpha adaptive and related to the population number is more convenient and matched to the notion of evolution in Equation (11). In the original EHO algorithm, the scale factor-alpha is a constant value. Now, α is varying with the generation number by this function:

$$\alpha_{new} = \alpha + \frac{\alpha_{\max} - \alpha_{\min}}{n}. \tag{11}$$

3.2. *Cultural-Based EHO (CEHO)*

By utilizing the space of the best prior members, the cultural-based algorithm aids in the improvement of the algorithm [26,52,53]. The cultural-based algorithm constructs a better community by considering a belief space comprised of selected population members by acceptance function, as shown in Figure 6. A new member can be generated by using the belief space. A cultural-based algorithm is used to generate new solutions among belief space boundaries in the separating operation.

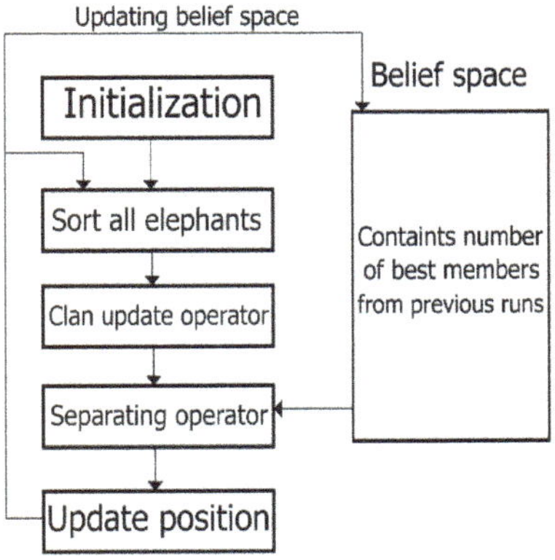

Figure 6. Belief space in cultural-based.

3.3. Biased Initialization EHO (BIEHO)

The main idea of the biased initialization algorithm is that the algorithm did not start evolving while the population's average fitness did not exceed a certain threshold. Therefore, the clan should be satisfied with its population's quality and ensure high-quality elephants. Start the generation with a population with functional fitness. The next step of evolution will not begin until the quality of the first generation reaches a suitable predetermined threshold. Biased algorithms are used in the initialization step by adding a rule or a limit [54]. Forcing the first generation of the population to have a good candidate solution may lead to another good production.

4. Computer Results and Simulations

EHO variants were tested using 57 mm diameter commercial silicon solar cells from the RTC Company of France to verify their performance against single- and double-diode models. The experiment is carried out under 1 sun (1000 W/m^2) at 33 °C [8,42,55]. A multi-crystalline PV solar module CS6P-240P is used to represent the three-diode model. CS6P-240P experimental data based on [56,57] are established for four irradiance levels (109.2, 246.65, 347.8, and 580.3 W/m^2) at temperatures (37.32, 40.05, 347.8, and 51.91 °C), respectively. Table 2 shows the manufacture specification for CS6P-240P under standard test conditions (STD). The basic EHO and its three variants are compared with the results of two algorithms from [42] called Artificial Bee Swarm Optimization algorithm (ABSO) and Harmony Search (HS) algorithm. The few adjustable parameters for EHO can be set as $\alpha = 0.9$, $\beta = 0.1$, number of clans = 4, population size = 32, and maximum iteration = 5000.

Table 2. Manufacture specification under standard test condition.

Maximum Power at STC	240 W
Optimum operating voltage	29.9 V
Optimum operating current	8.03 A
Open circuit voltage	37.0 V
Short circuit current	8.59 A
V Temperature coefficient *Voc*	−0.43%
I Temperature coefficient *Isc*	0.065%
Cell arrangement	60 (6 × 10)

Tables 3 and 4 present the optimal solar cell parameters and RMSE by EHO algorithms, Artificial Bee Swarm Optimization algorithm (ABSO), and Harmony Search (HS) for single- and double-diode modes. The single-diode model is considered the simplest model among all models with only five parameters. Table 3 shows that the four EHO algorithms obtained the same result due to the model's simplicity, but all four algorithms outperformed ABSO and HS. Table 4 shows the results for the double-diode model with seven parameters, showing differences between the extracted parameters and the RMSE. Compared to other algorithms, CEHO achieved the lowest RMSE. Figure 7 shows the convergence of the four EHO algorithms for the single-diode and double-diode model at the first 250 generations, respectively. In addition, it showed the fast convergence of the proposed EHO algorithms for obtaining good results.

Table 3. Comparison between EHO algorithms, ABS, and HS for single-diode solar cells.

Item	EHO	EHO Variants			ABSO	HS
		αEHO	CEHO	BIEHO		
I_{ph} (A)	0.76078	0.76077	0.76078	0.76077	0.7608	0.7607
I_d (μA)	0.32201	0.32143	0.32098	0.320479	0.30623	0.30495
R_s (Ω)	0.036388	0.036397	0.0364027	0.0364085	0.03659	0.03663
R_{sh} (Ω)	53.5851	53.58874	53.52479	53.49828	52.2903	53.5946
n	1.48086	1.48068	1.48054	1.48038	1.47583	1.47538
(RMSE)	9.861×10^{-4}	9.861×10^{-4}	9.861×10^{-4}	9.861×10^{-4}	9.9124×10^{-4}	9.9510×10^{-4}

Table 4. Comparison between EHO algorithms, ABS, and HS for double-diode solar cells.

Item	EHO	EHO Variants			ABSO	HS
		αEHO	CEHO	BIEHO		
I_{ph}(A)	0.76079	0.7607	0.76077	0.76079	0.76078	0.76176
I_{d1}(μA)	0.19895	0.23015	0.470885	0.294513	0.26713	0.12545
I_{d2}(μA)	0.25005	0.22753	0.258635	0.478734	0.38191	0.25470
R_s(Ω)	0.0367292	0.036594	0.036595	0.036591	0.03657	0.03545
R_{sh}(Ω)	53.47509	54.0848	54.85623	53.415705	54.6219	46.8269
$n1$	1.44596	1.45601	1.994023	1.47250	1.46512	1.49439
$n2$	1.69709	1.73558	1.462378	1.98067	1.98152	1.49989
$(RMSE)F$	9.876×10^{-4}	9.853×10^{-4}	9.830×10^{-4}	9.852×10^{-4}	9.834×10^{-4}	0.00126

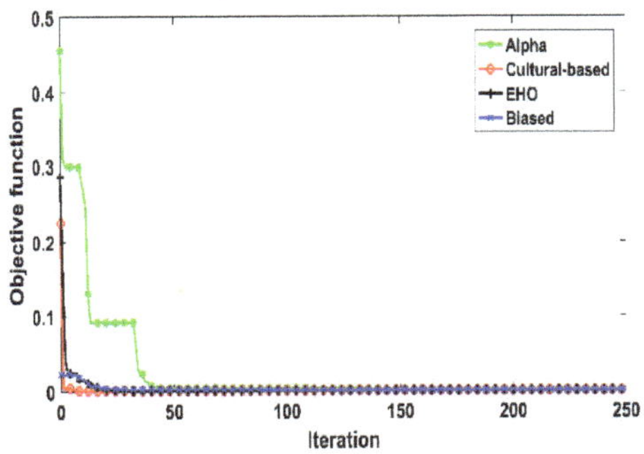

(a)

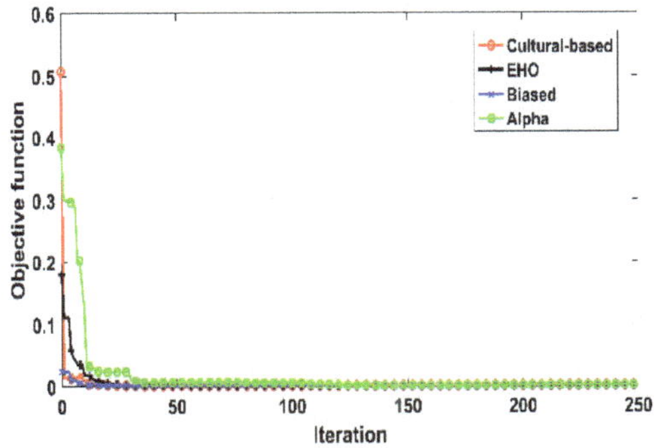

(b)

Figure 7. Convergence rates of EHO and its variants. (**a**) single diode. (**b**) double diode.

As demonstrated by Table 5, the measured current is very close to the calculated current. In addition, cultural-based EHO leads to outperformed results compared with other EHO variants.

Figures 8 and 9 show the power and current of the calculated and measured current from cultural-based EHO. Again, the measured and calculated curves are almost identical, while the relative error for the double-diode model for cultural-based EHO is presented in Table 6.

The previous results were for the PV panels at standard temperature and radiation. The four EHO algorithms were tested against three other algorithms at different irradiance levels and temperatures for more testing. Table 7 shows the extracted parameters for the seven algorithms at different irradiance levels and temperatures. Finally, the three-diode model is tested against three algorithms from [43] (Moth-Flame Optimizer (MFO), FPA, and Hybrid Evolutionary algorithm (DEIM)). The RMSEs for each algorithm at varying irradiance levels are listed in Table 8. Again, at low radiation with 109.2 W/m^2, CEHO outperforms EHO with a slightly small difference but a big difference compared to other algorithms. CEHO outperformed other algorithms at other radiations, and BIEHO's results were slightly different from CEHO's. The superiority of the CEHO algorithm is proven as the best compared with the other three variants and the other three algorithms for all irradiance levels. Figure 10 shows that calculated data fit the I-V curve of measured data for CEHO.

Table 5. The relative error for 26 measurements (single diode) with CEHO.

No.	V_t (v)	I_t (A) Measured	I_{ph} (A) Calculated	Relative Error
1	−0.2057	0.764	0.764104	−0.000104
2	−0.1291	0.762	0.762674	−0.000674
3	−0.0588	0.7605	0.761362	−0.000762
4	0.0057	0.7605	0.760156	0.000344
5	0.0646	0.76	0.759053	0.000947
6	0.1185	0.759	0.758037	0.000963
7	0.1678	0.757	0.757083	−0.000083
8	0.2132	0.757	0.756130	0.00087
9	0.2545	0.7555	0.755073	0.000427
10	0.2924	0.754	0.753649	0.000351
11	0.3269	0.7505	0.751377	−0.000877
12	0.3585	0.7465	0.747342	−0.000842
13	0.3873	0.7385	0.740110	−0.000161
14	0.4137	0.728	0.727382	0.000618
15	0.4373	0.7065	0.706981	−0.000481
16	0.459	0.6755	0.675295	0.000205
17	0.4784	0.632	0.630777	0.001223
18	0.496	0.573	0.571946	0.001054
19	0.5119	0.499	0.499618	−0.000618
20	0.5265	0.413	0.413650	−0.00065
21	0.5398	0.3165	0.317502	−0.001002
22	0.5521	0.212	0.212138	−0.000138
23	0.5633	0.1035	0.102232	0.0013
24	0.5736	−0.01	−0.008728	−0.001272
25	0.5833	−0.123	−0.125504	0.002504
26	0.59	−0.21	−0.208448	−0.007552

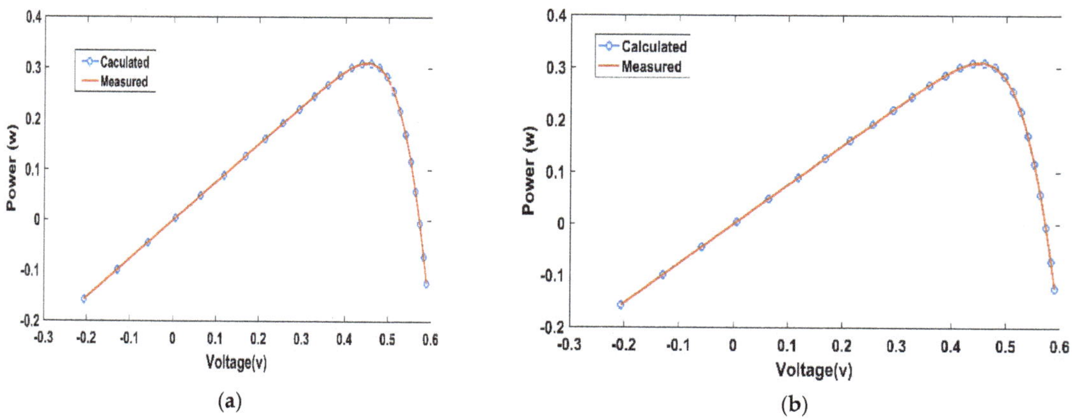

Figure 8. Measured power vs. calculated by CEHO. (**a**) single diode. (**b**) double diode.

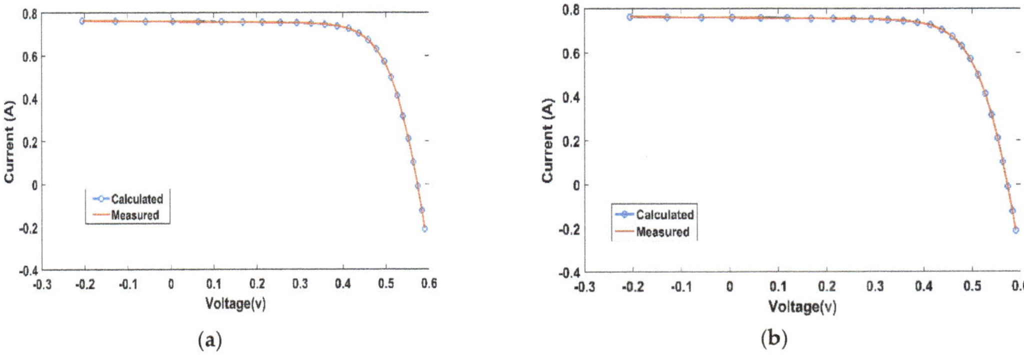

Figure 9. Measured current vs. calculated by CEHO. (**a**) single diode. (**b**) double diode.

Table 6. The relative error for 26 measurements (double diode) with CEHO.

No.	V_t (v)	I_t (A) Measured	I_{ph} (A) Calculated	Relative Error
1	−0.2057	0.764	0.764019	−0.000019
2	−0.1291	0.762	0.762623	−0.000623
3	−0.0588	0.7605	0.761343	−0.00843
4	0.0057	0.7605	0.760166	0.000334
5	0.0646	0.76	0.759088	0.006912
6	0.1185	0.759	0.758093	0.001093
7	0.1678	0.757	0.757154	−0.000154
8	0.2132	0.757	0.756208	0.000792
9	0.2545	0.7555	0.755147	0.000353
10	0.2924	0.754	0.753704	0.000296
11	0.3269	0.7505	0.751400	−0.000900
12	0.3585	0.7465	0.747325	−0.000825
13	0.3873	0.7385	0.740054	−0.001554
14	0.4137	0.728	0.727300	0.0007
15	0.4373	0.7065	0.706897	−0.000397
16	0.459	0.6755	0.675236	0.0003
17	0.4784	0.632	0.630758	0.001242
18	0.496	0.573	0.571968	0.001032
19	0.5119	0.499	0.499668	−0.000668
20	0.5265	0.413	0.413703	−0.000703
21	0.5398	0.3165	0.317536	−0.001036
22	0.5521	0.212	0.212140	−0.00014
23	0.5633	0.1035	0.102201	0.001299
24	0.5736	−0.01	−0.008761	−0.008761
25	0.5833	−0.123	−0.125531	0.002531
26	0.59	−0.21	−0.208418	−0.001582

Table 7. Comparison between different EHO algorithms among irradiance levels.

Irradiance	Algorithm	I_{ph}	I_{d1}	I_{d2}	I_{d3}	R_s	R_{sh}	n_3
580.3 W/m^2 51.91 °C	EHO	5.9992	1.7321×10^{-8}	6.7684×10^{-7}	1.1186×10^{-5}	0.39027	493.15	3.9986
	αEHO	5.9947	1.5744×10^{-8}	2.2779×10^{-7}	5.8365×10^{-5}	0.35637	4729.3	2.1169
	CEHO	5.9997	1.7336×10^{-8}	3.84148×10^{-7}	2.0015×10^{-10}	0.39055	478.607	2.7151
	BIEHO	5.9988	1.7292×10^{-8}	1.3773×10^{-6}	2.3406×10^{-8}	0.38977	502.29	3.048
	MFO	6.00066	1.7346×10^{-11}	9.210×10^{-7}	1.210×10^{-6}	0.38481	461.866	3.2135
	FBA	6.0075	1.7297×10^{-11}	8.7857×10^{-7}	9.0089×10^{-7}	0.39137	457.054	3.2926
	DEIM	6.0016	1.7363×10^{-11}	9.9751×10^{-7}	1.0234×10^{-6}	0.38524	457.282	3.2658
347.8 W/m^2 43.95 °C	EHO	3.0421	5.6398×10^{-9}	2.116×10^{-6}	2.0512×10^{-7}	0.41272	522.92	3.925
	αEHO	3.0328	4.5677×10^{-9}	7.2634×10^{-8}	3.1293×10^{-5}	0.26283	4927.8	2.001
	CEHO	3.0415	5.6216×10^{-9}	2.6116×10^{-6}	9.3642×10^{-9}	0.41118	540.79	3.9941
	BIEHO	3.0413	5.676×10^{-9}	2.1503×10^{-7}	7.9926×10^{-5}	0.41442	562.44	3.4295
	MFO	3.0457	5.6724×10^{-12}	9.985×10^{-7}	1.0234×10^{-6}	0.4163	461.524	3.2256
	FBA	3.04277	5.6211×10^{-12}	5.3506×10^{-7}	9.6777×10^{-7}	0.42925	517.401	3.1541
	DEIM	3.0454	5.6773×10^{-12}	9.9482×10^{-6}	1.3562×10^{-6}	0.41479	465.385	3.6897
246.65 W/m^2 40.05 °C	EHO	2.138	3.2543×10^{-9}	5.3294×10^{-6}	0.0005517	0.44377	4961.5	3.2946
	αEHO	2.135	2.4755×10^{-9}	9.8526×10^{-8}	4.3367×10^{-5}	0.28076	4977.9	2.0474
	CEHO	2.1379	3.2050×10^{-9}	8.1838×10^{-6}	0.0005982	0.43518	4999.86	3.4380
	BIEHO	2.1378	3.2494×10^{-9}	4.9161×10^{-6}	0.0004857	0.43959	4971.3	3.2026
	MFO	2.1435	3.3585×10^{-12}	1.62×10^{-6}	0.9391×10^{-3}	0.45333	4989.25	3.5252
	FBA	2.1484	3.453×10^{-12}	5.4342×10^{-7}	9.0503×10^{-7}	0.4923	4889.44	3.6523
	DEIM	2.1498	3.4402×10^{-12}	9.9684×10^{-7}	1.025×10^{-6}	0.8932	4746.08	3.5697
109.2 W/m^2 37.32 °C	EHO	0.99658	1.8992×10^{-9}	3.7627×10^{-7}	7.6187×10^{-7}	0.74613	469.11	3.9922
	αEHO	0.98919	1.7615×10^{-9}	6.7713×10^{-9}	3.1843×10^{-5}	0.58626	709.92	2.5034
	CEHO	0.99641	1.8939×10^{-9}	5.1732×10^{-7}	3.4151×10^{-8}	0.74203	472.73	3.8993
	BIEHO	0.99641	1.901×10^{-9}	2.5001×10^{-7}	1.8607×10^{-5}	0.74568	475.05	3.9983
	MFO	0.99853	2.2787×10^{-12}	1.0698×10^{-9}	9.9999×10^{-7}	0.7337	450.15	3.7173
	FBA	0.9978	2.2761×10^{-12}	1.0399×10^{-7}	5.8927×10^{-7}	0.7230	473.45	3.7569
	DEIM	0.9985	2.2658×10^{-12}	3.1652×10^{-8}	4.9986×10^{-7}	0.7351	449.34	3.3526

Table 8. Comparison between EHO algorithms, MFO, FBA, and DEIM for three-diode solar cells.

EHO	EHO Variants			MFO	FBA	DEIM
	αEHO	CEHO	BIEHO			
0.014598	0.026082	0.014591	0.014607	0.02455	0.02708	0.02807
0.0014236	0.02323	0.001337	0.001359	0.009927	0.016307	0.015864
0.0018575	0.006821	0.0017978	0.001821	0.012602	0.13287	0.012913
0.0009912	0.003721	0.00099094	0.0001923	0.001855	0.003607	0.0035508

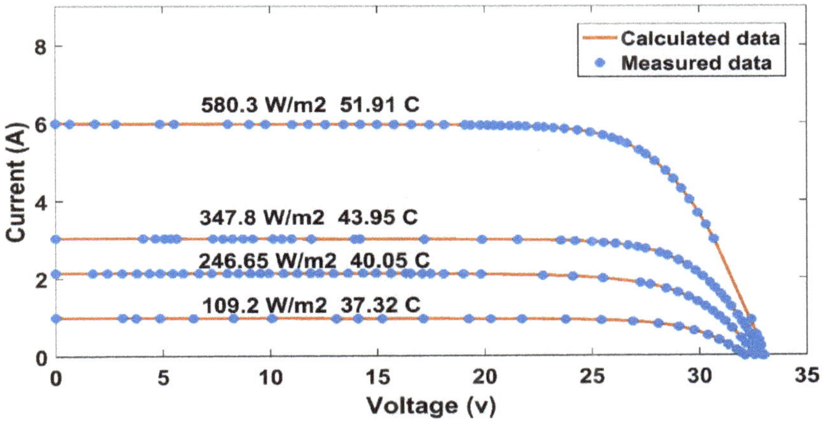

Figure 10. Measured current vs. calculated for three-diode model by CEHO.

5. Conclusions

This paper presents a new optimization algorithm based on elephant herding behavior called Elephant Herding Optimization (EHO) and three improved variants called αEHO, CEHO, and BIEHO. The EHO and its three variants are developed to estimate single, double, and three-diode solar cell models. The 57 mm diameter RTC Company of France commercial silicon solar cell with 26 points of measured data was chosen to present single and double models' problem under one irradiance level (25 °C and 1000 W/m^2). The EHO variants results are compared with two good algorithms (ABSO, HS). For presenting the three-diode model multi-crystalline PV solar module CS6P-240P under four irradiance levels (109.2, 246.65, 347.8, and 580.3 W/m^2) at temperature (37.32, 40.05, 347.8, and 51.91 °C) respectively. The EHO algorithms are compared with another three algorithms (MFO, FBA, and DEIM). The superiority of the four EHO algorithms is proven in the results. Cultural-based algorithms outperformed all algorithms used in the double- and three-diode models and ABSO, HS, and Biased in the single-diode model. Finally, it can be concluded from the results that EHO algorithms are very suitable for solving parameters extraction of solar cell problems for variant models.

Among the drawbacks of conventional EHO is its scale factor alpha being a constant value. Additionally, the behavior of EHO requires more attention to the solutions. Therefore, it would be helpful to employ more hybrid solutions, as this study recommends. Moreover, due to the practical nature of elephant herding, there are more processes involved than clan updating and separating. Thus, more models should be developed and incorporated into the EHO method that models elephant behavior. Finally, the main EHO was designed for solving continuous problems, so it must be validated for continuous and discrete problems [58].

Future work will include extracting parameters for more complex models for more accurate parameter extraction. In addition, the adaptive scaling factor is more promising than being a constant value in the range [0, 1]. Moreover, due to the superiority of the CEHO algorithm, we can do more enhancements to the CEHO algorithm to get more accurate results for more complex optimization problems. In addition, more behavior characteristics are recommended to investigate an advanced version of EHO accomplished with new hybrid algorithms.

Author Contributions: Conceptualization, Y.I.R. and M.A.E., R.A.E.-S.; methodology, A.M., A.A.M.; software, A.M., A.A.M.; validation, A.M., A.A.M.; formal analysis, M.A.E., R.A.E.-S.; investigation, A.M., A.A.M.; resources, A.M., A.A.M.; data curation, Y.I.R.; writing—original draft preparation, Y.I.R.; writing—review and editing, M.A.E., R.A.E.-S. and A.M.; visualization, A.M., A.A.M.; supervision, M.A.E., R.A.E.-S.; project administration, A.M., A.A.M. All authors have read and agreed to the published version of the manuscript.

Funding: This research received no external funding.

Institutional Review Board Statement: Not applicable.

Informed Consent Statement: Not applicable.

Data Availability Statement: Data available upon request.

Conflicts of Interest: The authors declare no conflict of interest.

Nomenclature

I_{ph}	The photogenerated current
I_d	The diode current
I_{d1}	The first diode current
I_{d2}	The second diode current
I_{d3}	The third diode current
V_t	The internal voltage
R_s	The series resistance
R_{sh}	The shunt resistance
n_1	The first diode ideality factor
n_2	The second diode ideality factor
n_3	The third diode ideality factor
k	Boltzmann's constant
T	Temperature
ql	The charge of an electron
N	Number of experimental data
$x_{n,c,j}$	Updated position for elephant j in clan c
$x_{c,j}$	Old position for elephant j in clan c
α	A scale factor ϵ [0, 1]
r	Random number ϵ [0, 1]
β	A scale factor ϵ [0, 1]
$x_{center,c}$	Centre of clan c
$x_{c,j,d}$	The dth of the elephant individual $x_{c,j}$
x_{man}	Upper bound of the position of elephant
x_{min}	Lower bound of the position of elephant
$x_{worst,c}$	Worst elephant individual in clan ci
α_{min}	Lower bound of permissible range of α
α_{max}	Upper bound of permissible range of α

References

1. Ali, E.S.; El-Sehiemy, R.A.; Abou El-Ela, A.A.; Mahmoud, K.; Lehtonen, M.; Darwish, M.M.F. An effective Bi-stage method for renewable energy sources integration into unbalanced distribution systems considering uncertainty. *Processes* **2021**, *9*, 471. [CrossRef]
2. Haider, H.T.; See, O.H.; Elmenreich, W. A review of residential demand response of smart grid. *Renew. Sustain. Energy Rev.* **2016**, *59*, 166–178. [CrossRef]
3. Shahabuddin, M.; Asim, M.; Sarwar, A. Parameter Extraction of a Solar PV Cell Using Projectile Search Algorithm. In Proceedings of the 2020 International Conference on Advances in Computing, Communication Materials (ICACCM), Dehradun, India, 21–22 August 2020; pp. 357–361.
4. El-Sayed, M.M.; Abou El-Ela, A.A.; El-Sehiemy, R.A. Effect of photovoltaic system on power quality in electrical distribution networks. In Proceedings of the 2016 18th International Middle-East Power Systems Conference, MEPCON 2016—Proceedings, Cairo, Egypt, 27–29 December 2017; pp. 1005–1012.
5. Gude, S.; Jana, K.C. Parameter extraction of photovoltaic cell using an improved cuckoo search optimization. *Sol. Energy* **2020**, *204*, 280–293. [CrossRef]
6. Chellaswamy, C.; Ramesh, R. Parameter extraction of solar cell models based on adaptive differential evolution algorithm. *Renew. Energy* **2016**, *97*, 823–837. [CrossRef]
7. Ali, S. New method for computing single diode model parameters of photovoltaic modules. *Renew. Energy* **2018**, *128*, 30–36. [CrossRef]
8. Chenouard, R.; El-Sehiemy, R.A. An interval branch and bound global optimization algorithm for parameter estimation of three photovoltaic models. *Energy Convers. Manag.* **2020**, *205*, 112400. [CrossRef]
9. Said, M.; Shaheen, A.M.; Ginidi, A.R.; El-Sehiemy, R.A.; Mahmoud, K.; Lehtonen, M.; Darwish, M.M.F. Estimating Parameters of Photovoltaic Models Using Accurate Turbulent Flow of Water Optimizer. *Processes* **2021**, *9*, 627. [CrossRef]
10. El-Sehiemy, R.A.; Shaheen, A.M.; Ginidi, A.; Sherif, S.M. Ghoneim A Forensic-Based Investigation Algorithm for Parameter Extraction of Solar Cell models. *IEEE Access* **2020**, *205*, 112400. [CrossRef]

11. Ginidi, A.; Ghoneim, S.M.; Elsayed, A.; El-Sehiemy, R.; Shaheen, A.; El-Fergany, A. Gorilla Troops Optimizer for Electrically Based Single and Double-Diode Models of Solar Photovoltaic Systems. *Sustainability* **2021**, *13*, 9459. [CrossRef]
12. Lun, S.X.; Wang, S.; Yang, G.H.; Guo, T.T. A new explicit double-diode modeling method based on Lambert W-function for photovoltaic arrays. *Sol. Energy* **2015**, *116*, 69–82. [CrossRef]
13. AlRashidi, M.R.; AlHajri, M.F.; El-Naggar, K.M.; Al-Othman, A.K. A new estimation approach for determining the I–V characteristics of solar cells. *Sol. Energy* **2011**, *85*, 1543–1550. [CrossRef]
14. Khalik, M.A.; Sherif, M.; Saraya, S.; Areed, F. Parameter identification problem: Real-coded GA approach. *Appl. Math. Comput.* **2007**, *187*, 1495–1501. [CrossRef]
15. Ishaque, K.; Salam, Z.; Mekhilef, S.; Shamsudin, A. Parameter extraction of solar photovoltaic modules using penalty-based differential evolution. *Appl. Energy* **2012**, *99*, 297–308. [CrossRef]
16. Askarzadeh, A.; Rezazadeh, A. Parameter identification for solar cell models using harmony search-based algorithms. *Sol. Energy* **2012**, *86*, 3241–3249. [CrossRef]
17. Oliva, D.; Cuevas, E.; Pajares, G. Parameter identification of solar cells using artificial bee colony optimization. *Energy* **2014**, *72*, 93–102. [CrossRef]
18. El-Naggar, K.M.; AlRashidi, M.R.; AlHajri, M.F.; Al-Othman, A.K. Simulated annealing algorithm for photovoltaic parameters identification. *Sol. Energy* **2012**, *86*, 266–274. [CrossRef]
19. Elsakaan, A.A.; El-Sehiemy, R.A.; Kaddah, S.S.; Elsaid, M.I. An enhanced moth-flame optimizer for solving non-smooth economic dispatch problems with emissions. *Energy* **2018**, *157*, 1063–1078. [CrossRef]
20. Rizk-Allah, R.M.; El-Sehiemy, R.A.; Wang, G.-G. A novel parallel hurricane optimization algorithm for secure emission/economic load dispatch solution. *Appl. Soft Comput.* **2017**, *63*, 206–222. [CrossRef]
21. Bayoumi, A.S.; El-Sehiemy, R.A.; Mahmoud, K.; Lehtonen, M.; Darwish, M.M.F. Assessment of an improved three-diode against modified two-diode patterns of MCS solar cells associated with soft parameter estimation paradigms. *Appl. Sci.* **2021**, *11*, 1055. [CrossRef]
22. Meena, N.K.; Parashar, S.; Swarnkar, A.; Gupta, N.; Niazi, K.R. Improved Elephant Herding Optimization for Multiobjective DER Accommodation in Distribution Systems. *IEEE Trans. Ind. Inf.* **2018**, *14*, 1029–1039. [CrossRef]
23. Zaky, A.A.; El Sehiemy, R.A.; Rashwan, Y.I.; Elhossieni, M.A.; Gkini, K.; Kladas, A.; Falaras, P. Optimal Performance Emulation of PSCs using the Elephant Herd Algorithm Associated with Experimental Validation. *ECS J. Solid State Sci. Technol.* **2019**, *8*, Q249–Q255. [CrossRef]
24. Zaky, A.A.; Ibrahim, M.N.; Rezk, H.; Christopoulos, E.; El Sehiemy, R.A.; Hristoforou, E.; Kladas, A.; Sergeant, P.; Falaras, P. Energy efficiency improvement of water pumping system using synchronous reluctance motor fed by perovskite solar cells. *Int. J. Energy Res.* **2020**. [CrossRef]
25. Wang, G.G.; Deb, S.; Gao, X.Z.; Dos Santos Coelho, L. A new metaheuristic optimisation algorithm motivated by elephant herding behaviour. *Int. J. Bio-Inspired Comput.* **2016**, *8*, 394–409. [CrossRef]
26. Elhosseini, M.A.; El Sehiemy, R.A.; Rashwan, Y.I.; Gao, X.Z. On the performance improvement of elephant herding optimization algorithm. *Knowl.-Based Syst.* **2019**, *166*, 58–70. [CrossRef]
27. Chin, V.J.; Salam, Z.; Ishaque, K. Cell modelling and model parameters estimation techniques for photovoltaic simulator application: A review. *Appl. Energy* **2015**, *154*, 500–519. [CrossRef]
28. Bai, J.; Liu, S.; Hao, Y.; Zhang, Z.; Jiang, M.; Zhang, Y. Development of a new compound method to extract the five parameters of PV modules. *Energy Convers. Manag.* **2014**. [CrossRef]
29. Alam, D.F.; Yousri, D.A.; Eteiba, M.B. Flower pollination algorithm based solar PV parameter estimation. *Energy Convers. Manag.* **2015**, *101*, 410–422. [CrossRef]
30. Kler, D.; Goswami, Y.; Rana, K.P.S.; Kumar, V. A novel approach to parameter estimation of photovoltaic systems using hybridized optimizer. *Energy Convers. Manag.* **2019**, *187*, 486–511. [CrossRef]
31. Premkumar, M.; Babu, T.S.; Umashankar, S.; Sowmya, R. A new metaphor-less algorithms for the photovoltaic cell parameter estimation. *Optik* **2020**, *208*, 164559. [CrossRef]
32. Elazab, O.S.; Hasanien, H.M.; Alsaidan, I.; Abdelaziz, A.Y.; Muyeen, S.M. Parameter estimation of three diode photovoltaic model using grasshopper optimization algorithm. *Energies* **2020**, *13*, 497. [CrossRef]
33. Zhang, Y.; Jin, Z.; Zhao, X.; Yang, Q. Backtracking search algorithm with Lévy flight for estimating parameters of photovoltaic models. *Energy Convers. Manag.* **2020**, *208*, 112615. [CrossRef]
34. Oliva, D.; Elaziz, M.A.; Elsheikh, A.H.; Ewees, A.A. A review on meta-heuristics methods for estimating parameters of solar cells. *J. Power Sources* **2019**, *435*, 126683. [CrossRef]
35. Bayoumi, A.S.A.; El-Sehiemy, R.A.; Abaza, A. Effective PV Parameter Estimation Algorithm Based on Marine Predators Optimizer Considering Normal and Low Radiation Operating Conditions. *Arab. J. Sci. Eng.* **2021**. [CrossRef]
36. Gnetchejo, P.J.; Ndjakomo Essiane, S.; Dadjé, A.; Ele, P. A combination of Newton-Raphson method and heuristics algorithms for parameter estimation in photovoltaic modules. *Heliyon* **2021**, *7*, e06673. [CrossRef]
37. Ibrahim, I.A.; Hossain, M.J.; Duck, B.C.; Nadarajah, M. An improved wind driven optimization algorithm for parameters identification of a triple-diode photovoltaic cell model. *Energy Convers. Manag.* **2020**, *213*, 112872. [CrossRef]
38. Ginidi, A.R.; Shaheen, A.M.; El-Sehiemy, R.A.; Elattar, E. Supply demand optimization algorithm for parameter extraction of various solar cell models. *Energy Rep.* **2021**, *7*, 5772–5794. [CrossRef]

39. Chin, V.J.; Salam, Z. A New Three-point-based Approach for the Parameter Extraction of Photovoltaic Cells. *Appl. Energy* **2019**, *237*, 519–533. [CrossRef]
40. Abdelghany, R.Y.; Kamel, S.; Sultan, H.M.; Khorasy, A.; Elsayed, S.K.; Ahmed, M. Development of an Improved Bonobo Optimizer and Its Application for Solar Cell Parameter Estimation. *Sustainability* **2021**, *13*, 3863. [CrossRef]
41. Kumar, C.; Raj, T.D.; Premkumar, M.; Raj, T.D. A new stochastic slime mould optimization algorithm for the estimation of solar photovoltaic cell parameters. *Optik* **2020**, *223*, 165277. [CrossRef]
42. Askarzadeh, A.; Rezazadeh, A. Artificial bee swarm optimization algorithm for parameters identification of solar cell models. *Appl. Energy* **2013**, *102*, 943–949. [CrossRef]
43. Chin, V.J.; Salam, Z. Coyote optimization algorithm for the parameter extraction of photovoltaic cells. *Solar Energy* **2019**, *194*, 656–670. [CrossRef]
44. Rezk, H.; Babu, T.S.; Al-Dhaifallah, M.; Ziedan, H.A. A robust parameter estimation approach based on stochastic fractal search optimization algorithm applied to solar PV parameters. *Energy Rep.* **2021**, *7*, 620–640. [CrossRef]
45. Naraharisetti, J.N.L.; Devarapalli, R.; Bathina, V. Parameter extraction of solar photovoltaic module by using a novel hybrid marine predators–success history based adaptive differential evolution algorithm. *Energy Sources Part A Recover. Util. Environ. Eff.* **2020**, 1–23. [CrossRef]
46. El-Dabah, M.A.; El-Sehiemy, R.A.; Becherif, M.; Ebrahim, M.A. Parameter estimation of triple diode photovoltaic model using an artificial ecosystem-based optimizer. *Int. Trans. Electr. Energy Syst.* **2021**, e13043. [CrossRef]
47. Nayak, B.; Mohapatra, A.; Mohanty, K.B. Parameter estimation of single diode PV module based on GWO algorithm. *Renew. Energy Focus* **2019**, *30*, 1–12. [CrossRef]
48. Ghani, F.; Duke, M. Numerical determination of parasitic resistances of a solar cell using the Lambert W-function. *Sol. Energy* **2011**, *85*, 2386–2394. [CrossRef]
49. Macabebe, E.Q.B.; Sheppard, C.J.; van Dyk, E.E. Parameter extraction from I–V characteristics of PV devices. *Sol. Energy* **2011**, *85*, 12–18. [CrossRef]
50. Khanna, V.; Das, B.K.; Bisht, D.; Singh, P.K. A three diode model for industrial solar cells and estimation of solar cell parameters using PSO algorithm. *Renew. Energy* **2015**, *78*, 105–113. [CrossRef]
51. Laudani, A.; Riganti Fulginei, F.; de Castro, F.; Salvini, A. Irradiance intensity dependence of the lumped parameters of the three-diodes model for organic solar cells. *Sol. Energy* **2018**, *163*, 526–536. [CrossRef]
52. Haikal, A.; Elhosseini, M. Modified cultural-based genetic algorithm for process optimization. *Ain Shams Eng. J.* **2011**, *2*, 173–182. [CrossRef]
53. Daneshyari, M.; Yen, G.G. Cultural-Based Multiobjective Particle Swarm Optimization. *IEEE Trans. Syst. Man Cybern. Part B* **2011**, *41*, 553–567. [CrossRef]
54. Ruiz, E.; Soto-Mendoza, V.; Ruiz Barbosa, A.E.; Reyes, R. Solving the open vehicle routing problem with capacity and distance constraints with a biased random key genetic algorithm. *Comput. Ind. Eng.* **2019**, *133*, 207–219. [CrossRef]
55. Liu, Y.; Chong, G.; Heidari, A.A.; Chen, H.; Liang, G.; Ye, X.; Cai, Z.; Wang, M. Horizontal and vertical crossover of Harris hawk optimizer with Nelder-Mead simplex for parameter estimation of photovoltaic models. *Energy Convers. Manag.* **2020**, *223*. [CrossRef]
56. Stropnik, R.; Stritih, U. Increasing the efficiency of PV panel with the use of PCM. *Renew. Energy* **2016**, *97*, 671–679. [CrossRef]
57. Islam, M.A.; Merabet, A.; Beguenane, R.; Ibrahim, H. Modeling solar photovoltaic cell and simulated performance analysis of a 250W PV module. In Proceedings of the 2013 IEEE Electrical Power Energy Conference, Halifax, NS, Canada, 21–23 August 2013; pp. 1–6.
58. Li, W.; Wang, G.-G.; Alavi, A.H. Learning-based elephant herding optimization algorithm for solving numerical optimization problems. *Knowl.-Based Syst.* **2020**, *195*, 105675. [CrossRef]

applied sciences

MDPI

Article

Performance Analysis of Modular Multilevel Converter with NPC Sub-Modules in Photovoltaic Grid-Integration

Suresh Mikkili [1,*], Raghu Vamsi Krishna [1], Praveen Kumar Bonthagorla [1] and Tomonobu Senjyu [2]

[1] Department of Electrical and Electronics Engineering, National Institute of Technology Goa, Ponda 403401, India; iamvamsi777@gmail.com (R.V.K.); praveennitgoa2017@gmail.com (P.K.B.)

[2] Department of Electrical and Electronics Engineering, Faculty of Engineering, University of the Ryukyus, Okinawa 9030213, Japan; b985542@tec.u-ryukyu.ac.jp

* Correspondence: mikkili.suresh@nitgoa.ac.in

Abstract: In this article, a three-phase modular multilevel converter (MMC) with three-level neutral point clamped converter (NPC) sub-modules (SMs) along with the placement of transformers in place of arm inductors is proposed for PV grid integration. Compared to the traditional MMCs, the proposed configuration reduces the voltage and power rating for the switches and the requirement of a high capacitor bank. In order to analyze the performance of the proposed converter arrangement, we have implemented four pulse width modulation schemes, such as Sine PWM with phase-level shifted carrier (SPWMLSC), Sine PWM with a phase-shifted carrier (SPWMPSC), Sine with the third harmonic injected level-shifted carrier (STHILSC), and Sine with the third harmonic injected phase-shifted carrier (STHIPSC). The proposed converter was simulated in the MATLAB/Simulink platform. Under normal and faulty operation, the results were presented with their performance indices of voltage and current harmonic distortion and sub-module capacitor voltage ripples at various modulation indices.

Keywords: modular multilevel converter (MMC); fault-tolerant; voltage source modular multilevel converter (VSMMC); SPWMLSC; SPWMPSC; STHILSC; STHIPSC

Citation: Mikkili, S.; Krishna, R.V.; Bonthagorla, P.K.; Senjyu, T. Performance Analysis of Modular Multilevel Converter with NPC Sub-Modules in Photovoltaic Grid-Integration. *Appl. Sci.* **2022**, *12*, 1219. https://doi.org/10.3390/app12031219

Academic Editor: Giovanni Petrone

Received: 6 December 2021
Accepted: 21 January 2022
Published: 24 January 2022

Publisher's Note: MDPI stays neutral with regard to jurisdictional claims in published maps and institutional affiliations.

Copyright: © 2022 by the authors. Licensee MDPI, Basel, Switzerland. This article is an open access article distributed under the terms and conditions of the Creative Commons Attribution (CC BY) license (https://creativecommons.org/licenses/by/4.0/).

1. Introduction

In order to achieve lower harmonic content in the output waveforms and lower filtering requirements at the grid side, multi-level inverters would have been the future trend in grid integration applications. To boost the efficiency and performance of grid-connected systems, multi-level inverters would have been integrated into renewable energy applications. Recently, a topology, that is, the modular multilevel converter (MMC) evolved from the multi-level family that has a profound application in photovoltaics, offshore, onshore wind energy, and medium voltage motor drives STATCOM, and UPQC, etc. The advantages of MMC are handled for high voltage applications, compact construction, stability, and reliability, whereas the PV is integrated into the grid. We could see some special characteristics like peak power capacity, fault ride-through, power quality, and higher redundancy capabilities popularly in MMC PV-based inverters [1]. One of the key topics that has been in current research is how to improve the efficiency and performance of photo-voltaic systems in the field of advanced power electronics for the energy conversion stages. There are various publications directed toward improvements and future implementation areas of before and after MMC studies. Many of those studies mostly focused on the various sub-module arrangements. The idea of a unidirectional sub-module has been explored in [2]. The efficiency of the MMC has been tested and compared to traditional two-level cells using various cell configurations, such as neutral point clamped and flying capacitor topologies. DC-MMC-based systems have been used as PV conversion systems with two stages with MMC-based HVDC [3,4], where high irradiance is available at longer distances from the consumption centers. Generally, in some MMC-based PV systems, it is required to provide each dc-dc converter without independent MPPT to supply for the SM capacitor balance, which in turn helps in constant production of output required voltage levels without any power fluctuations [5]. Having said that, by maintaining lossless energy conversion, we have had MV grid integration of PV using MMC [6], because the individual MPPT arrangement in MMC and multi-string central inverter have lower MPPT losses when compared to a central inverter. In addition, a high-power hybrid-MMC was used for PV grid integration to provide reliable performance, even though under partial shading condition, because of the presence of individual MPPTs in each SM [7]. Thus, some of the MMC-based PV-BESs (photovoltaic-battery energy storage systems) [8] and energy distribution areas [9] had also been potential options for handling power mismatches and smoothening the output power. In addition, for applications where high power and medium voltage requirements, we can have some topologies of inverters without any need for the line-frequency transformer are given in [10]. The organization of the paper is as follows: Section 1 deals with the introduction. Section 2 explains the MMC and its sub-module variants. Section 3 details the control of MMC. Section 4 illustrates the modulation strategies for MMC. Section 5 discusses the results. Finally, the paper is concluded in Section 6.

2. Modular Multilevel Converter and Sub Module Variants

Figure 1a is a three-phase MMC, and it consists of the top arm (Tarm) and the bottom arm (Barm). This arrangement generally could have an N-number of SM in each arm [11]. Each phase consists of arm inductors (Larm) and the submodules to provide better arm-currents without circulating harmonics and fault currents. Submodule (SM) or power cell represents a topological connection of IGBTs and could be of any voltage level. Generally, this SM is half-bridge (HBSM) or full-bridge (FBSM), but some researchers [12] started using SM with high levels. The SMs in the arms must be identical to get a symmetrical output wave. In the basic operation of the MMC, there is a chance of two events—sub-module insertion or by-passing state—whenever all the sub-modules are inserted in series and would result in arm voltages, respectively, and in turn produces a multilevel output. During the insertion processes, the SM capacitors were charged to their rated voltage magnitudes in the blocking mode based on the SM variants. The DC-link energy was equally divided among the SM capacitors; otherwise, there would be an unbalance of voltages and circulating currents within the phases. Figure 1b is a single-phase equivalent representation of a conventional MMC. The top and bottom arms' SM's total voltage magnitudes constitute V au and Val, respectively. For many applications, like motor drive, the number of SMs drastically increased day by day. The ongoing developer has structurally connected multi-level SM topologies to have dc fault handling capability, dimension reduction, and a higher number of voltage levels. Table 1 here provides us with the topological variants of SMs. Two-level SMs: Figure 2a shows the half-bridge submodule (HBSM) to produce the unipolar voltage by chopping the dc-link voltage. It consists of two IGBTs and a capacitor [13]. It can produce two levels of voltages 0, +VC1 when bypassed and inserted. In the HBSM, the dc voltage component will be present in the MMC arms [14]. Figure 2b is unipolar voltage FBSM [12], and this is similar to FBSM but with a slight replacement of a S3 switch with D3 of the FBSM (Figure 3a); it would be suitable for dc-fault handling capability. Another two-level variant is the unidirectional SM [15], and it has fewer semiconducting switches, as shown in Figure 2c. The clamp single submodule (CSSM) configuration is another configuration [16], which is shown in Figure 2d. The CSSM is derived from the three-level flying capacitor submodules (TLFCSM) for providing dc-fault handling capability during unidirectional operation.

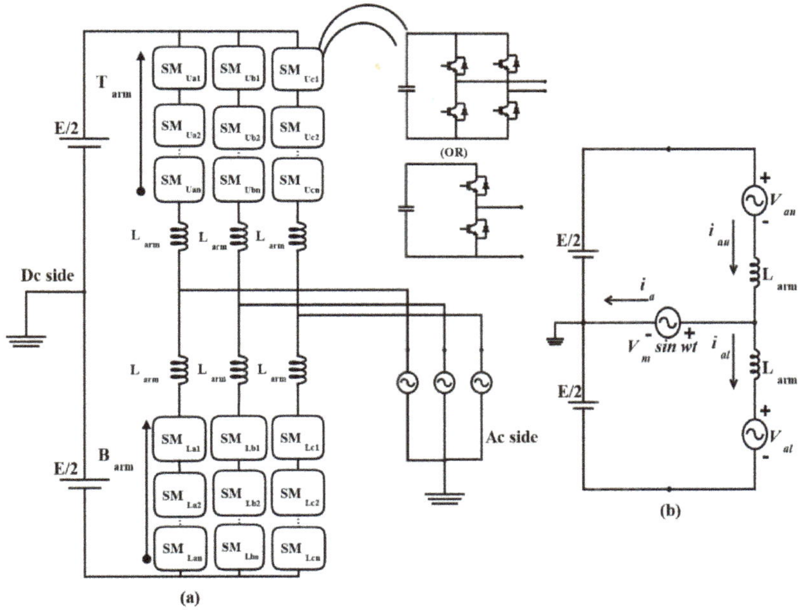

Figure 1. (**a**) Configuration of three-phase MMC, (**b**) single-phase equivalent circuit of conventional MMC.

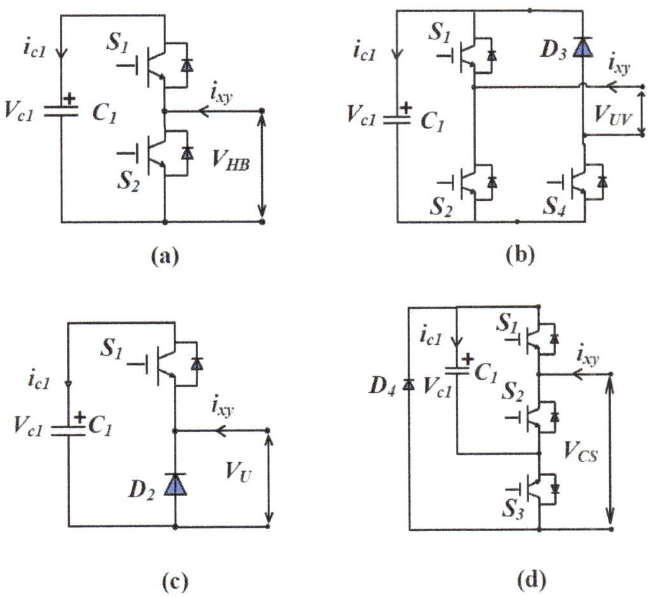

Figure 2. Two-level submodule variants (**a**) Half-bridge (**b**) Unipolar voltage full-bridge (**c**) Unidirectional SM (**d**) Clamp single SM.

Table 1. Comparison of different submodules [13–20].

Sub-Module (SM)	NVL	Voltage Level	NS	MNSC	MBV	Bipolar	DC-Fault Handling
Two-Level SMs							
HBSM	2	0, $-V_{C1}$	2	1	V_{C1}	No	No
Unipolar SM	2	0, $-V_{C1}$	3	1	V_{C1}	No	Yes
Unidirectional	2	0, $-V_{C1}$	1	1	V_{C1}	No	No
CSSM	2	0, $-V_{C1}$	3	2	V_{C1}	No	Yes
Three-Level SMs							
FBSM	3	0, $+V_{C1}$, $-V_{C1}$	4	2	V_{C1}	Yes	Yes
TLFCSM	3	0, $+V_{C2}$, $+(V_{C1} + V_{C2})$	4	2	$V_{C1} + V_{C2}$	No	No
TLNPC-1	3	0, $+V_{C2}$, $+(V_{C1} + V_{C2})$	4	2	$V_{C1} + V_{C2}$	No	No
TLNPC-2	3	0, $+V_{C2}$, $+(V_{C1} + V_{C2})$	6	2	$V_{C1} + V_{C2}$	No	No
TLNPC-3	3	0, $+V_{C2}$, $+(V_{C1} + V_{C2})$	4	2	$V_{C1} + V_{C2}$	No	No
DSM	3	0, $+V_{C2}$, $+(V_{C1} + V_{C2})$	8	4	$V_{C1} + V_{C2}$	No	Yes
TLCCSM	3	0, $+V_{C2}$, $+(V_{C1} + V_{C2})$	5	3	$V_{C1} + V_{C2}$	No	Yes
DBSM	3	0, $+V_{C1}$, $-V_{C2}$	2	1	V_{C1}	Yes	Yes
IHSM	3	0, $+V_{C1}$, $+(V_{C1} + V_{C2})$	5	3	$V_{C1} + V_{C2}$	No	Yes
Four-Level SMs							
CDSM-1	4	0, $+V_{C1}$, $+(V_{C1} + V_{C2})$	5	3	$V_{C1} + V_{C2}$	Yes	Yes
CDSM-2	4	0, $+V_{C2}$, $+(V_{C1} + V_{C2})$	7	3	$V_{C1} + V_{C2}$	Yes	Yes
Asymmetrical	4	0, $+V_{C2}$, $+(V_{C1} + V_{C2})$	4	2	$V_{C1} + V_{C2}$	Yes	Yes
Mixed SM	4	0, $+V_{C2}$, $+(V_{C1} + V_{C2})$	6	3	$V_{C1} + V_{C2}$	Yes	Yes
SDSM	4	0, $+V_{C2}$, $+(V_{C1} + V_{C2})$	5	3	$V_{C1} + V_{C2}$	Yes	Yes
Five-Level SMs							
FLCCSM	5	0, $+V_{C1}$, $+V_{C2}$, $\pm(VC_1 + V_{C2})$	6	3	$V_{C1} + V_{C2}$	Yes	Yes
CCSM	5	0, $+V_{C1}$, $+V_{C2}$, $\pm(VC_1 + V_{C2})$	8	4	$V_{C1} + V_{C2}$	Yes	Yes
PCSM	5	0, $+V_{C1}$, $+V_{C2}$, $\pm(VC_1 + V_{C2})$	8	4	$V_{C1} + V_{C2}$	Yes	Yes

Note: SM—Submodule, NVLs—Number of Voltage Levels, NSs—Number of Switches, MNSC—Maximum Number of Switches in the Conduction Path.

Three-level SMs: The well-known FBSM [16] is depicted in Figure 3a, and it generates bipolar output voltage levels, and we can use this for bipolar operation systems [17]. Moreover, if we connect two FBSMs in parallel, we could have a four-quadrant operation [18]. From Figure 3e, we can see that the three-level flying capacitor sub-module (TLFCSM) [14,15,18] has two capacitors—C1 which is twice in voltage rating with C2. This configuration does not have dc-fault handling and computationally complex control [19,20]. We see a three-level neutral point clamped submodules (TLNPC) in Figure 3b,d,g [14,18,21]. SMs in Figure 3d and TLFCSM are structurally close and produce 0, +VC2, and +(VC1 + VC2) magnitudes; SM in Figure 3b contains four switches, two capacitors, and two diodes, whereas SM in Figure 3g has six switches and two capacitors. Figure 3d, SM will be formed by the two-series combination of commutation circuits, while Figure 3b can be possible with the T-connection of switches; the midpoint switch here can block the voltages in both directions [22].

Even though it has more components, it lacks the dc-link short circuit handling capability, which results in severe control losses [17]. Two FBSMs are connected, as shown in Figure 3c, for improving the capacitor voltage performance, while keeping their SM's power level the same. This configuration [23] double-submodule (DSM) can improve the functionality of the SM without an increment in the semiconducting device's cost. With this capacitor, voltage ripples at low frequencies could also be reduced. However, it lacks the presence of bipolar operation. In [12], another three-level cross-connected SM (TL-CCSM,) formed by connecting two HBSMs back-to-back with switch and diode, is shown in Figure 3f. Making the connection like this, we could produce voltage magnitudes 0, VC1, and (VC1 + VC2). The diagonal-bridge SM (DBSM), shown in Figure 3i, is another variant that is similar to FBSM but with two diagonal switches that are replaced by diodes [24].

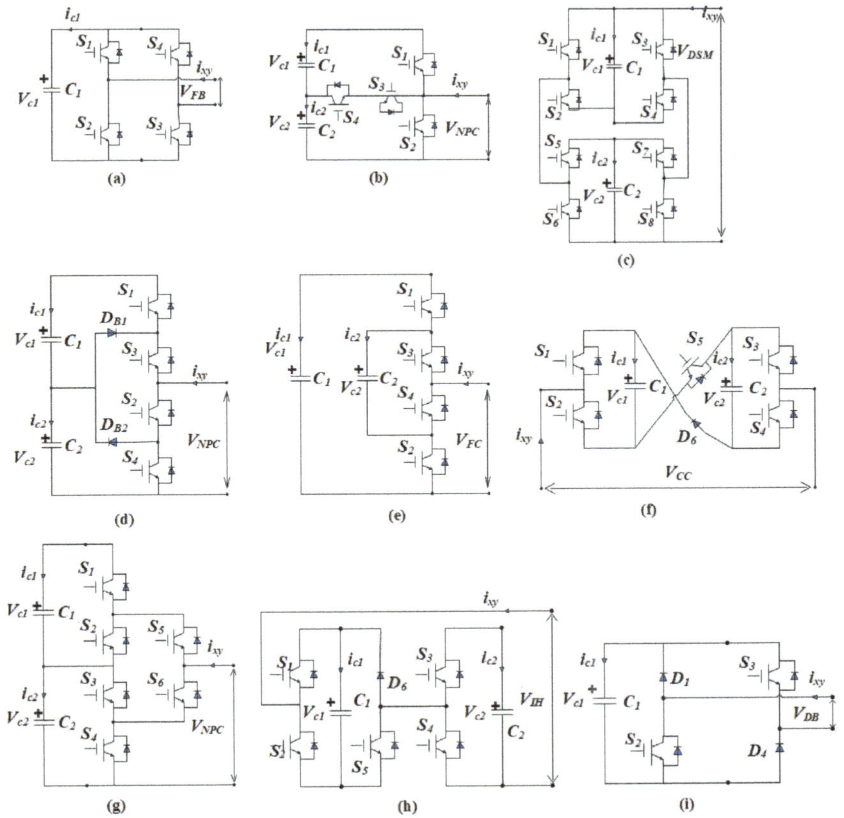

Figure 3. Three-level submodule variants (**a**) Full-bridge SM (**b**) Three-level flying capacitor submodule (TLFCSM) (**c**) TLNPC-1 (**d**) TLNPC-2 (**e**) TLNPC-3 (**f**) double-submodule (DSM) (**g**) three-level cross-connected (TLCCSM) (**h**) Improved hybrid SM(IHSM) (**i**) Diagonal-bridge SM (DBSM).

The last one is the three-level SM that is an improved hybrid SM (IHSM) [14], which has maximum voltage blocking of (VC1 + VC2), as depicted in Figure 3h. Four-level SMs: Figure 4a,b shows two kinds of double clamp submodules (CDSM) [12,14,21]. These configurations are the combination of two HBSMs connected in series. Here the switch S5 is on continuously, so it becomes two HBSMs connected in series under normal operating conditions. When the IGBTs are in blocking mode, both capacitors form either series or parallel connections in this CDSM. Further, this SM, operated as FBSM, has undergone different capacitor voltages as paralleled. To prevent this paralleling issue, diodes presented in that path could be replaced with IGBT switches. They would make another configuration, as shown in Figure 4b (CDSM); so far as losses are considered, CDSM is the combination of HBSM and FBSM. The alternate way of arranging commutation cells is shown in Figure 4c. This arrangement could be able to provide a dc-fault handling capability and can generate four voltage magnitudes. A combination of FBSM and HBSM would result in forming a hybrid submodule termed an asymmetrical SM [25], and it can perform both unipolar and bipolar operations, as shown in Figure 4d. The series connected double SM(SDSM) in shown in Figure 4e, which is obtained by connecting two HBSMs with S5 and D6 [26]. It has the capability of arc extension whenever a short circuit has taken place, and it could protect MMCs even when zero impedance current would have been taken place. Five-level SMs: As shown in Figure 5a, five-level cross-connected SM (FLCCSM) is the same as CDSM. Structurally FLCCSM is designed with two HBSMs in back-to-back cross-connection with the help of S5 and S6 [21,27]. Due to the two capacitors of the SM being connected in series, a blocking effect on the dc-fault current can be created. A different cross-connected SM (CCSM) [27] is shown in Figure 5b. It has a balanced bipolar voltage output, which could achieve greater voltage [28]. Similarly, if FBSMs are connected as parallel-connected SMs [29], which is shown in Figure 5c, they would give a reduced capacitor ripple.

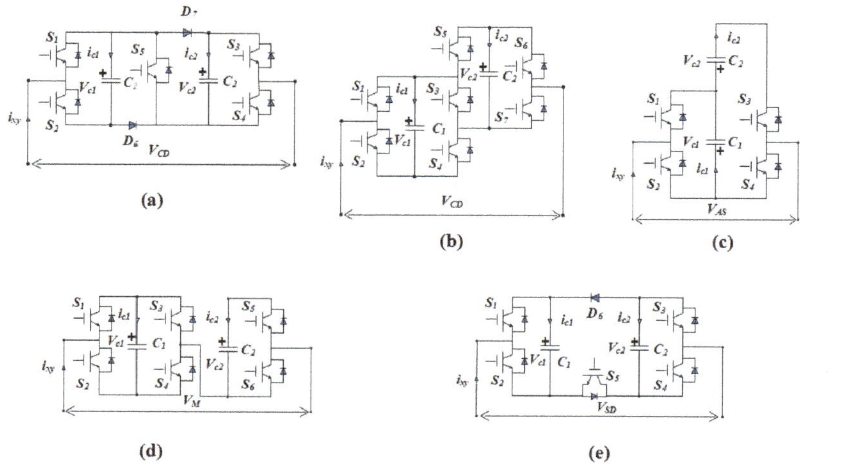

Figure 4. Four-level submodule variants (**a**) Clamp Double Submodules (CDSM), (**b**) Clamp Double Submodules (CDSM), (**c**) Mixed SM, (**d**) Asymmetrical SM, (**e**) Series Connected Double SM (SDSM).

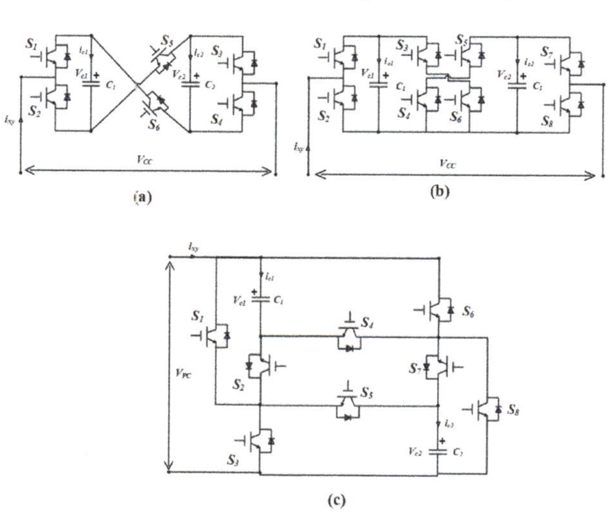

Figure 5. Five-level submodule variants (**a**) Five-Level Cross-Connected SM(FLCCSM), (**b**) Cross-Connected SM(CCSM), (**c**) Parallel-Connected SM.

3. Control Strategy of the Proposed MMC

The main part is the SM, which is may be a DC-DC or DC-AC power converter. Selection of better SM can provide fair control complexity, voltage blocking capability, and bipolar operation at a minimal cost. A new circuit performance of MMC with HB-SM is discussed in [2] by interchanging inductors with a transformer and also it offers twice the DC-AC voltage gain. The conventional MMC is having the advantage of power devices voltage ratings are halved, and the capacitor size of the SM is also lessened.

This paper uses neutral point clamped (NPC) SMs for the converter arrangement. The schematic (Figure 6) of the proposed MMC has slightly changed the SM with NPC SM with every single PV-panel, and the DC-DC converter and the DC-link to the MMC has been maintained by the PV array followed with the DC-DC converter. Figure 7 shows the controlling structure. For this controller implementation, initially, we are sensing the line-to-line three-phase voltages, and then these are transformed to two-phase voltage quantities by *alpha-beta* park's transformation. Using these *alpha-beta* voltages, PLL is implemented, *alpha-beta* voltages are then converted into *dq* voltages using Clark's transformation. Now from here, we are sensing the inverter side currents for the controller implementation. These currents are then transformed to the *alpha-beta* domain using park's transformation and then transformed to the *dq* domain using Clark's transformation. Here, *Id* corresponds to active current, and *Iq* corresponds to reactive current. Later *Id*&*Iq* are then subtracted from reference currents to find the error, and the error is fed to the *PI* controller to produce *Eq and Ed*. It is then transformed to *abc* voltages to get the references for PWM generation and, finally, we get the PWM generation block.

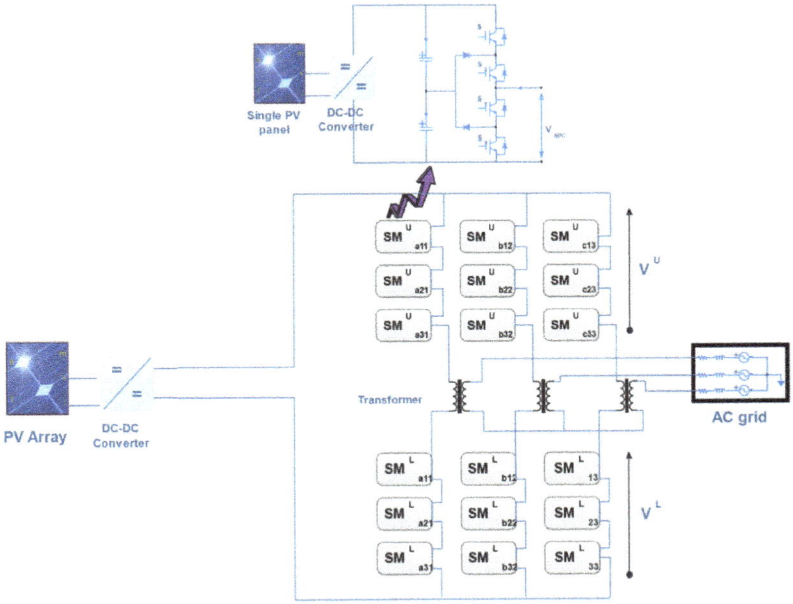

Figure 6. Schematic of the proposed modular multilevel converter.

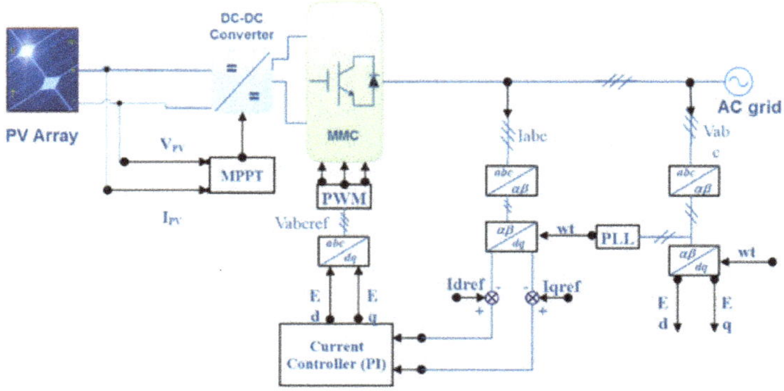

Figure 7. Control block diagram.

Generally, arm components in the traditional MMC are designed using the following equations.

$$L_{arm} \cdot C_{arm} = \frac{1}{\omega^2} \left\{ \frac{2(h^2 - 1) + m^2 h^2}{8h^2(h^2 - 1)} \right\} \tag{1}$$

$$C_{arm} = \frac{1}{L_{arm} \cdot \omega^2} \left\{ \frac{2(h^2 - 1) + m^2 h^2}{8h^2(h^2 - 1)} \right\} \tag{2}$$

where ω is the operating frequency, h is the harmonic order, and m is the modulation index.

However, in the proposed strategy, we use the transformers instead of arm inductors in the traditional MMC.

$$(L_{arm} + mutual\ inducance) \cdot C_{arm} = \frac{1}{\omega^2} \left\{ \frac{2(h^2 - 1) + m^2 h^2}{8h^2(h^2 - 1)} \right\} \tag{3}$$

From Equations (1) and (3), the inductance component is more in the proposed strategy, because of the mutual inductance of the transformer when compared to the traditional MMC for the same harmonic order and modulation index. Hence, from Equations (2) and (4), the proposed system requires a lesser value of capacitance when compared to the traditional MMC.

4. Modulation Strategies

Pulse width modulation (PWM) is generally used for regulating the power converter AC output voltage. The desired (reference) AC output voltage is achieved by regulating the duty cycle of the switching equipment. PWM methods were intended to eliminate harmonic components in the output voltage and increase the magnitude of the output voltage at any switching frequency. Figure 8 demonstrates the classification of various pulse width modulation techniques commonly used in multilevel converters. These are the PWM techniques adopted in VSC applications based on the switching frequency requirements. Figure 9 shows the simulation results with the (a) Sine PWM with the level-shifted carrier (SPWMLSC), (b) Sine PWM with the phase-shifted carrier (SPWMPSC), (c) Sine with the third harmonic injected level-shifted carrier (STHILSC), and (d) Sine with the third harmonic injected phase-shifted carrier (STHIPSC).

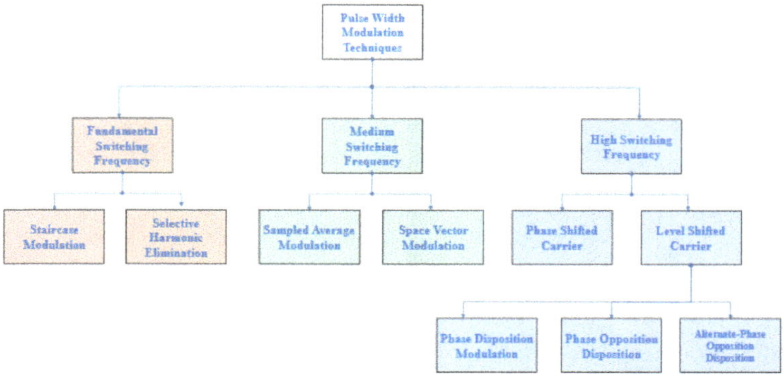

Figure 8. General pulse width modulation classification based on switching frequency.

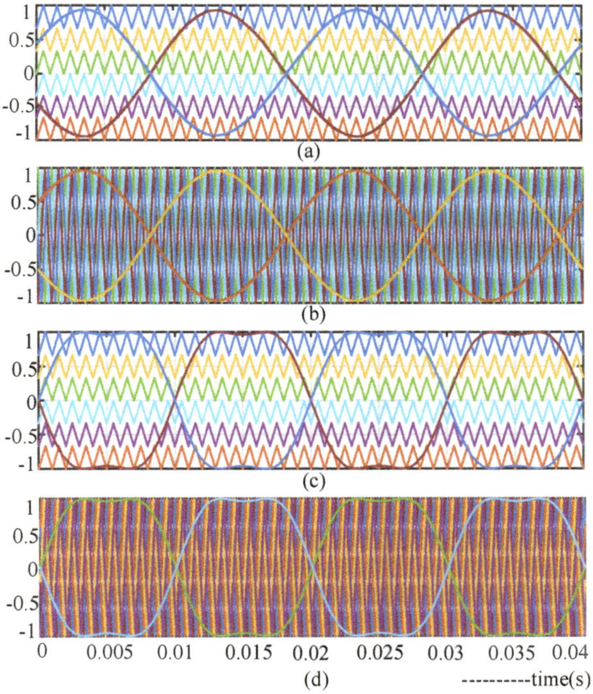

Figure 9. Modulation schemes conducted on this converter. (**a**) Sine PWM with level-shifted carrier (SPWMLSC); (**b**) Sine PWM with phase-shifted carrier (SPWMPSC); (**c**) Sine with third harmonic injected level-shifted carrier (STHILSC); and (**d**) Sine with third harmonic injected phase-shifted carrier (STHIPSC).

5. Results and Discussion

The simulation results of the proposed converter with various PWM schemes are clearly explained in this section. The parameters chosen for the simulation are given in Table 2. From Figure 10, we can see that the converter is operated with a modulation index (MI) of 0.5 from t = 0.8 s to t = 1.2 s, and it is operated with M = 0.95 from t = 0.8 s to t = 1.2 s. It is observed from Figure 10a,k that the THD in line voltage of the converter with d-q control and SPWMLSC is 2.58% at 50 Hz. respectively. In addition, from Figure 10b,m, the THD in line voltage of the converter with d-q control and SPWMPSC is 1.63%. Similarly, the

THD in line current with d-q control and SPWMLSC is 12.20% and with SPWMPSC is 6.64% at 50 Hz, as shown in Figure 10c,l, respectively. Figure 10e,g shows the capacitor voltages of upper arm phase-a SM-1 and lower arm phase-a SM-1 with SPWMLSC modulation scheme. Figure 10f,h shows the capacitor voltages of the upper arm phase-a SM-1 and lower arm phase-a SM-1 with SPWMPSC modulation scheme. Figure 10i,j shows the transformer primary and secondary phase-a currents for both PWM techniques, respectively.

Table 2. Simulated converter parameters.

Number of Cells in Each Arm	3
Ac line inductor	3 μH
Carrier frequency	1 KHz
Transformer resistance	74 m-ohm
Transformer inductor	3.48 mH
Cell capacitance	1000 μF

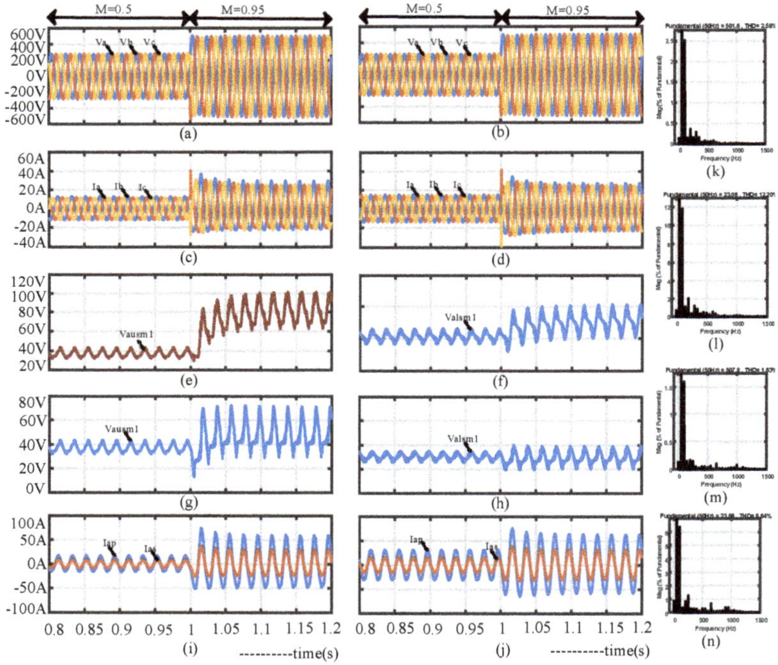

Figure 10. Change in modulation index. (**a**) Line voltages of converter with d-q control and SPWMLSC, (**b**) line voltages of converter with d-q control and SPWMPSC, (**c**) Line currents of converter with d-q control and SPWMLSC, (**d**) line currents of converter with d-q control and SPWMPSC, (**e**) SPWMLSC implemented upper arm phase-a SM-1 capacitor voltage, (**f**) SPWMPSC implemented upper arm phase-a SM-1 capacitor voltage, (**g**) SPWMLSC implemented lower arm phase-a SM-1 capacitor voltage, (**h**) SPWMPSC implemented lower arm phase-a SM-1 capacitor voltage, (**i**) primary and secondary phase-a transformer currents when SPWMLSC is implemented, (**j**) primary and secondary phase-a transformer currents when SPWMPSC is implemented, (**k**) SPWMLSC implemented line voltage THD, (**l**) SPWMLSC implemented line current THD, (**m**) SPWMPSC implemented line voltage THD, and (**n**) SPWMPSC implemented line current THD.

Further, the converter's dynamic performance is analyzed when there is a step-change from time t = 1 s to t = 1.2 s in the current magnitude of I = 7.25 A to I = 17.5 A at 50 Hz frequency. It is observed from Figure 11a,k that the THD in line voltage of the converter with d-q control and STHILSC is 1.26% at 50 Hz, respectively, and from Figure 11b,m the THD in line voltage of the converter with d-q control and STHIPSC is 1.18%. Similarly, the THD in line current with d-q control and STHILSC is 10.26% and with STHIPSC is 8.84% at 50 Hz, as is shown in Figure 11c,l and Figure 11d,n respectively. Figure 11e,g shows the capacitor voltages of upper arm phase-a SM-1 and lower arm phase-a SM-1 with STHILSC modulation scheme. Figure 11f,h shows the capacitor voltages of upper arm phase-a SM-1 and lower arm phase-a SM-1 with STHIPSC modulation scheme. Figure 11i,j shows the transformer primary and secondary phase-a currents for both STHILSC and STHIPSC PWM techniques, respectively.

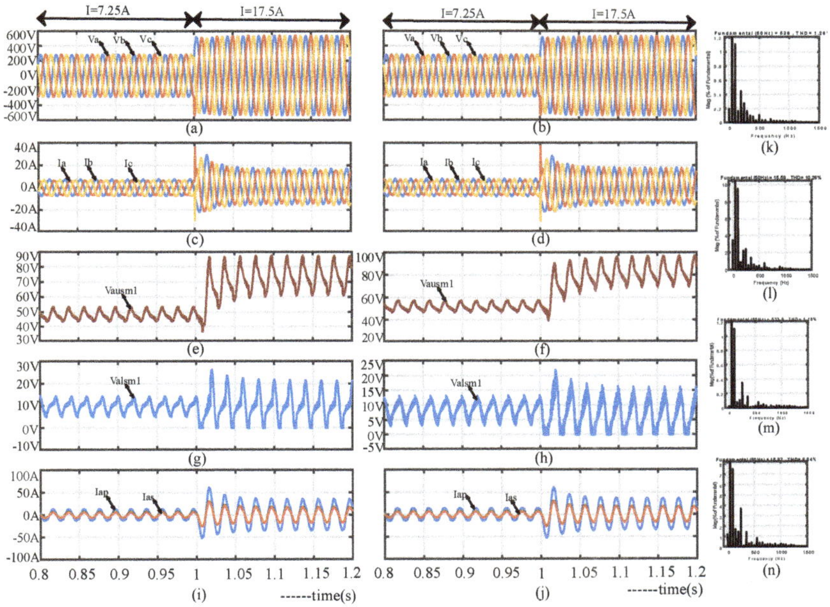

Figure 11. Change in current. (**a**) Line voltages of converter with d-q control and STHILSC, (**b**) line voltages of converter with d-q control and STHIPSC, (**c**) line currents of converter with d-q control and STHILSC, (**d**) line currents of converter with d-q control and STHIPSC, (**e**) STHILSC implemented upper arm phase-a SM-1 capacitor voltage, (**f**) STHIPSC implemented upper arm phase-a SM-1 capacitor voltage, (**g**) STHILSC implemented lower arm phase-a SM-1 capacitor voltage, (**h**) STHIPSC implemented lower arm phase-a SM-1 capacitor voltage, (**i**) primary and secondary phase-a transformer currents when STHILSC is implemented, (**j**) primary and secondary phase-a transformer currents when STHIPSC is implemented, (**k**) STHILSC implemented line voltage THD, (**l**) STHILSC implemented line current THD, (**m**) STHIPSC implemented line voltage THD, and (**n**) STHIPSC implemented line current THD.

Now, the performance of the MMC is analyzed when it is operating in open and short circuit conditions. The Figure 12a,b shows the line voltages with d-q control and SPWMLSC and SPWMPSC respectively, when one of the phases is open-circuited from time t = 1 s to t = 1.2 s at 50 Hz frequency. Figure 12c,d is the line currents with d-q control and SPWMLSC and SPWMPSC. During the faulty open-circuited operation, Figure 12e–h shows the capacitor voltages of the upper arm and lower arm of phase-a SM-1 with SPWMLSC and SPWMPSC modulation schemes, respectively. Further, the primary and secondary phase-a transformer currents, when SPWMLSC and SPWMPSC schemes are implemented, are shown, respectively, in Figure 12i,j. Similarly, Figure 13a,b shows the line voltages with d-q control and STHILSC and STHIPSC respectively, when one of the phases is open-circuited from time t = 1 s to t = 1.2 s at 50 Hz. Figure 13c,d shows the line currents with d-q control and STHILSC and STHIPSC respectively. During the faulty open circuit condition, Figure 13e–h shows the STHILSC implemented upper arm and lower arm phase-a SM-1 capacitor voltages, respectively, and STHIPSC implemented upper arm and lower arm phase-a SM-1 capacitor voltages, respectively. The primary and secondary phase-a transformer currents when implementing STHILSC STHIPSC are shown in Figure 13i,j. Further, the short circuit condition is created from time t = 1 s to t = 1.2 s. At this time interval, the converter performance is visualized, with simulation results shown in Figure 14. Figure 14a,b shows the line voltages with d-q control and SPWMLSC and SPWMPSC respectively. Furthermore, Figure 14c,d shows the line currents with d-q control and SPWMLSC and SPWMPSC respectively. During the faulty short circuit condition, Figure 14e–h shows the upper arm mad lower arm phase-a SM-1 capacitor voltages with SPWMPSC and SPWMLSC schemes, respectively. Furthermore, the primary and secondary phase-a transformer currents SPWMLSC and SPWMPSC modulation schemes are shown in Figure 14i,j, respectively.

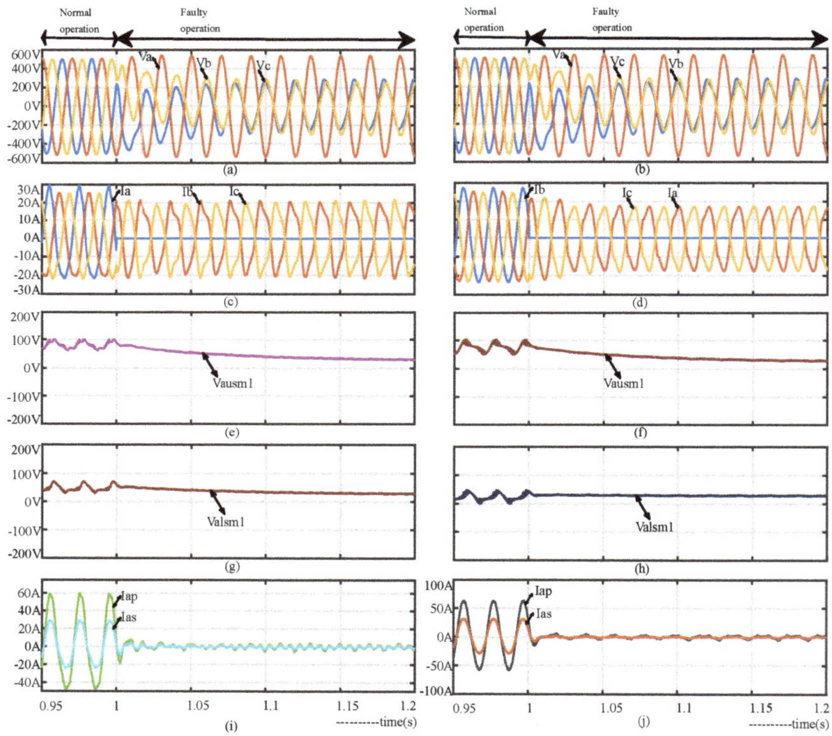

Figure 12. Converter when one phase is open circuited. (**a**) Line voltages of converter with d-q control and SPWMLSC, (**b**) line voltages of converter with d-q control and SPWMPSC, (**c**) line currents of converter with d-q control and SPWMLSC, (**d**) line currents of converter with d-q control and SPWMPSC, (**e**) SPWMLSC implemented upper arm phase-a SM-1 capacitor voltage, (**f**) SPWMPSC implemented upper arm phase-a SM-1 capacitor voltage, (**g**) SPWMLSC implemented lower arm phase-a SM-1 capacitor voltage, (**h**) SPWMPSC implemented lower arm phase-a SM-1 capacitor voltage, (i) primary and secondary phase-a transformer currents when SPWMLSC is implemented, (j) primary and secondary phase-a transformer currents when SPWMPSC is implemented.

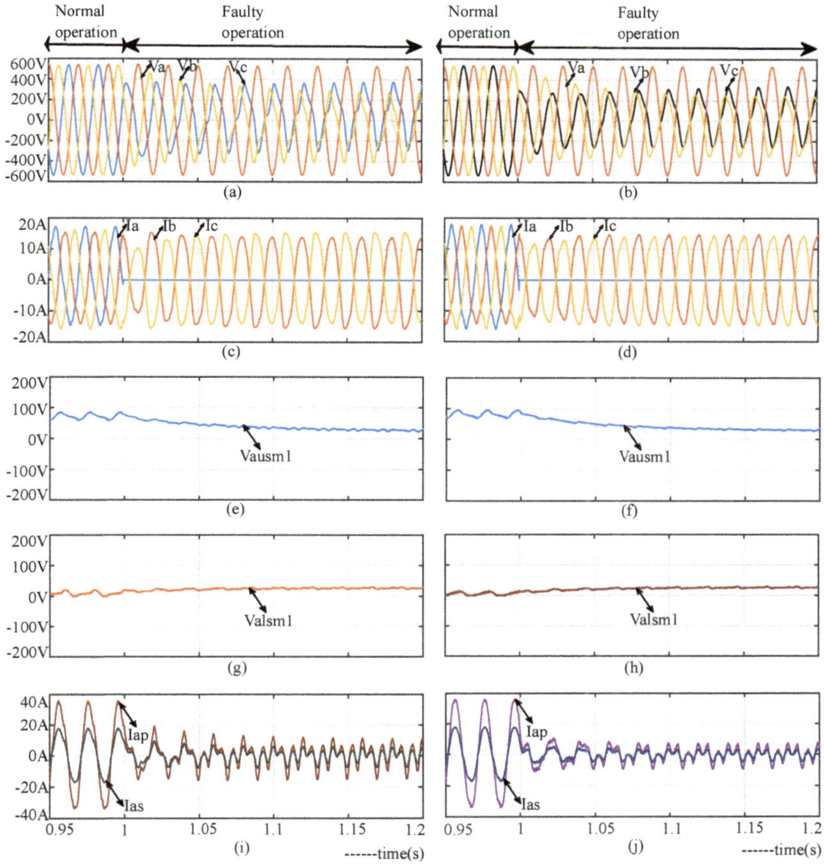

Figure 13. Converter when one phase is open circuited. (**a**) Line voltages of converter with d-q control and STHILSC, (**b**) line voltages of converter with d-q control and STHIPSC, (**c**) line currents of converter with d-q control and STHILSC, (**d**) line currents of converter with d-q control and STHIPSC, (**e**) STHILSC implemented upper arm phase-a SM-1 capacitor voltage, (**f**) STHIPSC implemented upper arm phase-a SM-1 capacitor voltage, (**g**) STHILSC implemented lower arm phase-a SM-1 capacitor voltage, (**h**) STHIPSC implemented lower arm phase-a SM-1 capacitor voltage, (**i**) primary and secondary phase-a transformer currents when STHILSC is implemented, (**j**) primary and secondary phase-a transformer currents when STHIPSC is implemented.

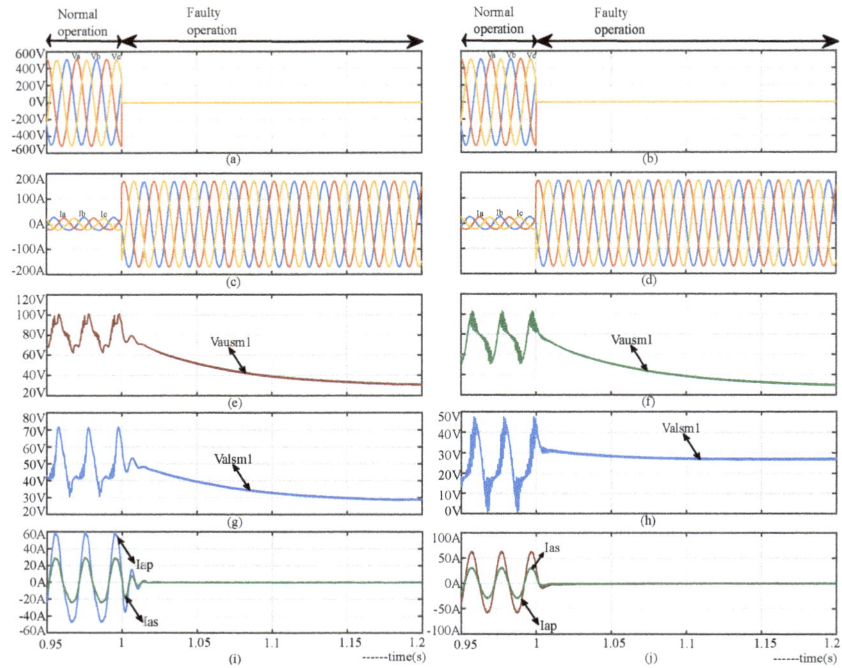

Figure 14. Converter during short circuited condition. (**a**) Line voltages of converter with d-q control and SPWMLSC, (**b**) line voltages of converter with d-q control and SPWMPSC, (**c**) line currents of converter with d-q control and SPWMLSC, (**d**) line currents of converter with d-q control and SPWMPSC, (**e**) SPWMLSC implemented upper arm phase-a SM-1 capacitor voltage, (**f**) SPWMPSC implemented upper arm phase-a SM-1 capacitor voltage, (**g**) SPWMLSC implemented lower arm phase-a SM-1 capacitor voltage, (**h**) SPWMPSC implemented lower arm phase-a SM-1 capacitor voltage, (**i**) primary and secondary phase-a transformer currents when SPWMLSC is implemented, (**j**) primary and secondary phase-a transformer currents when SPWMPSC is implemented.

Here, the proposed converter's performance is analyzed with sin third harmonic injection for both levels—shifted and phase-shifted PWM schemes—when the converter has been short-circuited from t = 1 s to t = 1.2 s at 50 Hz. Figure 15a,b shows the line voltages with d-q control and STHILSC and STHIPSC respectively. Figure 15c,d shows the line currents with d-q control and STHILSC and STHIPSC respectively. During short-circuited operation, Figure 15e–h shows the capacitor voltages of the upper arm and lower arm phase-a SM-1 with STHILSC modulation scheme. The primary and secondary phase-a transformer currents with STHILSC and STHIPSC schemes are shown in Figure 15i,j respectively.

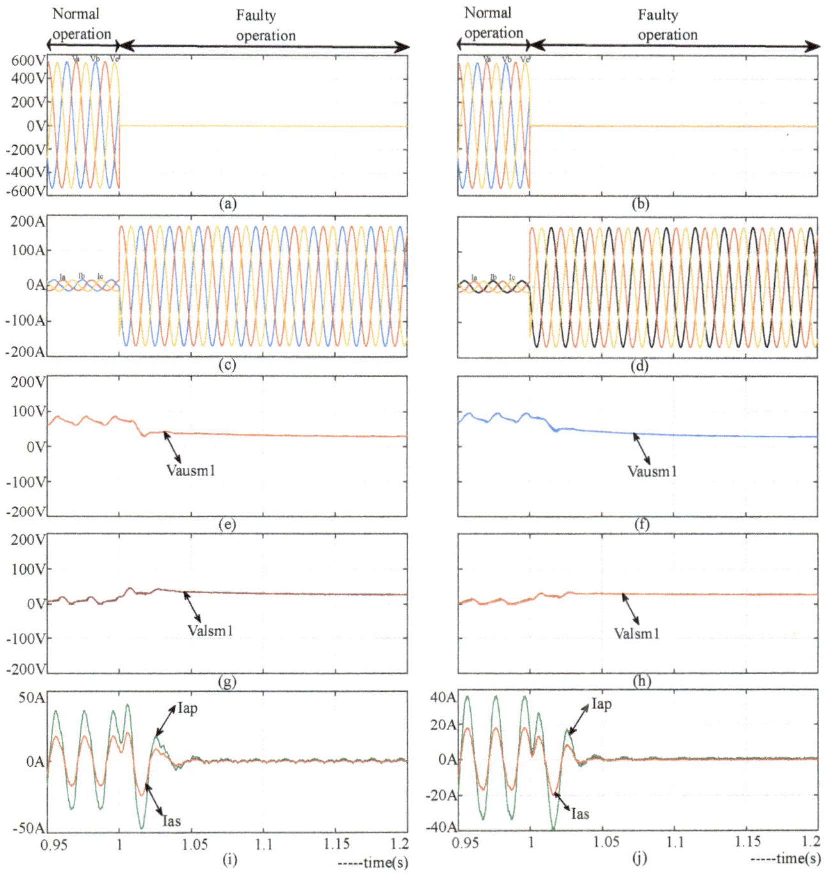

Figure 15. Converter during short circuited condition. (**a**) Line voltages of converter with d-q control and STHILSC, (**b**) line voltages of converter with d-q control and STHIPSC, (**c**) line currents of converter with d-q control and STHILSC, (**d**) line currents of converter with d-q control and STHIPSC, (**e**) STHILSC implemented upper arm phase-a SM-1 capacitor voltage, (**f**) STHIPSC implemented upper arm phase-a SM-1 capacitor voltage, (**g**) STHILSC implemented lower arm phase-a SM-1 capacitor voltage, (**h**) STHIPSC implemented lower arm phase-a SM-1 capacitor voltage, (**i**) primary and secondary phase-a transformer currents when STHILSC is implemented, (**j**) primary and secondary phase-a transformer currents when STHIPSC is implemented.

6. Conclusions

In this paper the performance of the proposed MMC with NPC sub-modules in PV grid-connected applications was investigated under steady and transient conditions by employing various PWM techniques. The proposed MMC with NPC sub-modules with various SPWMLSC, SPWMPSC, STHILSC, and STHIPSC modulation techniques reports the voltage THDs were 2.58%, 1.63%, 1.26%, and 1.18%, and the current THDs were 12.20%, 6.64%, 10.26%, and 8.84%, respectively. The phase-shifted carrier modulation scheme has features like a low computational burden, can manage fault tolerance, and provides superior voltage balance under abnormal conditions when it comes to the digital controllers. From the results it is clearly revealed that the modulation schemes, such as STHILSC and STHIPSC, give superior performance in terms of lower output voltage and current harmonic distortions. The proposed arrangement achieves any volage level with lower rating of power switches when compared to the conventional neutral point clamped converter. This strategy can also significantly reduce the requirement of the high rating capacitor bank due to the presence of the mutual inductance component in the transformers, which is not present in the traditional MMC. Hence it can used for many high-power and medium-voltage grid-connected applications.

Author Contributions: Conceptualization—S.M. and R.V.K; Methodology, Software, Validation—S.M., R.V.K., P.V.B., and T.S.; Investigation, Resources, Writing—Original Draft Preparation, Writing—Review and Editing, Visualization; Supervision; Project Administration, S.M., R.V.K., P.V.B., and T.S. All authors have read and agreed to the published version of the manuscript.

Funding: The authors gratefully acknowledge the support offered by the Science and Engineering Research Board (SERB), Department of Science and Technology, Government of India, under the Grant number: ECR/2017/000316 for this research work.

Institutional Review Board Statement: Not applicable.

Informed Consent Statement: Not applicable.

Data Availability Statement: Not applicable.

Conflicts of Interest: The authors declare no conflict of interest.

References

1. Mei, J.; Xiao, B.; Shen, K.; Tolbert, L.M.; Zheng, J.Y. Modular multilevel inverter with new modulation method and its application to photovoltaic grid-connected generator. *IEEE Trans. Power Electron.* **2013**, *28*, 5063–5073. [CrossRef]
2. Nademi, H.; Das, A.; Burgos, R.; Norum, L.E. A new circuit performance of modular multilevel inverter suitable for photovoltaic conversion plants. *IEEE J. Emerg. Sel. Top. Power Electron.* **2015**, *4*, 393–404. [CrossRef]
3. Rojas, C.A.; Kouro, S.; Perez, M.A.; Echeverria, J. DC–DC MMC for HVDC grid interface of utility-scale photovoltaic conversion systems. *IEEE Trans. Ind. Electron.* **2017**, *65*, 352–362. [CrossRef]
4. Wu, P.; Huang, W.; Tai, N. Novel grid connection interface for utility-scale PV power plants based on MMC. *J. Eng.* **2019**, *2019*, 2683–2686. [CrossRef]
5. Bayat, H.; Yazdani, A. A power mismatch elimination strategy for an MMC-based photovoltaic system. *IEEE Trans. Energy Convers.* **2018**, *33*, 1519–1528. [CrossRef]
6. Acharya, A.B.; Ricco, M.; Sera, D.; Teoderscu, R.; Norum, L.E. Performance analysis of medium-voltage grid integration of PV plant using modular multilevel converter. *IEEE Trans. Energy Convers.* **2019**, *34*, 1731–1740. [CrossRef]
7. Basu, T.S.; Maiti, S. A hybrid modular multilevel converter for solar power integration. *IEEE Trans. Ind. Appl.* **2019**, *55*, 5166–5177. [CrossRef]
8. Bayat, H.; Yazdani, A. A hybrid MMC-based photovoltaic and battery energy storage system. *IEEE Power Energy Technol. Syst. J.* **2019**, *6*, 32–40. [CrossRef]
9. Verdugo, C.; Candela, J.I.; Blaabjerg, F.; Rodriguez, P. Three-phase isolated multi-modular converter in renewable energy distribution systems. *IEEE J. Emerg. Sel. Top. Power Electron.* **2019**, *8*, 854–865. [CrossRef]
10. Zhang, X.; Wang, M.; Zhao, T.; Mao, W.; Hu, Y.; Cao, R. Topological comparison and analysis of medium-voltage and high-power direct-linked PV inverter. *CES Trans. Electr. Mach. Syst.* **2019**, *3*, 327–334. [CrossRef]
11. Solas, E.; Abad, G.; Barrena, J.A.; Aurtenetxea, S.; Carcar, A.; Zając, L. Modular multilevel converter with different submodule concepts—Part I: Capacitor voltage balancing method. *IEEE Trans. Ind. Electron.* **2012**, *60*, 4525–4535. [CrossRef]
12. Solas, E.; Abad, G.; Barrena, J.A.; Aurtenetxea, S.; Carcar, A.; Zajac, L. Modular Multilevel Converter With Different Submodule Concepts—Part II: Experimental Validation and Comparison for HVDC Application. *IEEE Trans. Ind. Electron.* **2012**, *60*, 4536–4545. [CrossRef]
13. Saad, H.; Dennetière, S.; Mahseredjian, J.; Delarue, P.; Guillaud, X.; Peralta, J.; Nguefeu, S. Modular multilevel converter models for electromagnetic transients. *IEEE Trans. Power Deliv.* **2013**, *29*, 1481–1489. [CrossRef]
14. Dekka, A.; Wu, B.; Fuentes, R.L.; Perez, M.; Zargari, N.R. Evolution of topologies, modeling, control schemes, and applications of modular multilevel converters. *IEEE J. Emerg. Sel. Top. Power Electron.* **2017**, *5*, 1631–1656. [CrossRef]
15. Jacobson, B.; Karlsson, P.; Asplund, G.; Harnefors, L.; Jonsson, T. VSC-HVDC Transmission with Cascaded Two-Level Converters. Proceeding of CIGRE SC B4 Session, Paris, France, 22 August 2010; pp. 1–8.
16. Qin, J.; Saeedifard, M.; Rockhill, A.; Zhou, R. Hybrid design of modular multilevel converters for HVDC systems based on various submodule circuits. *IEEE Trans. Power Deliv.* **2014**, *30*, 385–394. [CrossRef]
17. de Sousa, G.J.M.; Dias, A.D.S.; Alves, J.A.; Heldwein, M.L. June. In Modeling and control of a modular multilevel converter for medium voltage drives rectifier applications. In Proceedings of the 2015 IEEE 24th International Symposium on Industrial Electronics (ISIE), Buzios, Brazil, 3–5 June 2015; pp. 1080–1087.
18. Xu, J.; Zhao, P.; Zhao, C. Reliability analysis and redundancy configuration of MMC with hybrid submodule topologies. *IEEE Trans. Power Electron.* **2015**, *31*, 2720–2729. [CrossRef]

19. Perez, M.A.; Bernet, S.; Rodriguez, J.; Kouro, S.; Lizana, R. Circuit topologies, modeling, control schemes, and applications of modular multilevel converters. *IEEE Trans. Power Electron.* **2014**, *30*, 4–17. [CrossRef]
20. Nami, A.; Liang, J.; Dijkhuizen, F.; Demetriades, G.D. Modular multilevel converters for HVDC applications: Review on converter cells and functionalities. *IEEE Trans. Power Electron.* **2014**, *30*, 18–36. [CrossRef]
21. Debnath, S.; Qin, J.; Bahrani, B.; Saeedifard, M.; Barbosa, P. Operation, control, and applications of the modular multilevel converter: A review. *IEEE Trans. Power Electron.* **2014**, *30*, 37–53. [CrossRef]
22. Soeiro, T.B.; Kolar, J.W. Novel 3-level hybrid neutral-point-clamped converter. In Proceedings of the IECON 2011-37th Annual Conference of the IEEE Industrial Electronics Society, Melbourne, VIC, Australia, 7–10 November 2011; pp. 4457–4462.
23. Ilves, K.; Taffner, F.; Norrga, S.; Antonopoulos, A.; Harnefors, L.; Nee, H.P. A submodule implementation for parallel connection of capacitors in modular multilevel converters. *IEEE Trans. Power Electron.* **2014**, *30*, 3518–3527. [CrossRef]
24. Yu, X.; Wei, Y.; Jiang, Q.; Xie, X.; Liu, Y.; Wang, K. A novel hybrid-arm bipolar MMC topology with DC fault ride-through capability. *IEEE Trans. Power Deliv.* **2016**, *32*, 1404–1413. [CrossRef]
25. Mao, M.; Ding, Y.; Chang, L.; Hatziargyriou, N.D.; Chen, Q.; Tao, T.; Li, Y. Multi-objective power management for EV fleet with MMC-based integration into smart grid. *IEEE Trans. Smart Grid* **2017**, *10*, 1428–1439. [CrossRef]
26. Zhang, J.; Zhao, C. The research of SM topology with DC fault tolerance in MMC-HVDC. *IEEE Trans. Power Deliv.* **2015**, *30*, 1561–1568. [CrossRef]
27. Nami, A.; Wang, L.; Dijkhuizen, F.; Shukla, A. September. In Five level cross connected cell for cascaded converters. In Proceedings of the 2013 15th European conference on power electronics and applications (EPE), Lille, France, 2–6 September 2013; pp. 1–9.
28. Abd Razak, A.B. Modulation Strategy for New Asymmetrical Multilevel Inverter Topology Using Nearest Dc Level. Ph.D. Thesis, Universiti Teknologi Malaysia, Skudai, Malaysia, 2019.
29. Chaudhuri, T. Cross connected multilevel voltage source inverter topologies for medium voltage applications. Ph.D. Dissertation, EPFL, Lausanne, Switzerland, 2008.

applied sciences

MDPI

Article

A Comparison between Particle Swarm and Grey Wolf Optimization Algorithms for Improving the Battery Autonomy in a Photovoltaic System

Habib Kraiem [1,2], Flah Aymen [2], Lobna Yahya [2], Alicia Triviño [3], Mosleh Alharthi [4] and Sherif S. M. Ghoneim [4,*]

1 Department of Electrical Engineering, College of Engineering, Northern Border University, Arar 73222, Saudi Arabia; habib.kraiem@yahoo.fr
2 Processes, Energy, Environment and Electrical Systems, National Engineering School of Gabès, University of Gabès, Gabes 6072, Tunisia; flahaymening@gmail.com (F.A.); lobnayahya2@gmail.com (L.Y.)
3 Department of Electrical Engineering, University of Málaga, 29071 Málaga, Spain; atc@uma.es
4 Department of Electrical Engineering, College of Engineering, Taif University, Taif 21944, Saudi Arabia; m.harthi@tu.edu.sa
* Correspondence: s.ghoneim@tu.edu.sa

Citation: Kraiem, H.; Aymen, F.; Yahya, L.; Triviño, A.; Alharthi, M.; Ghoneim, S.S.M. A Comparison between Particle Swarm and Grey Wolf Optimization Algorithms for Improving the Battery Autonomy in a Photovoltaic System. *Appl. Sci.* **2021**, *11*, 7732. https://doi.org/10.3390/app11167732

Academic Editors: Marcos Tostado-Véliz and Rodolfo Dufo-López

Received: 18 July 2021
Accepted: 20 August 2021
Published: 22 August 2021

Publisher's Note: MDPI stays neutral with regard to jurisdictional claims in published maps and institutional affiliations.

Copyright: © 2021 by the authors. Licensee MDPI, Basel, Switzerland. This article is an open access article distributed under the terms and conditions of the Creative Commons Attribution (CC BY) license (https://creativecommons.org/licenses/by/4.0/).

Abstract: This research focuses on a photovoltaic system that powers an Electric Vehicle when moving in realistic scenarios with partial shading conditions. The main goal is to find an efficient control scheme to allow the solar generator producing the maximum amount of power achievable. The first contribution of this paper is the mathematical modelling of the photovoltaic system, its function and its features, considering the synthesis of the step-up converter and the maximum power point tracking analysis. This research looks at two intelligent control strategies to get the most power out, even with shading areas. Specifically, we show how to apply two evolutionary algorithms for this control. They are the "particle swarm optimization method" and the "grey wolf optimization method". These algorithms were tested and evaluated when a battery storage system in an Electric Vehicle is fed through a photovoltaic system. The Simulink/Matlab tool is used to execute the simulation phases and to quantify the performances of each of these control systems. Based on our simulation tests, the best method is identified.

Keywords: optimization algorithm; control system; renewable energy; PSO; GWO; battery storage energy; electric vehicle

1. Introduction

Due to its availability and the good conversion factor, solar energy technology has advanced at an exponential rate in the last few years [1]. As a renewable energy source, generation from solar energy eliminates pollution caused by traditional energy industries by lowering air nocive emissions [2]. Moreover, electricity generation from this resource is quite viable for a variety of uses. In particular, the rapid development of solar energy instruments gives a complete kit of tools that can be directly applied into the field of Electric Vehicles (EV) [3,4]. Several studies have suggested that photovoltaic cells can be used to cover EVs' surfaces to store a significant amount of electricity in the storage system [5,6]. This would increase EVs' autonomy, which will in turn increase the use of EVs. Some additional benefits are also associated to solar-powered Electric Vehicles [7,8]. First, the load peaks may be reduced so that the grid management is easier. Second, a decrement of the costs of charging the EVs would also be perceived by the drivers.

In [9], researchers provided some statistics that prove that there is a huge free space to place photovoltaic (PV) cells in the car. As pointed out, these components can be used to help feed the car with electricity. It is possible to have 6 kW of electrical power in some buses or trucks with this kind of installation [10].

In order to benefit from this technology, the solar cells must be installed in a particular location on the vehicle to enhance the average autonomy of the vehicle. In the previously cited references, it is stated that the Audi car model can get around 600 W from solar equipment. If the EV model is pure electric, the PV cells can help to feed the vehicle with 1% of the total electric power when it is moving. In addition, the energy can be obtained while the EV is parked. It is proved that the PV solution is interesting for heavy trucks because there is more vacant area that can be used to generate renewable electricity. The PV system can provide 2% of their consumption when moving along a long road. However, the statistics are different in the city, as there are many shading areas where the efficiency of the PV system may be affected. The way to estimate the energy return is challenging since it depends on the vehicle's relative situation compared to the sun's light and the presence of road obstructions. Because the car can quickly change from one type of shading scenario to another, obtaining the maximum amount of energy in all these conditions is quite tough. This is even more complex if we take into account that the efficiency of PV energy production mainly depends on the dynamic conditions associated to solar energy. The most relevant features are radiation, temperature, and the state of the PV field surface (dirt, deterioration). These factors directly influence its photon absorption and therefore affect the productivity of PV panels. In addition, the phenomenon of partial shading is one of the problems that impair the proper functioning of a PV plant. Partial shading is a non-uniform distribution of illumination on photovoltaic modules, which is due to several reasons. Indeed, there are two types of shading [11,12] The distant shade corresponds to the disappearance of the sun behind the horizon line. Alternatively, close shading is often due to unavoidable obstacles such as power lines, trees, neighboring buildings or dirt.

This undesirable phenomenon affects the conversion efficiency and the ability to extract the maximum available power from the PV field by generating multiple local maxima in the PV curves. In addition, shading also disrupts the operation of PV cells, causing two problems. The first problem is the mismatch, which is due to the fact that the total current in a PV field is limited by the current of the shaded module (low power) if the current *Isc* (the current flowing through the photovoltaic cell in short-circuit) of the shaded module is greater than the *Isc* of the uniformly lit modules. The second problem is the onset of hot spot. This problem occurs when the *Isc* current of the shaded module is less than the *Isc* of the uniformly illuminated modules, so that the shaded module behaves like an energy receiver extracting energy from the other PV modules. This effect can be noticed in the PV curve of a shaded PV panel, depicted in Figure 1. In partially shaded conditions, when the PV system does not receive uniform irradiation, the P-V characteristics become more complex, with multiple local peak displayed and a single global peak. These points are referred to as Maximum Power Points (MPP).

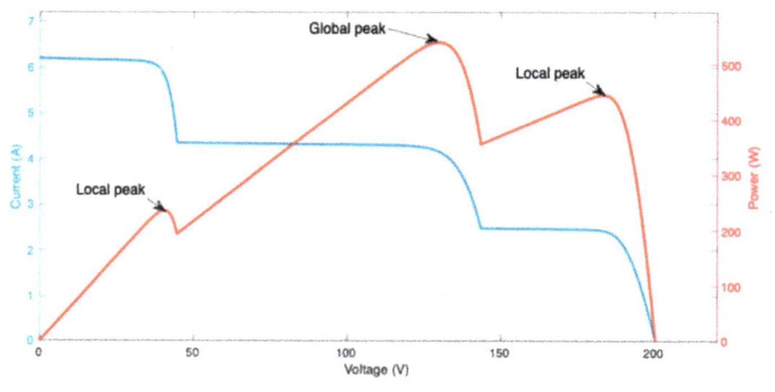

Figure 1. Typical curve of a shaded PV panel.

In addition to the deformation of the I-V curve, shading may lead to the heating of this module by dissipation of this energy. If the operating point of the shaded module reaches the breakdown voltage, the module will be destroyed by the avalanche effect [13,14].

Thus, it is crucial to find an optimal operating voltage to extract the maximum amount of PV power when designing the PV system while avoiding the aforementioned problems. A Maximum Power Point Tracking (MPPT) management tool is a required control loop that helps to get the most power out of a PV system. The basic performance of this algorithm is to adjust the duty cycle of the power converter connecting the PV modules to the DC-link. Concerning this tracking technique, various methods were carried out and help improve the energetic performances of the system. The traditional process relies on the incremental method's principle (IC) [15,16]. It is based on the principle of the zero derivative of the output power P of the PV panel, with respect to its output voltage V at the point maximum MPP power. In the maximum point, it is positive on the left and negative on the right, so the algorithm tries to find the voltage where this condition holds [17].

The Perturb and Observe (P&O) approach was the focus of further studies [18,19]. Because of its simplicity and convenience of use, it is frequently implemented. The primary benefit of this method is its direct torque design and lesser rate of monitored parameters. However, it has a significant flaw in terms of chattering on the supplied power form.

Classic MPPT methods such as P&O and IC are based on moving the next operating point (OP) in the direction of increasing PV power. However, when partial shading is present, the P–V curve is no longer monastically increasing, as shown in Figure 1. Thus, these conventional methods can only achieve a local MPP and may not reach the global maximum [20,21]. Therefore, it is necessary to develop an appropriate MPPT algorithm that can get to the global maximum power regardless of the state of illumination on the modules.

To overcome this limitation, other strategies based on intelligent optimization, such as the fuzzy logic technique or the neural network method, have been designed in the same sector [22,23]. The two techniques lead to higher profitability, but their main issue is the database required to adapt these algorithms to PV systems. Optimization algorithms were also used to help resolve this issue of partial shading of the photovoltaic system [24,25]. In particular, the evolutionary algorithms aim to have an adaptable MPPT tracking method, based on the animal behavior to find food [26,27].

There is a variety of swarm algorithms, which have been applied in multiple systems such as in [28–31]. Among them, Particle Swarm Optimization (PSO) [25] and Grey Wolf Optimization (GWO) [32] have shown their reliability to solve real optimization problems where the objective function is not linear. In particular, the works in [17,33] only considered these two algorithms to configure a DC/DC power converter. The review presented in [34] show that PSO algorithms are still investigated to tune the power converters of microgrids. Moreover, the study elaborated by Mirjalili in [32] presents a comparison between multiple swarm algorithms. As a conclusion, they state that the better results were found for the PSO and the GWO algorithms. Indeed, the two algorithms are inspired by natural competence to reach high speed and precision. Based on these previous works, in this paper we evaluate their relative feasibility and performance of employing the swarm algorithms to configure the power converter of the PV panels in order to cope with different shading conditions.

These two algorithms have already been applied and evaluated separately in PV systems. Particle Swarm Optimization algorithm can help to calculate the duty cycle of the power converter in the PV connection dynamically. Several works tested this solution for this application, as in [35] where the authors proved that this solution could be efficient if it is running offline. On the other side, the Grey Wolf Optimization (GWO) algorithm appeared as a useful solution for extracting energy from the PV system with maximum efficiency [36]. However, the two algorithms have several parameters and constants, which must be fixed initially to start the algorithms correctly.

The contribution of the paper is to apply and evaluate these two optimization algorithms for the same PV system, considering different partial shading conditions. The PV system model considered in this work is based on a commercial PV panel. So we can compare both performances to study their suitability in realistic implementations. The evaluation of each of these algorithms is based on the precision and the speed for tracking the global MPP with different partial shading conditions. Specifically, we have studied the two swarm-intelligence based algorithms for four shading conditions in a four-module PV system.

This research paper is organized to initially present an introduction section, which describes the objective of the paper and explains the state-of-the-art in this application field. In the second section, the PV system is modeled with the necessary equations that regroup all the parameters and constants that define this physical system. In the third section, the two optimization algorithms are explained. Their flowcharts are exposed and the principle running of each one is described for the MPPT algorithm. In the next section, the simulation conditions and the obtained results are shown for each irradiance case. In the end, a conclusion section is formatted for resuming the paper and giving some perspectives of this work.

2. Model of the PV System

In this section, we first describe the model for a solar PV cell. Then, we integrate it into the model of a PV system.

A. Solar PV Cell Model

A PV module consists of several solar cells connected in series and in parallel to achieve the desired voltage and current levels. A solar panel cell is essentially a p-n semiconductor junction. When exposed to light, a direct current is generated. For simplicity, the single diode model of Figure 2 is used in this document [37]. This model offers a good compromise between simplicity and precision with the basic structure.

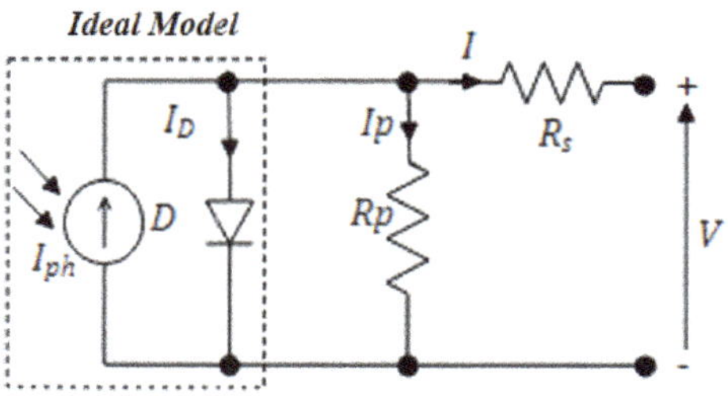

Figure 2. Simplified equivalent circuit of solar cell.

The equivalent circuit of the general model consists of a photo current (*Iph*), a diode, a parallel resistance (*Rp*) expressing a leakage current, and a series resistance (*Rs*) due to the contacts between the semiconductors and the metal parts. This equivalent circuit is depicted in Figure 2.

In Figure 2, we apply Kirchhoff's law. The current will be obtained by the following Equation:

$$I = I_{ph} - I_D - I_p \tag{1}$$

where I_{ph} the current generated by light or photocurrent and I_p the current flowing in the parallel resistor, which can be computed as:

$$I_p = \frac{V + R_s I}{R_p} \tag{2}$$

I_D is the current of the diode, which is proportional to the saturation current. The following equation expresses the value of this magnitude

$$I_D = I_{sd}\left(exp\left(\frac{q.(V+R_s.I)}{n.K.T}\right)-1\right) \tag{3}$$

being I_{sd} the reverse saturation current in amperes (A), q the electron charge (1.6×10^{-19} C), K the Boltzmann constant (1.38×10^{-23} J/K), T the cell temperature in Kelvin (K) and n the ideal factor.

We replace the voltage-current characteristic equation of a solar cell in Equation (1). So we derive that:

$$I = I_{ph} - I_{sd}\left(exp\left(\frac{q.(V+R_s.I)}{n.K.T}\right)-1\right) - \frac{V+R_s I}{R_p} \tag{4}$$

The photocurrent depends mainly on the solar radiation and the operating temperature of the cell, which is described by the following Equation:

$$I_{ph} = \left[I_{sc} + K_i\left(T - T_{ref}\right)\right]\frac{G}{G_{ref}} \tag{5}$$

where:

I_{sc}: is the short-circuit current of the cell at 25 °C and 1000 W/m^2

K_i: the temperature coefficient of the cell short-circuit current,

T_{ref}: is the reference temperature of the cell, in Kelvin (K) (=25 °C + 273),

G: is the solar radiation in watt/square meter (W/m^2),

G_{ref}: is the reference insolation of the cell (=1000 W/m^2),

On the other hand, the cell saturation current varies with the temperature of the cell, which is described as follows:

$$I_{sd} = I_{rs}\left(\frac{T}{T_{ref}}\right)^3 .exp\left(\frac{q.E_g\left(\frac{1}{T_{ref}}-\frac{1}{T}\right)}{K.n}\right) \tag{6}$$

being:

Irs: the reverse saturation current of the PV cell

Eg: is the gap energy of the semiconductor used in the PV cell in electron-volt (eV)

n: is the ideal factor, which depends on the PV technology listed in Table 1.

Table 1. Ideal Factor n dependence on PV cell technology.

Technology	N
Monocrystalline Silicon (Si-mono)	1.2
Polycrystalline Silicon (Si-poly)	1.3
Hydrogenated Amorphous Silicon (a-Si:H)	1.8
Hydrogenated Amorphous Silicon tande (a-Si:H tandem)	3.3
Hydrogenated Amorphous Silicon triple (a-Si:H triple)	5
Cadmium Telluride (CdTe)	1.5
Copper Indium Selenide (CIS)	1.5
Gallium arsenide (GaAs)	1.3

The reverse saturation current is given by the following Equation (7):

$$I_{rs} = \frac{I_{SC}}{exp\left(\frac{q.V_{oc}}{N_s.n.K.T}\right)-1} \tag{7}$$

B. Solar PV Module Model

A typical PV cell produces less than 2 W at about 0.5 V so that cells must be connected in series-parallel on a module to produce enough power [38]. A PV array, like presented in Figure 3, is a group of several PV modules electrically connected in series (N_s cells) and in parallel (N_p columns) to generate the required current and voltage.

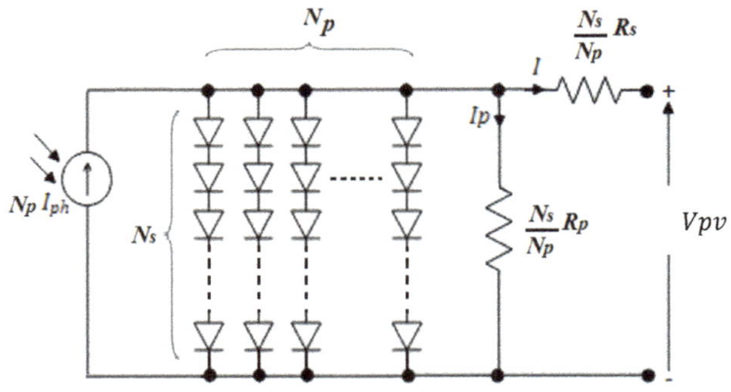

Figure 3. A general model of a Photovoltaic Module.

The voltage-current characteristic equation of a PV module becomes as follows:

$$I = N_p.I_{ph} - N_p.I_{sd}\left[exp\left(\frac{q\left(\frac{Vpv}{N_s} + \frac{I.R_s}{N_p}\right)}{K.T.n}\right) - 1\right] - \frac{\left(\frac{N_p.Vpv}{N_s} + I.R_s\right)}{R_p} \tag{8}$$

The following Table 2 gives the parameters of the PV module used in this work.

Table 2. Characteristics of PV Module "Tata Power Solar Systems TP250MBZ".

Maximum Power (Pmax)	249 W
Voltage at maximum power point (Vmpp)	30 V
Current at maximum power point (Impp)	8.3 A
Open Circuit Voltage (Voc)	36.8 V
Short Circuit Current (Isc)	8.83 A
Number of cell per module	60
Temperature coefficient of Voc(%/deg.c)	−0.33
Temperature coefficient of Isc(%/deg.c)	0.063805
Nominal Voltage $V_{battery}$	100 V
Rated Capacity Battery	10 Ah
Initial Stat of Charge	50%
Battery response time	0.001 s

Figure 4 shows the proposed block diagram of the PV system. In this model. We include the intelligent MPPT algorithm (based on GWO and PSO), which will be described in Sections 3 and 4. The proposed algorithms aim to compensate for the limitations of conventional maximization algorithms, which cannot find the overall maximum, especially during a partial shading phenomenon. Regardless of the partial shading profile, it can find the maximum power point on the power-voltage characteristic (*Ppv*) of a photovoltaic system. This characteristic is usually referred to as *Ppv* = f (*Vpv*). The algorithm is applied to the semiconductor switch of the boost converter installed to connect the PV panels to the DC-link. Specifically, the duty cycle of this device is adjusted according to the output of the Intelligent MPPT algorithm. In our implementation, V_{out} is the voltage that will be used for feeding the battery with the required power for its charge.

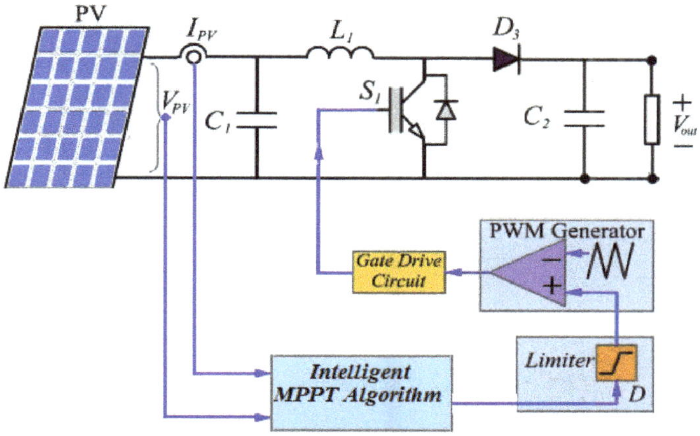

Figure 4. Block diagram of a PV system using an intelligent MPPT algorithm.

3. Optimization Algorithms

This paper focuses on two swarm-intelligence based algorithms. Specifically, they are the Particle Swarm Optimization and the Grey-Wolf Optimization algorithms. They have been selected for this PV system due to its demonstrated capability to adjust power converters.

A. The PSO algorithm

This algorithm was one of the solutions that was used for resolving a huge mathematical optimization problem. It is classified as an evolutionary metaheuristic that belongs to the larger class of evolutionary algorithms.

It was exposed by Russel Eberhart (electrical engineer) and James Kennedy (sociopsychologist) in 1995 [39]. It was originally inspired by the social behavior of animals evolving in swarms, such as schools of fish and flight groups of birds. We can see highly complex movement dynamics in these organisms, despite the fact that each individual has minimal intelligence and only local knowledge of his place in the swarm. [40]. Therefore, local information and the memory of each individual are used to decide their displacement. Simple rules, such as "stay close to another candidate", "go in the same direction, "or" to go at the same speed, "are sufficient to maintain the cohesion of the swarm and allow the implementation of complex adaptative collective behaviors [26].

The particle swarm is a population of simple agents called particles. Each particle is considered as a solution to the problem, where it has a position and a speed. In addition, each particle has a memory allowing it to remember its best performance and the best performance achieved by "neighboring" particles (informants). In addition, each particle has a group of informants, historically called its neighborhood [41,42]. A swarm of particles, which are potential solutions to the optimization problem, will look for the global optimum with their movements.

The following three components that influence the movement of a particle are:

1. An inertia component: the particle tends to follow its current direction of movement;
2. A cognitive component: the particle tends to move towards the best site through which it has already passed;
3. A social component: particle tends to rely on the congeners' experience and, thus, to go to the best site already reached by its neighbors.

The strategy for moving a particle is shown in Figure 5, where the three previous trends are illustrated.

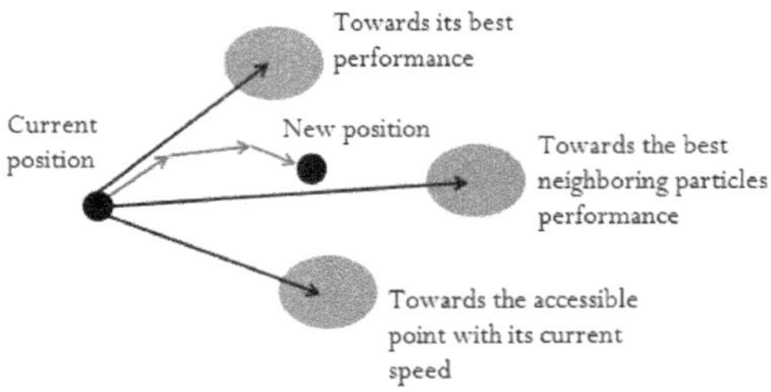

Figure 5. Displacement of a particle in the PSO algorithm.

In an N-dimensional search space, particle i of the swarm is modeled by its position vector $\vec{x}_i = (x_{i1}, x_{i2}, \ldots, x_{iN})$ and by its speed vector $\vec{v}_i = (v_{i1}, v_{i2}, \ldots, v_{iN})$. The value of the objective function determines the quality of its position at this point. The particle keeps in memory the best position through which it has ever passed, which we denote by $\vec{P}_{best\_i}(p_{best\_i1}, p_{best\_i2}, \ldots, p_{best\_iN})$. The best position reached by the particles of the swarm is denoted by $\vec{G}_{best} = (g_{best\_1}, g_{best\_2}, \ldots, g_{best\_N})$.

At the start of the algorithm, the swarm particles are randomly initialized in the search space of the problem. Then, at each iteration, each particle moves, linearly combining the three components mentioned above. Indeed, at iteration t + 1, the speed vector and the position vector are calculated from Equations (9) and (10), respectively, as follows:

$$v_{i,j}^{t+1} = \omega v_{i,j}^{t} + c_1 r_{1,t}^{t}\left[p_{best}{}_{i,j}^{t} - x_{i,j}^{t}\right] + c_2 r_{2i,j}^{t}\left[g_{best}{}_{j}^{t} - x_{i,j}^{t}\right] \quad (9)$$

$$x_{i,j}^{t+1} = x_{i,j}^{t} + v_{i,j}^{t+1} \quad j \in \{1, 2 \ldots, N\} \quad (10)$$

where ω is a constant, called the coefficient of inertia; c1 and c2 are two constants, called acceleration coefficients; r1 and r2 are two random numbers drawn uniformly in the interval [0 1], at each iteration t and for each dimension j.

The three components mentioned above (i.e., inertia, cognitive and social) are represented in Equation (9) by the following terms:

$\omega v_{i,j}^{t}$ corresponds to the inertia component of the displacement, where the parameter ω controls the influence of the direction of displacement on the future movement.

$c_1 r_{1,t}^{t}\left[p_{best}{}_{i,j}^{t} - x_{i,j}^{t}\right]$ corresponds to the cognitive component of the displacement, where the parameter c1 controls the cognitive behavior of the particle.

$c_2 r_{2i,j}^{t}\left[g_{best}{}_{j}^{t} - x_{i,j}^{t}\right]$ corresponds to the social component of the displacement, where the parameter c2 controls the social aptitude of the particle.

Once the particles have moved, the new positions are evaluated. The two vectors P_{best_i} and G_{best} are updated, at iteration t + 1, according to the two Equations (11) and (12). This procedure is presented in Algorithm 1, where M is the number of particles in the swarm.

$$\vec{P}_{best\_i}(t+1) = \begin{cases} \vec{P}_{best\_i}(t), \text{iff}\left(\vec{x}_i(t+1)\right) \geq \vec{P}_{best\_i}(t) \\ \vec{x}_i(t+1), \text{iff}\left(\vec{x}_i(t+1)\right) < \vec{P}_{best\_i}(t) \end{cases} \quad (11)$$

$$\vec{G}_{best}(t+1) = arg\ min_{P_{best\_i}} f\left(\vec{P}_{best\_i}(t+1)\right),\ 1 \leq i \leq M \quad (12)$$

The corresponding flowchart that describes this organization can be found in Figure 6 and explained as follows:

Algorithm 1. The procedure of corresponding flowchart

Randomly initialize M particles: position and speed.
Evaluate the positions of the particles
For each particle i, $\vec{P}_{best\ i} = \vec{x}_i$
Calculate $\vec{G}_{best}$ according to Equation (12)
While the stop criterion is not satisfied, do
Move the particles according to Equations (9) and (10)
Evaluate the positions of the particles
Update $\vec{P}_{best\ i}$ and $\vec{G}_{best}$ according to Equations (11) and (12)
end

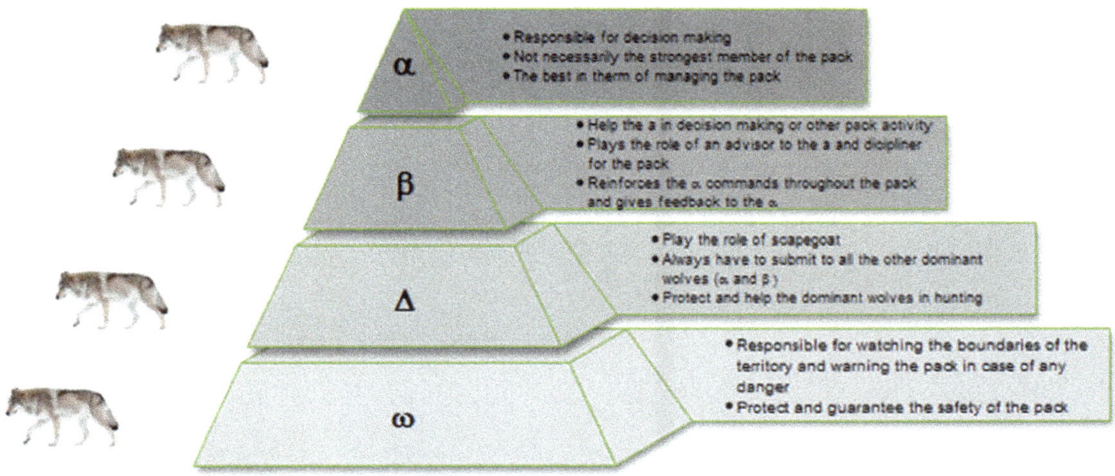

Figure 6. Hierarchical levels of grey wolves and their tasks.

B. The GWO Algorithm

Similarly to the particle swarm Optimization algorithm, one of the new algorithms, which are based on the metaheuristic principle is the Grey Wolf Optimization GWO algorithm. The researcher, Mirjalili, was one of the first researchers who developed this algorithm and exposed its running principle in 2014 [32,43]. To obtain the optimum solution of the problem to be optimized, the algorithm principle uses social authority, which is represented by the behaviour of the wolves when surrounding a victim. During the operation of hunting for the victim, this algorithm simulates the hierarchical supremacy of grey wolves until their movements end. It works in a similar way to population-based algorithms in which it simulates the natural behavior of grey wolves foraging for food in their social lives. Four types of grey wolf groups can be used to compose hierarchical commands. Figure 6 shows this hierarchy, with the following three levels:

a. The first level represents the group's command. A wolf in this level is called (alpha) (α). The alpha is responsible for deciding to hunt and orders the other wolves in the pack. Therefore, it can be considered as the correct solution.
b. The next level in the chain is called (β). Wolves in this level help the alpha to make decisions and monitor the actions of other groups. They can replace the alphas when they die or get older.
c. The lower level contains the delta and omega, which are the lowest ranks and who eat last after the wolves of the upper levels have finished.

The hunting process of the wolf pack involves three main steps: chasing, surrounding the prey, and attacking the prey. The algorithm starts with a given number of grey wolves with their positions generated arbitrarily.

The following equations determine the encircling behavior of each group of the pack:

$$\vec{D} = \left|\vec{C}.\vec{X}_p(t) - \vec{X}(t)\right| \tag{13}$$

$$\vec{X}(t+1) = \left|\vec{X}_p(t) - \vec{A}.\vec{D}\right| \tag{14}$$

where $\vec{X}(t)$ is the vector position of the grey wolf, $\vec{X}_p(t)$ is the vector prey position and $\vec{A}$ and $\vec{C}$ are the vectors gives by the following Equations:

$$\begin{cases} \vec{A} = 2.\vec{a}.\vec{r}_1 - \vec{a} \\ \vec{C} = 2.\vec{r}_2 \end{cases} \quad \text{with} : \; a = 2.\left(1 - \frac{t}{T_{max}}\right) \tag{15}$$

being t the current iteration, T_{max} the total number of iterations and r_1 and r_2 are random vectors chosen in the interval [0,1].

The prey position *Xp (t + 1)* update is calculated by averaging the positions of grey wolves α, β and Δ (three temporarily optimal solutions). The following average function is used for this purpose

$$\vec{X}_p(t+1) = \frac{\vec{X}_1(t) + \vec{X}_2(t) + \vec{X}_3(t)}{3} \tag{16}$$

where:

$$\begin{cases} \vec{X}_1(t) = \vec{X}_\alpha(t) - \vec{A}_1.\vec{D}_\alpha \\ \vec{X}_2(t) = \vec{X}_\beta(t) - \vec{A}_2.\vec{D}_\beta \\ \vec{X}_3(t) = \vec{X}_\Delta(t) - \vec{A}_3.\vec{D}_\Delta \end{cases} \text{and} \begin{cases} \vec{D}_\alpha = \left|\vec{C}_1\vec{X}_\alpha(t) - \vec{X}(t)\right| \\ \vec{D}_\beta = \left|\vec{C}_2\vec{X}_\beta(t) - \vec{X}(t)\right| \\ \vec{D}_\Delta = \left|\vec{C}_3\vec{X}_\Delta(t) - \vec{X}(t)\right| \end{cases}$$

Equation (13) represents the distance from the current position, which should be minimized as much as possible so that the next position represented by Equation (14) gets closer and closer to the position of the prey. This will imply that the algorithm will get to the correct solution of the problem *Xp(t)*.

The parameter "*a*" used in this algorithm decreases linearly in the interval [2,0] for the successive iterations using Equation (15). Thus, it will model the behavior of the wolves when approaching the victim (exploration phase). For this phase, if the condition | A | < 1 is verified, the wolves attack the victim.

The alpha group are said to have the best possible knowledge of the location of prey. Once the position of the prey is determined, the hunt will be guided by the alpha group followed by the beta and delta wolves. The latter two groups participate in the hunt occasionally. The rest of the group is limited to take care of the injured wolves of the pack. When the prey stops moving, the wolves attack and finish the hunt [43].

The flowchart of this algorithm is illustrated in Figure 7.

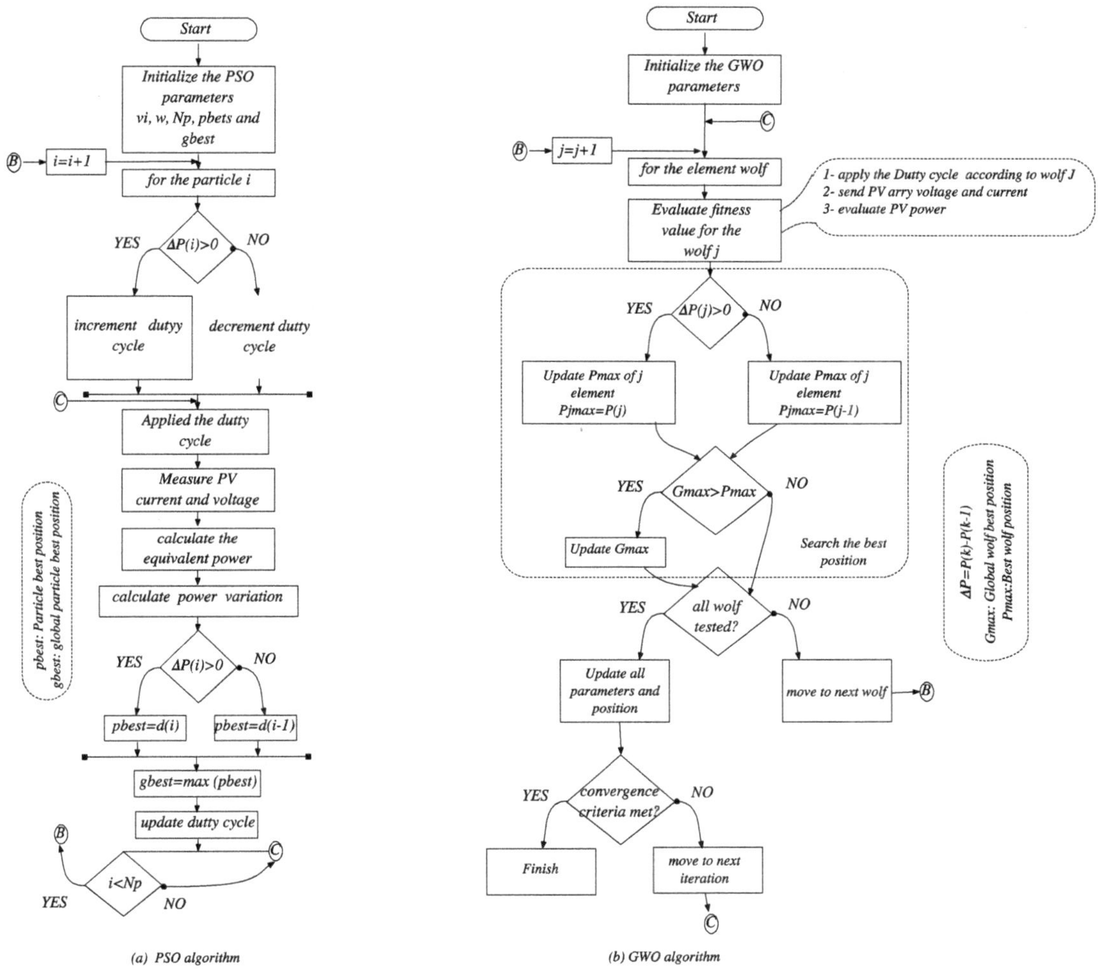

Figure 7. PSO and GWO algorithms flowchart.

C. The PSO and GWO MPPT controllers

To apply the previous evolutionary algorithms, we need to define the functions and parameters for the proposed problem. The objective is to maximize the energy extraction from the PV cells installed on EVs, even when there are shadowing areas involved in the scenario to consider.

For the PSO algorithm, the function to optimize is expressed in Equation (17).

$$\max(Ppv) = f\left(x_{i,j}^{t},\ v_{i,j}^{t+1}\right) \tag{17}$$

In the GWO, each wolf position corresponds to the duty cycle factor applied on the MPPT. Thus, the *Gmax* represents the global best position of all of the wolves and *Pmax* is the corresponding best position of the corresponding wolf. If no partial shading exists, the global best wolf best position and the local wolf best position will be the same. However, for the partial shading case, there is several (*Pmax*) wolf best positions, and the global best position (*Gmax*) will be related to the max of the best wolf position *Gmax* = *max (Pmax)*.

The proposed GWO algorithm aims to compensate for the limitations of conventional maximization algorithms, which cannot find the overall maximum, especially during a partial shading phenomenon. Regardless of the partial shading profile, this algorithm can find the maximum power point on the power-voltage characteristic of a photovoltaic system. Towards this goal, the objective function for this algorithm is defined as it is in Equation (18).

$$\max(Ppv) = f(\alpha,\ \beta,\ \Delta, \omega) \quad (18)$$

The proposed MPPT algorithm is based on the application of GWO to control the duty cycle. The expected consequence is that this will turn into decreasing the steady-state oscillations presented by the conventional MPP tracking algorithms. Consequently, the power loss due to oscillation is reduced, which increases the photovoltaic system efficiency.

For the implementation of GWO MPPT algorithm, duty cycle d is defined as a grey wolf. Therefore, the Equation (14) can be rewritten as follows:

$$d_i(k+1) = d_i(k) - A.d \quad (19)$$

The fitness function of the GWO algorithm is formulated to have the objective of function 20. It is for the two optimization algorithms.

$$P\left(d_j^k\right) > P\left(d_j^{k-1}\right) \quad (20)$$

where P represents the PV power, d is the duty cycle of the boost converter, j is the current grey wolves number, and k is the iteration number.

The flowchart and the parameters of the PSO and GWO MPPT algorithms are detailed in Figure 7 and Table 3, respectively.

Table 3. Parameter of the PSO and GWO MPPT algorithms.

	PSO	GWO
Inertia weight (w)	0.1	-
Personal Learning Coefficient $c1$	1.7	-
Global Learning Coefficient $c2$	1.5	-
Constant a	-	Self adaption
Coefficient r_1 and r_2	-	Random numbers in [0,1]
Number of iterations N	100	100
Number of particles P	10	10
Sampling time	0.0001s	0.0001s
Initial duty cycle	0.4	

The choice of these parameters was fixed after several online simulation tests, where the goal is to find the best combination in terms of the algorithm running speed and the best performances. Specifically, we have varied the number of particles/wolves and the maximum number of iterations. The first test had the configuration of 150 iterations and 10 particles. The corresponding simulation time was evaluated to 30 min when using an I5 laptop with 8-GB as RAM memory. The resulting performances were found perfect. The second configuration was fixed to 50 iteration and 8 particles, and then the corresponding simulation time was evaluated to 24 min, but the resulting energetic performances were not so good. There were some problems with the stability of the output power.

Many other tests (more than 5 combinations) were also applied, and the best combination was found as it is indicated in Table 3. For the selected configuration, the simulation time was 26 min, and we get a good performance in terms of extracted power and stability.

4. Implementation and Simulation Results

The algorithms previously studied are applied to a PV system composed of 4 photovoltaic modules. They are connected in a combination of serial and parallel cells and the characteristics are given in Table 2. Considering the size of the PV modules, this system could be implemented in electric vehicle application or even an isolated grid for feeding some isolating area using a solar energy. Some recharge stations can also benefit from this solution and the proposed method can improve energy yield.

The simulation bloc can be illustrated as it is Figure 8, all the PV cells are connected to the DC bus, which can give information about the global voltage and current, in relation to the solar radiation factor. Then the MPPT algorithm can extract the maximum of power using the PSO or the GWO algorithm. A DC converter is used for having information about the outputted power delivered to the load.

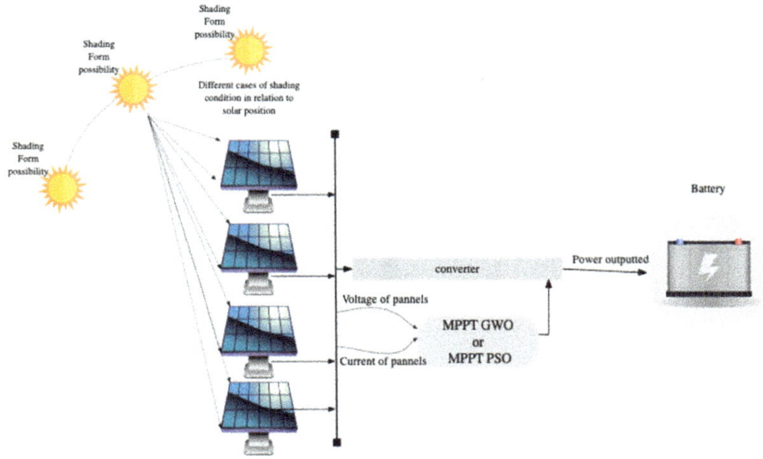

Figure 8. Overall system design in the simulation phase.

Indeed, to compare the efficiency of MPPT PSO and GWO we considered four simulation tests so that a wide variety of conditions are modelled, Table 4, gives the maximum power, voltage and current parameters for each shading condition. During the first test, Figure 9a, the irradiation was kept constant and uniform for the four modules (1000 W/m^2). The second test is characterized by partial shading on two of the four modules of around 40%, represented in Figure 9b. It corresponds to the following lighting distribution on the four modules (600 W/m^2, 600 W/m^2, 1000 W/m^2, 1000 W/m^2). The third test (in Figure 9c) resembles the second one, but the partial shading on the two modules is varied in the order of 50% on the first and 20% for the second (500 W/m^2, 800 W/m^2, 1000 W/m^2, 1000 W/m^2). Three modules operate under partial shade during the fourth Test, in Figure 9d. This corresponds to the following lighting distribution (200 W/m^2, 300 W/m^2, 700 W/m^2, 1000 W/m^2)

Table 4. PV characteristics study for a different type of shading.

	P_{mpp} (W)	V_{mpp} (V)	I_{mpp} (A)
Uniform irradiation	996	120	8.3
First case of Partial shading	646.3	126.2	5.12
Second case of Partial shading	637.9	93.04	6.85
Third case of Partial shading	363.4	60.74	5.98

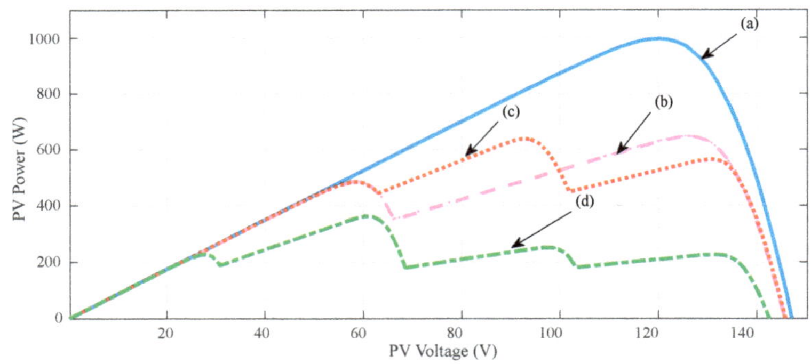

Figure 9. PV characteristics under different levels of partial shading.

Figures 10–13 show the main electrical measurements of the PV tested for different shadowing conditions. For these results, we have used the Matlab simulation tool.

A. Uniform irradiation

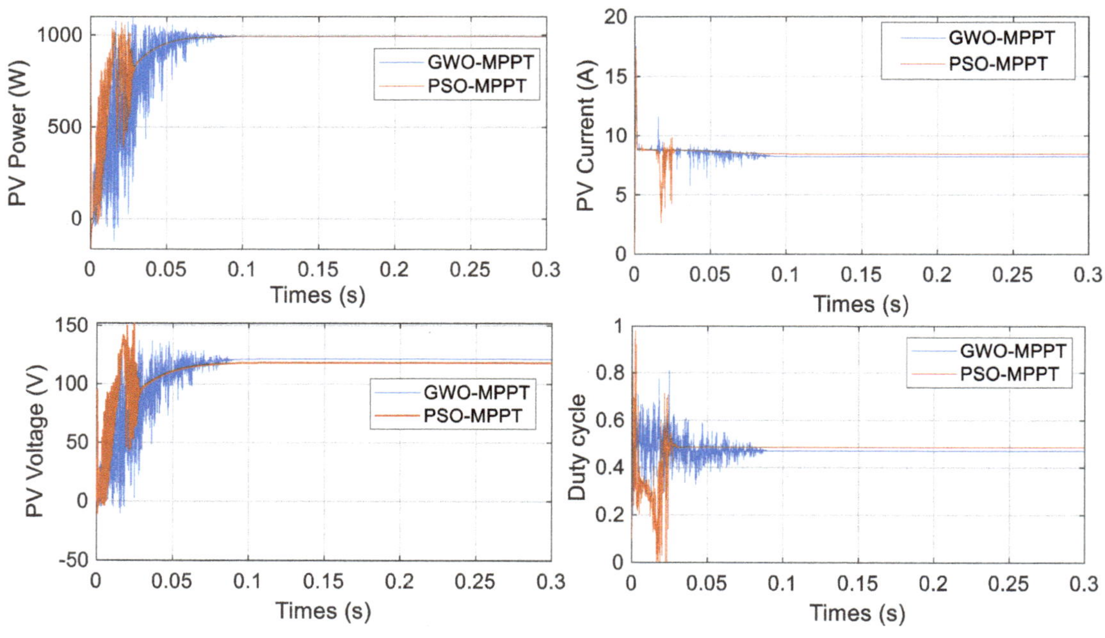

Figure 10. PV characteristics under uniform irradiation: 0% partial shading.

B. The first case of partial shading

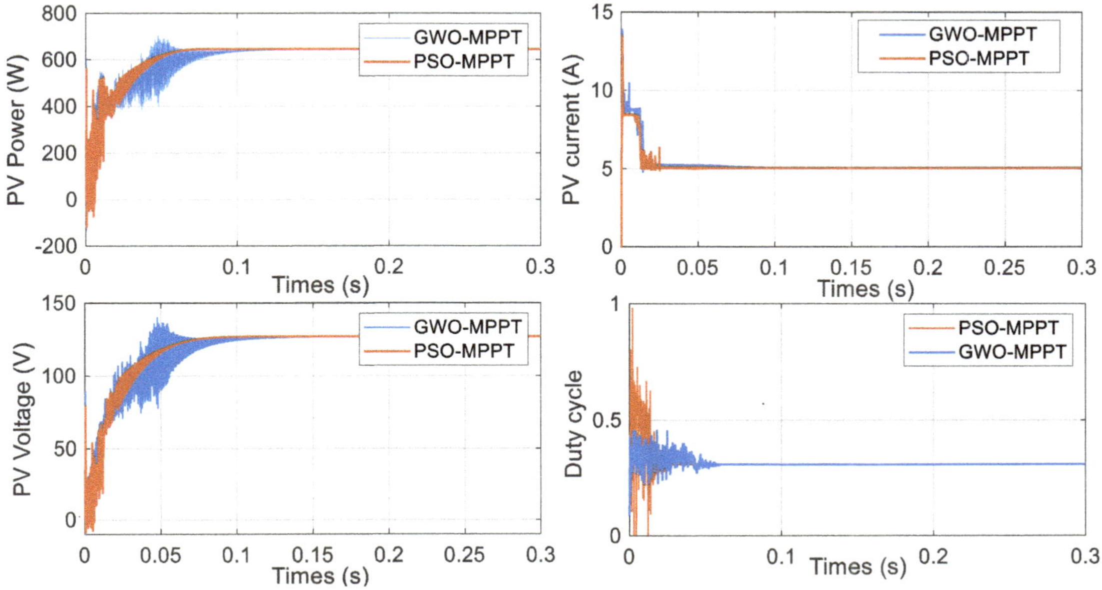

Figure 11. PV characteristics under partial shading: (600 W/m ^{2}, 600 W/m ^{2}, 1000 W/m ^{2}, 1000 W/m ^{2}).

C. Second case of partial shading.

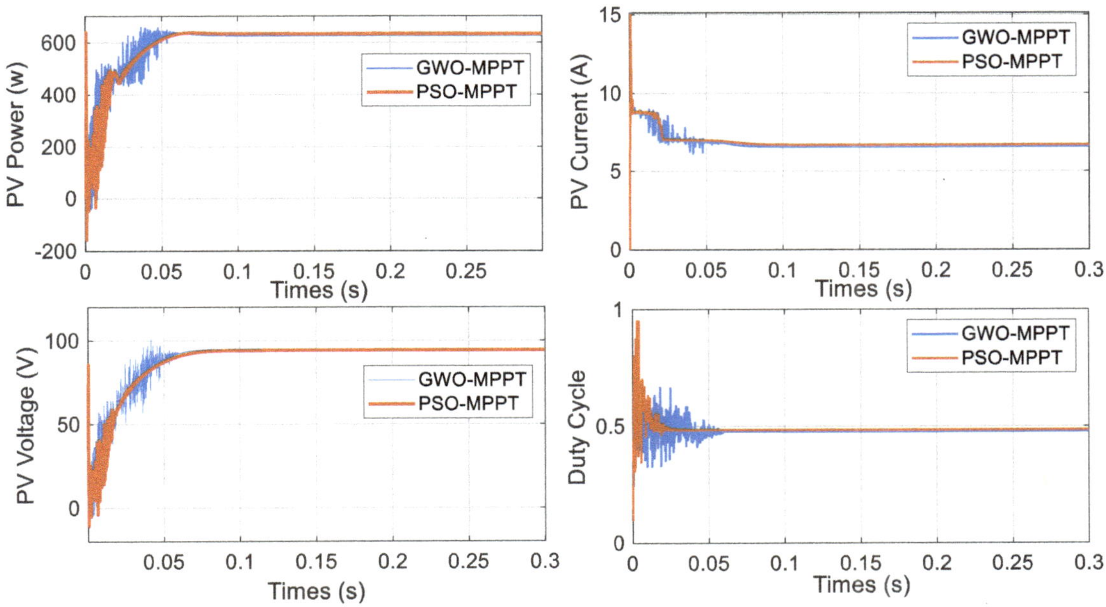

Figure 12. PV characteristics under partial shading: (500 W/m^2, 800 W/m^2, 1000 W/m^2, 1000 W/m^2).

D. Third case of partial shading

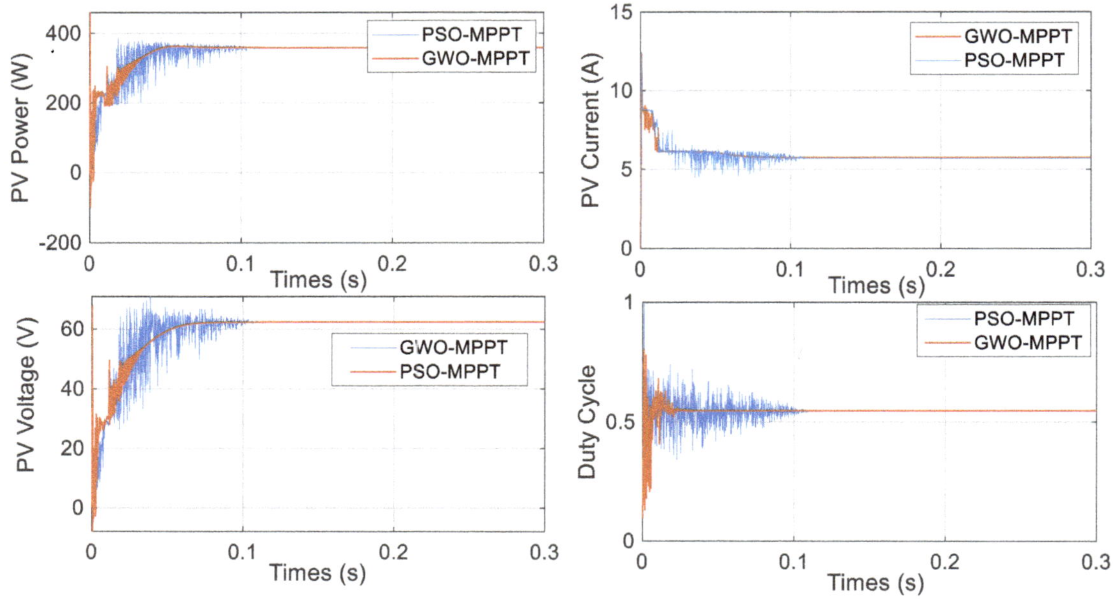

Figure 13. PV characteristics under partial shading: (200 W/m^2, 300 W/m^2, 700 W/m^2, 1000 W/m^2).

Table 5 presents a comparative study between the two MPPT algorithms for different types of shading distribution on the panels. The essential criteria for a judicious comparison between the two algorithms are:

- Track of the point of maximum power under the different levels of partial shading.
- Rapid convergence towards the point of the global maximum.

As can be observed, the simulation results from Figure 10 to Figure 13 show the evolution of the power, current, voltage and duty cycle of the PV system for the four types of shading distribution we have tested on the panels. From these results, it can be observed that the two solutions ensure a good MPP tracking. The advantage of the PSO MPPT over the GWO MPPT is related to two issues: (i) the amplitude of the oscillations at the transient state and (ii) the accuracy to track the point of maximum power. A high oscillation exists for the case of GWO, which can be one the weaknesses of this algorithm. There is also a small oscillation when executing the PSO at the beginning, when the radiation form changes. However, this will not cause a problem as in the real situation, the modification of the radiation comes very slow. So, we a look a better performance in a real situation.

Table 5. Comparative study between PSO-MPPT and GWO-MPPT.

MPPT Techniques	Uniform Irradiation		First Case of Partial Shading		Second Case of Partial Shading		Third Case of Partial Shading	
	PSO	GWO	PSO	GWO	PSO	GWO	PSO	GWO
Time to reach the MPP (s)	0.081	0.096	0.071	0.106	0.079	0.091	0.0561	0.105
Extracted Power at MPP (W)	993.2	994.7	645.6	645.1	633.9	628	359.1	357.5
Tracking Efficiency (%)	99.71	99.86	99.89	99.81	99.37	98.44	98.81	98.37

The application of these algorithms in real time requires the use of the high-speed processor given the large number of operations to be carried out in one second (processing and control measurement). Therefore, the time needed to converge towards the best response depends on the speed of the algorithm used and the material available (essentially the speed of the processor). In addition, the presence of high amplitude oscillations during the transient phase is a harmful phenomenon for electrical systems and can cause a variety of problems. According to the simulation results and the criteria indicated above, the PSO MPPT algorithm shows itself well for the real-time application.

We have also studied how these algorithms impact on the battery State-of-Charge (SoC). In particular, we have studied the effects on a representative lithium-ion battery characterized by the parameters in Table 6. Figure 14 shows the efficiency of PSO MPPT versus GWO MPPT in terms of storage charge in the battery under uniform irradiation. A gain of about 0.0014% SoC for 1 s corresponds to almost 5.04% of battery charge for one hour.

Table 6. Parameters of lithium-ion battery.

Nominal Voltage $V_{battery}$	**100 V**
Rated Capacity Battery	10 Ah
Initial Stat of Charge	50%
Battery response time	0.001 s

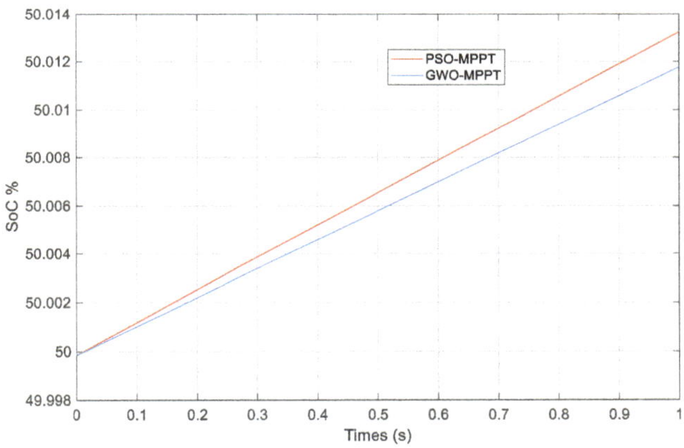

Figure 14. SoC using the PSO MPPT and GWO MPPT.

5. Conclusions

This work attempts to study and discuss two MPPT techniques based on two meta-heuristic optimization algorithms, i.e., PSO and GWO. These new techniques of MPPT overcome the problems of classic MPPT strategies (e.g., perturb and observe and incremental) when tracking the maximum power point, even in the presence of sudden changes of irradiation and shadows on the photovoltaic modules. The MPPT techniques studied show good behavior and better performance. A comparative study of simulation results for a different type of shading shows PSO-MPPT effectiveness compared to GWO-MPPT from the point of view of speed and oscillation during the transient state. In addition, a simulation test shows the efficiency of PSO MPPT versus GWO MPPT in terms of storage charge in the battery under uniform irradiation. As a future work, we would like to analyze how to adapt the algorithm's parameters (which are now constant) to the dynamic lighting conditions.

Moreover, one of the future endeavors of this work is to compare more algorithms performances and search the best combination that can be used for such an optimization problem. Therefore, incremental algorithm, perturb and observe algorithm, the fuzzy solution and other population-based metaheuristic algorithms as bio-inspired algorithms, evolutionary algorithm and physics-based algorithm will be studied, tested and evaluated.

Author Contributions: Conceptualization, H.K. and F.A.; Data curation, H.K.; Formal analysis, H.K. and F.A.; Funding acquisition, M.A.; Investigation, H.K. and L.Y.; Methodology, H.K. and F.A.; Project administration, H.K., A.T. and S.S.M.G.; Resources, H.K.; Software, H.K. and L.Y.; Supervision, H.K. and M.A.; Validation, H.K. and F.A.; Visualization, H.K., F.A., L.Y., M.A. and S.S.M.G.; Writing—original draft, H.K., F.A., L.Y., A.T., M.A. and S.S.M.G.; Writing—review & editing, H.K., F.A., L.Y., A.T., M.A. and S.S.M.G. All authors have read and agreed to the published version of the manuscript.

Funding: This research was funded by Taif University Researchers Supporting Project, grant number "TURSP-2020/122" and by II Plan Propio Smart Campus, project "Smart and Secure EV Urban Lab II" by the University of Málaga (Spain).

Institutional Review Board Statement: Not applicable.

Informed Consent Statement: Not applicable.

Data Availability Statement: Not applicable.

Acknowledgments: The authors would like to acknowledge the financial support received from Taif University Researchers Supporting Project Number (TURSP-2020/122), Taif University, Taif, Saudi Arabia and the financial support received from II Plan Propio Smart Campus, project "Smart and Secure EV Urban Lab II" by the University of Málaga (Spain).

Conflicts of Interest: The authors declare no conflict of interest.

References

1. Xie, W.T.; Dai, Y.J.; Wang, R.Z.; Sumathy, K. Concentrated solar energy applications using Fresnel lenses: A review. *Renew. Sustain. Energy Rev.* **2011**, *15*, 2588–2606. [CrossRef]
2. Mokarram, M.; Mokarram, M.J.; Khosravi, M.R.; Saber, A.; Rahideh, A. Determination of the optimal location for constructing solar photovoltaic farms based on multi-criteria decision system and Dempster–Shafer theory. *Sci. Rep.* **2020**, *10*, 8200. [CrossRef] [PubMed]
3. Lv, Z.; Qiao, L.; Cai, K.; Wang, Q. Big Data Analysis Technology for Electric Vehicle Networks in Smart Cities. *IEEE Trans. Intell. Transp. Syst.* **2021**, *22*, 1807–1816. [CrossRef]
4. Mohamed, N.; Aymen, F.; Ali, Z.M.; Zobaa, A.F.; Aleem, S.H.E.A. Efficient Power Management Strategy of Electric Vehicles Based Hybrid Renewable Energy. *Sustainability* **2021**, *13*, 7351. [CrossRef]
5. Habib, A.K.M.A.; Hasan, M.K.; Mahmud, M.; Motakabber, S.M.A.; Ibrahimya, M.I.; Islam, S. A review: Energy storage system and balancing circuits for electric vehicle application. *IET Power Electron.* **2021**, *14*, 1–13. [CrossRef]
6. Flah, A.; Majed, A.; Bajaj, M.; Naveen, K.S.; Mishra, S.; Sharma, S.K. Electric Vehicle Model Based on Multiple Recharge System and a Particular Traction Motor Conception. *IEEE Access* **2021**, *9*, 49308–49324.
7. Arun, P.; Mohanrajan, S. Effect of Partial Shading on Vehicle Integrated PV System. In Proceedings of the 2019 3rd International Conference on Electronics, Communication and Aerospace Technology (ICECA), Coimbatore, India, 12–14 June 2019; pp. 1262–1267.
8. Mobarak, M.H.; Kleiman, R.N.; Bauman, J. Solar-Charged Electric Vehicles: A Comprehensive Analysis of Grid, Driver, and Environmental Benefits. *IEEE Trans. Transp. Electrif.* **2021**, *7*, 579–603. [CrossRef]
9. Al-Shammaa, A.A.; Stocker, A. A Novel Method for Event Detection using Wireless Sensor Networks. In Proceedings of the ICCSW, London, UK, 20–21 September 2018.
10. Kraiem, H.; Flah, A.; Mohamed, N.; Alowaidi, M.; Bajaj, M.; Mishra, S.; Sharma, N.K.; Sharma, S.K. Increasing Electric Vehicle Autonomy Using a Photovoltaic System Controlled by Particle Swarm Optimization. *IEEE Access* **2021**, *9*, 72040–72054. [CrossRef]
11. Lin, G.; Bimenyimana, S.; Tseng, M.-L.; Wang, C.-H.; Liu, Y.; Li, L. Photovoltaic Modules Selection from Shading Effects on Different Materials. *Symmetry* **2020**, *12*, 2082. [CrossRef]
12. Picault, D.; Raison, B.; Bacha, S.; de la Casa, J.; Aguilera, J. Forecasting photovoltaic array power production subject to mismatch losses. *Sol. Energy* **2010**, *84*, 1301–1309. [CrossRef]
13. Haque, A.; Bharath, K.V.S.; Khan, M.A.; Khan, I.; Jaffery, Z.A. Fault diagnosis of Photovoltaic Modules. *Energy Sci. Eng.* **2019**, *7*, 622–644. [CrossRef]
14. Pandian, A.; Bansal, K.; Thiruvadigal, D.J.; Sakthivel, S. Fire Hazards and Overheating Caused by Shading Faults on Photo Voltaic Solar Panel. *Fire Technol.* **2016**, *52*, 349–364. [CrossRef]
15. Safari, A.; Mekhilef, S. Incremental conductance MPPT method for PV systems. In Proceedings of the 2011 24th Canadian Conference on Electrical and Computer Engineering (CCECE), Niagara Falls, ON, Canada, 8–11 May 2011; pp. 345–347.
16. Banu, I.V.; Beniugă, R.; Istrate, M. Comparative analysis of the perturb-and-observe and incremental conductance MPPT methods. In Proceedings of the 2013 8th International Symposium on Advanced Topics In Electrical Engineering (Atee), Bucharest, Romania, 23–25 May 2013; pp. 1–4.

17. Águila-León, J.; Chiñas-Palacios, C.D.; Vargas-Salgado, C.; Hurtado-Perez, E.; García, E.X.M. Optimal PID Parameters Tunning for a DC-DC Boost Converter: A Performance Comparative Using Grey Wolf Optimizer, Particle Swarm Optimization and Genetic Algorithms. In Proceedings of the 2020 IEEE Conference on Technologies for Sustainability (SusTech), Santa Ana, CA, USA, 23–25 April 2020; pp. 1–6.
18. Algarín, C.R.; Álvarez, O.R.; Castro, A.O. Data from a photovoltaic system using fuzzy logic and the P&O algorithm under sudden changes in solar irradiance and operating temperature. *Data Br.* **2018**, *21*, 1618–1621.
19. Ahmed, J.; Salam, Z. An improved perturb and observe (P&O) maximum power point tracking (MPPT) algorithm for higher efficiency. *Appl. Energy* **2015**, *150*, 97–108.
20. Liu, C.; Luo, Y.; Huang, J.; Liu, Y. A PSO-based MPPT algorithm for photovoltaic systems subject to inhomogeneous insolation. In Proceedings of the the 6th International Conference on Soft Computing and Intelligent Systems, and the 13th International Symposium on Advanced Intelligence Systems, Kobe, Japan, 20–24 November 2012; pp. 721–726.
21. Miyatake, M.; Veerachary, M.; Toriumi, F.; Fujii, N.; Ko, H. Maximum Power Point Tracking of Multiple Photovoltaic Arrays: A PSO Approach. *IEEE Trans. Aerosp. Electron. Syst.* **2011**, *47*, 367–380. [CrossRef]
22. Awais, M.; Khan, L.; Ahmad, S.; Mumtaz, S.; Badar, R. Nonlinear adaptive NeuroFuzzy feedback linearization based MPPT control schemes for photovoltaic system in microgrid. *PLoS ONE* **2020**, *15*, e0234992. [CrossRef]
23. Viloria-Porto, J.; Robles-Algarín, C.; Restrepo-Leal, D. Code and data from an ADALINE network trained with the RTRL and LMS algorithms for an MPPT controller in a photovoltaic system. *Data Br.* **2020**, *32*, 106296. [CrossRef] [PubMed]
24. Premkumar, M.; Subramaniam, U.; Babu, T.S.; Elavarasan, R.M.; Mihet-Popa, L. Evaluation of Mathematical Model to Characterize the Performance of Conventional and Hybrid PV Array Topologies under Static and Dynamic Shading Patterns. *Energies* **2020**, *13*, 3216. [CrossRef]
25. Díaz Martínez, D.; Trujillo Codorniu, R.; Giral, R.; Vázquez Seisdedos, L. Evaluation of particle swarm optimization techniques applied to maximum power point tracking in photovoltaic systems. *Int. J. Circuit Theory Appl.* **2021**, *49*, 1849–1867. [CrossRef]
26. Ab Wahab, M.N.; Nefti-Meziani, S.; Atyabi, A. A Comprehensive Review of Swarm Optimization Algorithms. *PLoS ONE* **2015**, *10*, e0122827. [CrossRef]
27. Chao, K.-H.; Rizal, M.N. A Hybrid MPPT Controller Based on the Genetic Algorithm and Ant Colony Optimization for Photovoltaic Systems under Partially Shaded Conditions. *Energies* **2021**, *14*, 2902. [CrossRef]
28. Min-Shui, H.; Mustafa, G.; Hong-Ping, Z. Vibration-Based Structural Damage Identification under Varying Temperature Effects. *J. Aerosp. Eng.* **2018**, *31*, 4018014.
29. Huang, M.; Lei, Y.; Li, X. Structural Damage Identification Based on l_1Regularization and Bare Bones Particle Swarm Optimization with Double Jump Strategy. *Math. Probl. Eng.* **2019**, *2019*, 5954104. [CrossRef]
30. Gupta, S.; Abderazek, H.; Yıldız, B.S.; Yildiz, A.R.; Mirjalili, S.; Sait, S.M. Comparison of metaheuristic optimization algorithms for solving constrained mechanical design optimization problems. *Expert Syst. Appl.* **2021**, *183*, 115351. [CrossRef]
31. Huang, M.; Li, X.; Lei, Y.; Gu, J. Structural damage identification based on modal frequency strain energy assurance criterion and flexibility using enhanced Moth-Flame optimization. *Structures* **2020**, *28*, 1119–1136. [CrossRef]
32. Mirjalili, S.; Mirjalili, S.M.; Lewis, A. Grey Wolf Optimizer. *Adv. Eng. Softw.* **2014**, *69*, 46–61. [CrossRef]
33. Nadweh, S.; Khaddam, O.; Hayek, G.; Atieh, B.; Haes Alhelou, H. Steady state analysis of modern industrial variable speed drive systems using controllers adjusted via grey wolf algorithm & particle swarm optimization. *Heliyon* **2020**, *6*, e05438.
34. Jumani, T.A.; Mustafa, M.W.; Alghamdi, A.S.; Rasid, M.M.; Alamgir, A.; Awan, A.B. Swarm Intelligence-Based Optimization Techniques for Dynamic Response and Power Quality Enhancement of AC Microgrids: A Comprehensive Review. *IEEE Access* **2020**, *8*, 75986–76001. [CrossRef]
35. Çetinbaş, İ.; Tamyürek, B.; Demirtaş, M. Sizing optimization and design of an autonomous AC microgrid for commercial loads using Harris Hawks Optimization algorithm. *Energy Convers. Manag.* **2021**, *245*, 114562. [CrossRef]
36. Li, Y.; Lin, X.; Liu, J. An improved gray wolf optimization algorithm to solve engineering problems. *Sustainability* **2021**, *13*, 3208. [CrossRef]
37. Preiser, K. Photovoltaic Systems. *Handb. Photovolt. Sci. Eng.* **2003**, 753–798.
38. Chang, W.J.; Lee, K.-H.; Ha, H.; Jin, K.; Kim, G.; Hwang, S.-T.; Lee, H.-M.; Ahn, S.-W.; Yoon, W.; Seo, H.; et al. Design Principle and Loss Engineering for Photovoltaic-Electrolysis Cell System. *ACS Omega* **2017**, *2*, 1009–1018. [CrossRef]
39. Eberhart, J.K.R. Particle swarm optimization. In Proceedings of the IEEE International Conference on Neural Networks, Perth, WA, Australia, 27 November–1 December 1995; pp. 1942–1948.
40. Kraiem, H.; Shaaban, S.M. Energy optimization of an electric car using losses minimization and intelligent predictive torque control. *J. Algorithm. Comput. Technol.* **2020**, *14*, 1748302620966698. [CrossRef]
41. Flah, A.; Oussama, B.; Sbita, L.; Mohamed, N.S. BLDC Control Method Optimized by PSO Algorithm. *Int. Symp. Adv. Electr. Commun. Technol.* **2019**, 1–5.
42. Kariem, H.; Touti, E.; Fetouh, T. The efficiency of PSO-based MPPT technique of an electric vehicle within the city. *Meas. Control.* **2020**. [CrossRef]
43. Saxena, A.; Sharma, A.; Shekhawat, S. Parameter extraction of solar cell using intelligent grey wolf optimizer. *Evol. Intell.* **2020**, *11*, 1–7. [CrossRef]

applied sciences

MDPI

Article

Cooperative Multi-Objective Optimization of DC Multi-Microgrid Systems in Distribution Networks

Zhiwen Xu, Changsong Chen *, Mingyang Dong, Jingyue Zhang, Dongtong Han and Haowen Chen

State Key Laboratory of Advanced Electromagnetic Engineering and Technology, School of Electrical and Electronic Engineering, Huazhong University of Science and Technology, Wuhan 430074, China; xuzhiwen@hust.edu.cn (Z.X.); dongmingyang@hust.edu.cn (M.D.); m202071581@hust.edu.cn (J.Z.); m202071541@hust.edu.cn (D.H.); m201971407@hust.edu.cn (H.C.)

* Correspondence: ccsfm@hust.edu.cn

Featured Application: A cooperative multi-objective optimization model of a DC multi-microgrid that considers across-time-and-space energy transmission of EVs is established to improve the economy of the system, decrease the loss of the distribution network, and reduce carbon emissions.

Abstract: By constructing a DC multi-microgrid system (MMGS) including renewable energy sources (RESs) and electric vehicles (EVs) to coordinate with the distribution network, the utilization rate of RESs can be effectively improved and carbon emissions can be reduced. To improve the economy of MMGS and reduce the network loss of the distribution network, a cooperative double-loop optimization strategy is proposed. The inner-loop economic dispatching reduces the daily operating cost of MMGS by optimizing the active power output of RESs, EVs, and DC/AC converters in MMGS. The outer-loop reactive power optimization reduces the network loss of the distribution network by optimizing the reactive power of the bidirectional DC/AC converters. The double-loop, which synergistically optimizes the economic cost and carbon emissions of MMGS, not only improves the economy of MMGS and operational effectiveness of the distribution network but also realizes the low-carbon emissions. The Across-time-and-space energy transmission (ATSET) of the EVs is considered, whose impact on economic dispatching is analyzed. Particle Swarm Optimization (PSO) is applied to iterative solutions. Finally, the rationality and feasibility of the cooperative multi-objective optimization model are proved by a revised IEEE 33-node system.

Keywords: DC multi-microgrid system; carbon emissions; economic dispatch; across-time-and-space energy transmission; cooperative multi-objective optimization

Citation: Xu, Z.; Chen, C.; Dong, M.; Zhang, J.; Han, D.; Chen, H. Cooperative Multi-Objective Optimization of DC Multi-Microgrid Systems in Distribution Networks. *Appl. Sci.* **2021**, *11*, 8916. https://doi.org/10.3390/app11198916

Academic Editor: Marcos Tostado-Véliz

Received: 7 August 2021
Accepted: 21 September 2021
Published: 24 September 2021

Publisher's Note: MDPI stays neutral with regard to jurisdictional claims in published maps and institutional affiliations.

Copyright: © 2021 by the authors. Licensee MDPI, Basel, Switzerland. This article is an open access article distributed under the terms and conditions of the Creative Commons Attribution (CC BY) license (https://creativecommons.org/licenses/by/4.0/).

1. Introduction

Since the national carbon neutrality and carbon peak requirements have been put forward [1], low carbon emissions and new energy have become hot research topics [2,3]. It is a trend to replace petrol vehicles with electric vehicles (EVs) and replace regional large-scale power grids with microgrids (MGs) containing renewable energy sources (RESs) [4,5]. As the output of RESs is intermittent and uncertain, the MGs need to coordinate with the distribution network to centrally regulate the RESs, which is a challenge to the operation mode of the traditional power system. With the popularity of EVs, the burden of the distribution network will greatly be increased. Additionally, the safe operation of the distribution network will be threatened if EVs are charged in the distribution network without control.

The research on the charging and discharging dispatching strategy of EVs is mainly from the view of the economy [6,7]. Many studies have considered charging/discharging strategies of EVs but overlooked the energy storage characteristics of EVs. Through the bidirectional Vehicle-to-grid (V2G) technology, EVs can also deliver electrical energy to the grid by discharging, and improve the operation of the grid [8–10].

Some research combines EVs with distributed RESs in the MG. In [11], an optimization method for the operation route and charging/discharging time of EVs is proposed, which uses the timely charging/discharging of EVs to consume the output of RESs and reduce the volatility of the equivalent load. In [12], the MG energy management strategy is discussed from the perspective of system operating cost and the consumption efficiency of RESs, and V2G technology has been applied. In [13], the structure and parameter design of the system have been discussed, and the actual MG system using V2G technology has been studied. However, most of the energy scheduling in MGs and the distribution network is to adjust the output of active power.

Some research focuses on the economic dispatch of the multi-microgrid system. In [14], an interconnected multi-microgrids (IMMGs) system using various complementary power sources effectively coordinates the energy sharing/trading among the MGs and the main grid to improve energy efficiency. In [15], A probabilistic modeling of both small-scale energy resources (SSERs) and load demand at each microgrid (MGs) is performed to determine the optimal economic operation of each MG with minimum cost based on the power transaction between the MGs and the main grid. The above does not consider the reactive power exchange between MMG and the main grid.

In the current research on reactive power exchange and network loss, most studies focus on the reactive power of a single distribution network. In [16], the trend of reactive power demand in the distribution network is evaluated. Reactive power demand management plays an important role in the cost-effectiveness and stable operation of the distribution network. A multi-objective planning algorithm for reactive power compensation of radial distribution networks is proposed in [17], which uses unified power quality conditioner (UPQC) compensation for load reactive power to reduce network loss. In [18], the solid-state transformer (SST) is used to supply the load reactive power demand and inject reactive power into the grid, which reduces network losses in a radial distribution network.

Some research focuses on the impact of reactive power optimization on the loss of MG. In [19], a distributed, leaderless and randomized algorithm is proposed, which controls the microgenerators in the island-operated MG system to compensate for reactive power and reduce power distribution loss in MG. A generalized approach for probabilistic optimal reactive power planning is proposed in [20], which can reduce the annual energy losses of the grid-connected MG system.

These papers mentioned above give less consideration to the collaborative optimization of MG clusters and the distribution network. To solve the above problems, a cooperative multi-objective optimization model of a DC multi-microgrid system (MMGS) including RESs, EVs, and DC/AC converters is established. The goal of the model is to obtain the optimal MMGS economic cost and the network loss of the distribution network. The main contributions of this paper are as follows:

1. A grid-connected MMGS containing RESs and EVs is constructed, where RESs, EVs, MGs and distribution networks are combined, bidirectional V2G technology is used and the across-time-and-space energy transmission (ATSET) of EVs is thoroughly considered. The effect of the across-time-and-space energy transmission on MMGS economic operation is analyzed to state the potential benefits of cooperative multi-objective optimization.
2. A cooperative multi-objective optimization model is established, including the dynamic economic dispatch of RESs, EVs, DC/AC converters, and the reactive power optimization of DC/AC converters in MMGS. The cooperative multi-objective optimization model consists of two loops. The inner-loop model uses the active power output of RESs, EVs, DC/AC converters as variables, and the daily operating cost of MMGS is used as the optimization objective. The outer-loop model uses the reactive power output of the DC/AC converters as the variable to optimize the network loss of the distribution network, thereby reducing network loss cost and carbon emissions

cost. The ultimate goal of the cooperative multi-objective is to obtain the optimal daily economic cost.

3. The concepts of carbon neutrality and carbon peaking are combined. Through the cooperative multi-objective optimization model, the carbon emissions generated by the operation of the MMGS and the distribution network are effectively reduced. The cooperative multi-objective optimization model not only improves the economy but also reduces the total carbon emissions of MMGS and the distribution network.

2. System Structure

2.1. Structure of the DC Multi-Microgrid System

The MMGS discussed in this paper includes multiple relatively independent MGs in space. The DC multi-microgrid energy management system (MMGEMS) manages all energy transactions in MMGS. Each MG is integrated into the distribution network through power electronic devices and exchanges energy with the distribution network. Each MG contains RESs and EVs charging/discharging infrastructures (EVCDIs). There are two main types of MGs in the MMGS: MGs located in residential areas (RMG) and MGs located in office areas (OBMG). The structure of the MMGS is shown in Figure 1.

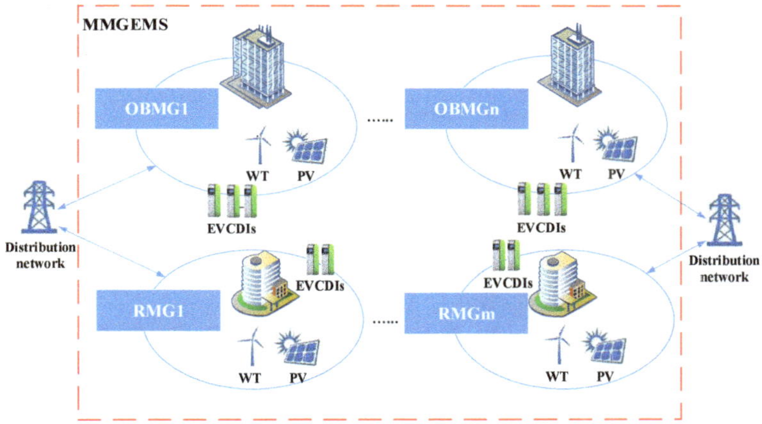

Figure 1. Structure of a DC multi-microgrid system.

The control of the system is mainly conducted by the collaboration of the MG energy management system (MEMS) and the EVs management system (EVMS). The MEMS is responsible for the energy dispatching of photovoltaics (PVs), wind turbines (WTs), and EVs in MGs, and the EVMS manages the charging and discharging behaviors of EVs. A DC multi-microgrid control system is shown in Figure 2.

2.2. DC Microgrid

The basic structure of the DC microgrid is shown in Figure 3. Each MG is connected to the distribution network through a transformer and a DC/AC converter, which can exchange energy with the distribution network. A connection switch is installed in the grid-connected circuit, which can switch the MG between island operation mode and grid-connected operation mode.

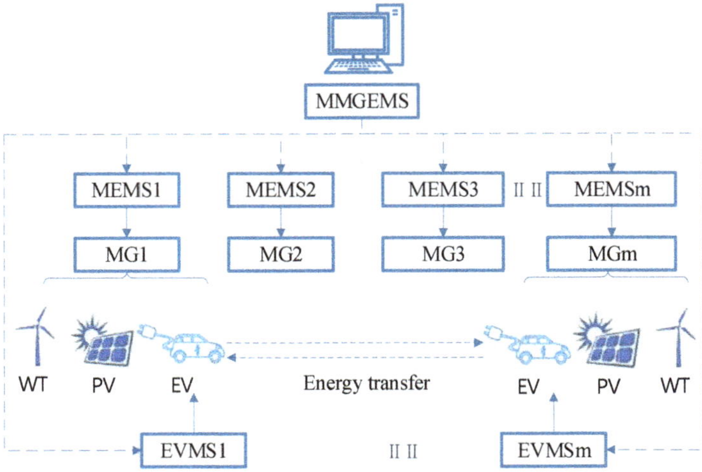

Figure 2. Structure of a DC multi-microgrid control system.

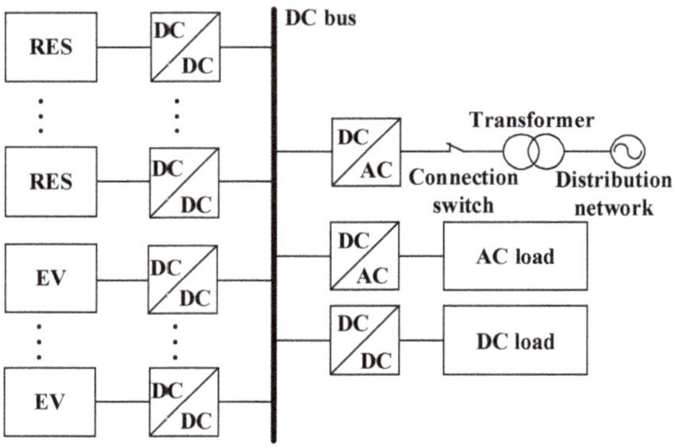

Figure 3. Structure of a DC microgrid.

2.3. *Bidirectional DC/AC Converter*

The bidirectional DC/AC converters are used to connect the MMGS and the distribution network, which can output active and reactive power with the distribution network. DC/AC converters use power factor correction (PFC) to obtain the unity power factor [21]. Therefore, the DC/AC converters are set to the unity power factor in this paper when the reactive power is not optimized [22]. However, by using the appropriate pulse-width modulation (PWM) switching technique, the power factor is adjusted to control the reactive power output of the DC/AC converters to the distribution network [22]. This is the basis for reactive power optimization.

3. Mathematical Model

3.1. *Renewable Power Generation*

3.1.1. Photovoltaic Module

In this paper, the power prediction module based on artificial neural networks (ANNs) [23] is applied to the economic dispatching of MMGS. The weather data are from the numerical weather forecast (NWP).

3.1.2. Wind Turbine

The output of WTs is mainly affected by wind speed [24]. The ANN is still used to predict wind power [25]. The inputs are the wind speed and wind direction from NWP.

3.2. Electric Vehicles

3.2.1. EVs Model

The EVs in the MMGS are all commuter vehicles, and the residents of the residential area are the workers in the office area. As 77.95% of EVs' users will reach the working area at 7:30–9:30 [26], the standard parking time slots in OBMG and RMG are assumed to be 9:00–17:00 and 19:00–7:00 [27,28]. The capacity of EV at t-th is

$$SOC_{EVm,n,t} = SOC_{EVm,n,t-1}(1-\sigma) + P^{EV}_{m,n,t} \bullet \Delta t \bullet \eta_{CEV} \ , \quad if \quad P^{EV}_{m,n,t} \geq 0 \tag{1}$$

$$SOC_{EVm,n,t} = SOC_{EVm,n,t-1}(1-\sigma) + P^{EV}_{m,n,t} \bullet \Delta t / \eta_{DEV} \ , \quad if \quad P^{EV}_{m,n,t} < 0 \tag{2}$$

The power output of EV in (1) and (2) is measured on the MMGS side. Where $SOC_{EVm,n,t}$ is the remaining power capacity of the n-th EV in the m-th MG in the t-th hour, σ is the self-discharge coefficient. $P^{EV}_{m,n,t}$ is the charging or discharging power in the t-th hour of the n-th EV in the m-th MG. If $P^{EV}_{m,n,t} \geq 0$, EVs are charged. If $P^{EV}_{m,n,t} < 0$, EVs release energy; Δt = 1 h. η_{DEV} and η_{CEV} are the discharging and charging efficiency of EVs to calculate the power actual charging or discharging power of EVs.

3.2.2. Across-Time-and-Space Energy Transmission of EV

In the same MG, the EV is used as an energy storage unit, and its charging/discharging power can be dispatched for the operation of the MG. When the EV is connected to the MG and the power is sufficient, MMGS controls the EV to charge during the low charging price or when the system has excess energy, and discharge during the peak discharging price or when the system is short of power. The EV is charged and discharged in the same MG to realize energy transfer over time, thereby reducing the cost of MMGS purchasing electricity directly from the distribution network. At the same time, it also allows the user of the EV to profit by selling part of the electricity, which enables both parties to obtain a certain amount of economic benefit.

On the other hand, EVs not only have energy storage characteristics but also can move between different locations. In the case of differences in the electricity price of the distribution network within a region, benefits can be obtained through the cross-space transfer of energy. For example, the electricity prices of RMG and OBMG for electricity trading with the distribution network are quite different. Most of the time, the electricity price of OBMG purchasing electricity from the distribution network is higher than RMG. Therefore, the electric energy charged in the RMG at a low charging price is sold to the MMGS at a high discharging price in the OBMG, and the electric energy is transferred between different spaces and times through charging and discharging.

EVs realize the across-time energy transmission in the same MG and realize the across-time-and-space energy transmission in different MGs, which can transfer the lower-priced electric energy in RMG to OBMG at a higher price. Under the right conditions, both MMGS and EV users can benefit. This characteristic of EVs for energy transfer between different times and different spaces is called the across-time-and-space energy transmission.

3.3. EV Charging/Discharging Infrastructures

The charging/discharging behaviors of EVs are carried out through the EVCDIs.

$$P^{EVCDIs}_{m,n,t} = P^{EV}_{m,n,t} \bullet \eta_{CEV} \ , \quad if \quad P^{EV}_{m,n,t} \geq 0 \tag{3}$$

$$P^{EVCDIs}_{m,n,t} = P^{EV}_{m,n,t} / \eta_{DEV} \ , \quad if \quad P^{EV}_{m,n,t} < 0 \tag{4}$$

where $P^{EVCDIs}_{m,n,t}$ is the power of the EVCDIs of the n-th EV in the m-th MG in the t-th hour.

4. Cooperative Multi-Objective Optimization Model

4.1. Description of the Optimization Model

The EVMS collects the dispatchable capacity forecast data of EVs and outputs the dispatching plan of the EVs. The MEMS collects the output of predicted RESs, the predicted load data, and the energy price of the distribution network. Based on this information, MEMS outputs the active power of RESs, EVs, and DC/AC converters in MMGS, and transmits it to the reactive power optimization module in MMGEMS. The reactive power optimization module outputs the optimal reactive power of the DC/AC converters according to the data. The two modules coordinate and output the optimal result.

4.2. Double-Loop Optimization Process

The process is shown in Figure 4. The inner-loop is the dynamic economic dispatch which is used to optimize the active power output of RESs, EVs, and DC/AC converters to obtain the optimal total operating cost of the MMGS. The outer-loop optimizes the reactive power output of the DC/AC converters according to the active power output of the inner-loop, to make the network loss of the distribution network minimum, thereby reducing the network loss cost and carbon emissions of MMGS and the distribution network. The inner-loop and the outer-loop work together to obtain the optimal active power output plan in MMGS and reactive power output of the DC/AC converters, which makes the economic cost of MMGS minimum.

4.3. Cooperative Multi-Objective Optimization Objective Function

The main goal of optimization is to reduce the daily economic total cost of MMGS. MMGS discussed in this model consists of multiple MGs, which are assumed to be owned by a single operator. Another goal of the model is the lowest network loss of the distribution network, which can be obtained through the outer-loop model. Therefore, the objective function to minimize the total economic cost of the entire system can be expressed as:

$$f = C_{ETC} = C_{OTC} + C_{WTC} \tag{5}$$

where f is the main goal of the cooperative optimization, C_{ETC} is the economic total cost of MMGS. C_{OTC} is the operating total cost of MMGS, C_{WTC} is the energy loss cost of the MMGS that is obtained from the outer-loop model, where

$$C_{WTC} = C_{il} + C_{co} \tag{6}$$

$$C_{co} = E_{CO} \bullet k_c \tag{7}$$

$$E_{CO} = (W_S^G - W_S^B) \bullet \Delta t \bullet e_c \tag{8}$$

$$C_{il} = (W_S^G - W_S^B) \bullet \Delta t \bullet k_{il} \tag{9}$$

$$E_C = \sum_{m=1}^{M} E_{CIm} + E_{CO} \tag{10}$$

C_{il} and C_{co} are the network loss cost and carbon emissions cost caused by the increase in the distribution network loss in the outer-loop model, respectively. E_{CO} is the carbon emissions generated by the distribution network. W_S^G is the total daily operating network loss of the distribution network when MMGS is integrated into the distribution network and runs. W_S^B is the total daily operating network loss when there is no MMGS access, which is a fixed value also called the original baseline loss. k_{il}, e_c, k_c are fixed factors, k_{il} is the loss cost coefficient of the distribution network, e_c is the carbon emissions factor, k_c is the carbon cost factor. Δt = 1 h. E_C is the total carbon emissions of MMGS and the distribution network, E_{CIm} is the carbon emissions generated by m-th MG in the inner-loop model, M is the number of MGs in the MMGS.

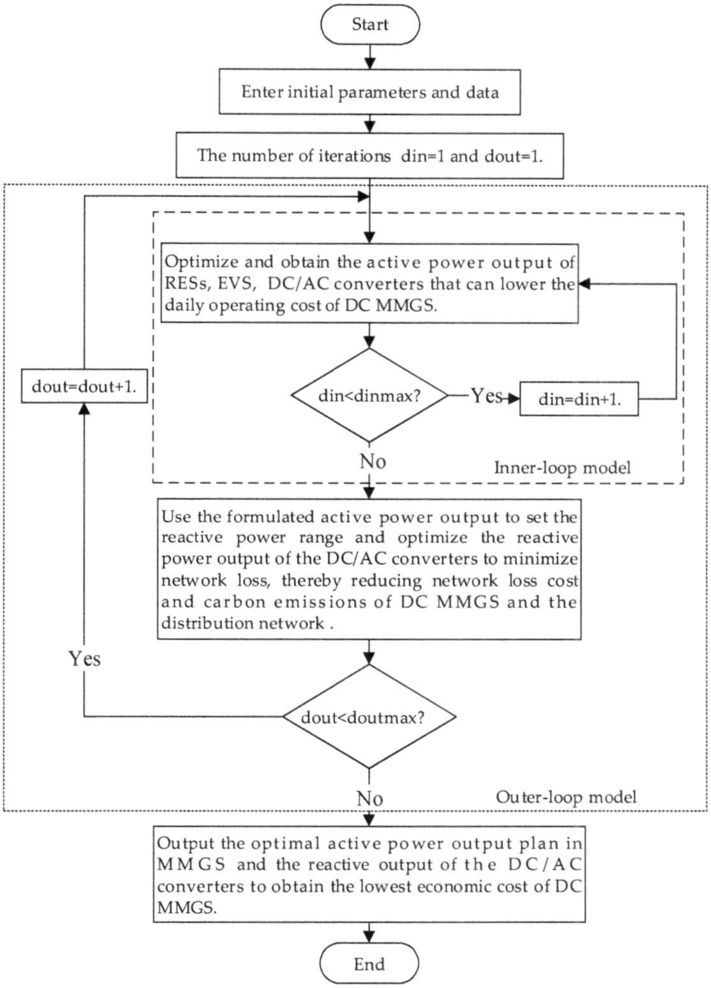

Figure 4. Process of the cooperative multi-objective optimization.

Since C_{OTC} and W_S^G are the optimization targets of the inner-loop model and the outer-loop model, respectively, the objective functions of the inner-loop model and the outer-loop model are set as follows:

$$f_1 = \min C_{OTC} \tag{11}$$

$$f_2 = \min W_S^G \tag{12}$$

where f_1 and f_2 are the objective functions of the inner-loop model and the outer-loop model, respectively.

Through (5)–(12), f can be expressed as:

$$f = f_1 + (f_2 - W_S^B) \bullet \Delta t \bullet [e_c \bullet k_c + k_{il}] \tag{13}$$

4.4. Inner-Loop Optimization

The goal of the inner-loop optimization model is to minimize the daily operating cost of the MMGS. The daily operating cost is mainly composed of system energy transaction cost and carbon emissions cost. The objective function is as follows:

$$f_1 = \min C_{OTC} \tag{14}$$

$$C_{OTC} = \sum_{m=1}^{M} C_{OCm} \tag{15}$$

$$C_{OCm} = C_{exm} + C_{cim} \tag{16}$$

$$C_{cim} = E_{CIm} \bullet k_c \tag{17}$$

C_{OCm} is the operating cost of the m-th MG that is obtained from the inner-loop model. C_{exm} is the energy transaction cost in the m-th MG. C_{cim} is the carbon emissions cost due to energy exchange in the inner-loop model.

4.4.1. Energy Transaction Cost

The energy transaction cost is the sum of RESs cost, energy exchange cost between MMGS and EVs, MMGS and distribution network, and the additional cycle cost of EV batteries. $P_{m,t}^{EV}$, $P_{m,t}^{PV}$, $P_{m,t}^{WT}$, and $P_{m,t}^{G}$ are the optimization variables.

$$C_{exm} = C_{resm} - C_{evm} + C_{gm} + C_{cym} \tag{18}$$

$$C_{resm} = C_{PVm} + C_{WTm} \tag{19}$$

$$C_{PVm} + C_{WTm} = \sum_{t=1}^{T} P_{m,t}^{PV} C_{m,t}^{PV} \Delta t + \sum_{t=1}^{T} P_{m,t}^{WT} C_{m,t}^{WT} \Delta t \tag{20}$$

$$C_{evm} = \sum_{t=1}^{T} \sum_{n=1}^{N} P_{m,n,t}^{EV} \bullet \eta_{CEV} C_{m,t}^{CEV} \Delta t \ , \quad if \quad P_{m,n,t}^{EV} \geq 0 \tag{21}$$

$$C_{evm} = \sum_{t=1}^{T} \sum_{n=1}^{N} P_{m,n,t}^{EV} C_{m,t}^{DEV} \Delta t \ , \quad if \quad P_{m,n,t}^{EV} < 0 \tag{22}$$

$$C_{gm} = \sum_{t=1}^{T} P_{m,t}^{G} C_{m,t}^{G} \Delta t \tag{23}$$

$$C_{cym} = \sum_{n=1}^{N} k_{cy} C_{cyn}^{EV} \tag{24}$$

where C_{resm} is the cost of RESs of the m-th MG in a day, C_{PVm}, and C_{WTm} are the cost of PVs and WTs. $P_{m,t}^{PV}$ is the power output of PVs in the m-th MG, at t-th hour, $C_{m,t}^{PV}$ is the PV power generation cost, $P_{m,t}^{WT}$ is the power output of WTs, $C_{m,t}^{WT}$ is the WT power generation cost. C_{evm} is the cost of energy exchange between MMGS and EVs, $C_{m,t}^{CEV}$ is the charging price of EVs in m-th MG, $C_{m,t}^{DEV}$ is the discharging price, Δt = 1 h, T = 24 h. C_{gm} is the energy exchange cost between the MG and the distribution network through the DC/AC converters, $P_{m,t}^{G}$ is the active power output between the MG and the distribution network through the DC/AC converters. If $P_{m,t}^{G} \geq 0$, MG purchases electricity from the distribution network. If $P_{m,t}^{G} < 0$, MMGS sells electricity to the distribution network. $C_{m,t}^{G}$ is the electricity price that MG purchases/sells to the distribution network. C_{cym} is the additional cycle cost of EV batteries, C_{cyn}^{EV} is the additional battery charging/discharging cycle cost of n-th EV, k_{cy} is the number of additional charging/discharging cycles, N is the number of EVs.

4.4.2. Carbon Emissions and Cost

The electricity of the distribution network mainly comes from thermal power generation. When MG exchanges energy with the distribution network, the distribution network emits more CO_2. To reduce carbon emissions as much as possible and increase the use of RESs, in this paper, the cost of carbon emissions is used as the penalty cost of CO_2 generated by the energy exchange between the MMGS and the distribution network [29].

$$C_{cim} = E_{CIm} \bullet k_c \tag{25}$$

$$E_{CIm} = \sum_{t=1}^{T} P_{m,t}^{G} \Delta t \bullet e_c \tag{26}$$

4.5. Constraints of the Inner-Loop Model

4.5.1. EVs Power Constraint

The charging/discharging power of EVs cannot exceed the rated power of EVCDIs.

$$\left| P_{m,n,t}^{EV} \right| \leq P_{m,n,R}^{EVCDIs} \tag{27}$$

where $P_{m,n,R}^{EVCDIs}$ is the rated power of the EVCDI serving the n-th EV.

4.5.2. EVs Capacity Constraint

The remaining power of EVs must meet the constraints of rated capacity.

$$SOC_{EVm,n,\min} \leq SOC_{EVm,n,t} \leq SOC_{EVm,n,\max} \tag{28}$$

where $SOC_{EVm,n,\min,}$ and $SOC_{EVm,n,\max}$ are the minima and maximum capacity, respectively, of n-th EV in m-th MG.

4.5.3. RESs Output Constraint

Considering the performance limitations of renewable energy, the output of RESs in m-th MG has a certain upper limit.

$$0 \leq P_{m,t}^{WT} \leq P_{m,\max}^{WT} \tag{29}$$

$$0 \leq P_{m,t}^{PV} \leq P_{m,\max}^{PV} \tag{30}$$

4.5.4. System Power Balance Constraint

For MMGS, the active power output should meet the power balance constraint.

$$P_{m,t}^{EV} + P_{m,t}^{G} + P_{m,t}^{WT} + P_{m,t}^{PV} = P_{m,t}^{L} \tag{31}$$

where $P_{m,t}^{L}$ is the total load of the m-th MG at time t.

4.6. Outer-Loop Optimization

The outer-loop optimization model takes the network loss as the optimization goal. By optimizing the reactive power output of the DC/AC converters $Q_{m,t}^{G}$, the daily network loss of the distribution network is minimized, thereby reducing network loss cost and carbon emissions of MMGS and the distribution network [30]. This paper assumes that the m-th MG is connected to node i of the distribution network.

$$f_2 = \min W_S^G \tag{32}$$

$$W_S^G = \sum_{t=1}^{T} \sum_{i,j=1}^{N_{br}} k_i R_{ij} \frac{P_{ij,t}^2 + Q_{ij,t}^2}{V_{ij,t}^2} \tag{33}$$

$$W_S^I = W_S^G - W_S^B \tag{34}$$

$$P_{ij,t} = P_{ij,t}^0 + P_{m,t}^G \tag{35}$$

$$Q_{ij,t} = Q_{ij,t}^0 + Q_{m,t}^G \tag{36}$$

$$W_S^I = f_2 - W_S^B \tag{37}$$

$$C_{il} + C_{co} = W_S^I \bullet \Delta t \bullet k_{il} + W_S^I \bullet \Delta t \bullet e_c \bullet k_c \tag{38}$$

W_S^I is the daily operating increased network loss of the distribution network. N_{br} is the number of branches. i, j are the nodes, k_i is the state variable of the i-th branch switch, 1 means closed, 0 means open; R_{ij} is the resistance of branch ij, $P_{ij,t}$, $Q_{ij,t}$ are the active and reactive power of branch ij in the t-th hour, $V_{ij,t}$ is the voltage, $P_{ij,t}^0$, $Q_{ij,t}^0$ are initially active, reactive power when connected without MMGS. $P_{m,t}^G$, $Q_{m,t}^G$ are the active and reactive power through the DC/AC converters injected into node i by the m-th MG connected to node i. To facilitate the calculation of network loss, a day is divided into 12 small periods, with a time interval of 2 h.

4.6.1. Network Loss Cost

The operation of MMGS connected to the distribution network will cause increased network loss in the distribution network. Therefore, the distribution network will sign a contract with the operator of MMGS, and the operator needs to pay a certain network loss fee for the daily operating increased network loss in the distribution network.

$$C_{il} = W_S^I \bullet \Delta t \bullet k_{il} \tag{39}$$

4.6.2. Carbon Emissions and Cost

When the network loss of the distribution network increases by the operation of MMGS, more CO_2 will be emitted. MMGS will still incur a penalty cost for carbon emissions by the increasing network loss, which differs from the carbon emissions cost due to energy exchange in the inner-loop model.

$$C_{co} = W_S^I \bullet \Delta t \bullet e_c \bullet k_c \tag{40}$$

4.7. Constraints of the Outer-Loop Model

The model takes the actual power flow of the power grid as the constraints.

4.7.1. Node Power Flow Constraint

$$P_{m,t}^G + P_{i,t}^0 = P_{Li,t} + V_{i,t} \sum_{j=1}^{N_n} V_{j,t} (G_{ij} \cos \delta_{ij} + B_{ij} \sin \delta_{ij}) \tag{41}$$

$$Q_{m,t}^G + Q_{i,t}^0 = Q_{Li,t} + V_{i,t} \sum_{j=1}^{N_n} V_{j,t} (G_{ij} \sin \delta_{ij} + B_{ij} \cos \delta_{ij}) \tag{42}$$

where $P_{i,t}^0$ and $Q_{i,t}^0$ are the initial input active and reactive power of node i in the t-th hour, $P_{Li,t}$ and $Q_{Li,t}$ are the active and reactive load, $V_{i,t}$ and $V_{j,t}$ are the voltage of node i *and* j, G_{ij}, B_{ij}, and δ_{ij} are the conductance, susceptance, and phase angle difference of branch ij.

4.7.2. Node Voltage Constraint

$$V_i^{\min} \le V_{i,t} \le V_i^{\max} \tag{43}$$

$V_i^{\min}$ and $V_i^{\max}$ are lower and upper limits of the node i voltage amplitude.

4.7.3. Branch Power Constraint

$$\left|P_{ij,t}\right| \leq P_{ij,\max} \tag{44}$$

$$\left|Q_{ij,t}\right| \leq Q_{ij,\max} \tag{45}$$

$P_{ij,\max}$, $Q_{ij,\max}$ are the maximum active and reactive power of the branch ij.

4.7.4. Branch Current Constraint

$$I_{ij} \leq I_{ij}^{\max} \tag{46}$$

where $I_{ij}^{\max}$ is the upper limit of branch ij current carrying capacity.

4.7.5. Reactive Output Constraint of DC/AC Converter

The reactive power output of the DC/AC converters must satisfy the constraint:

$$\left|Q_{m,t}^{G}\right| \leq \sqrt{S_m^2 - \left(P_{m,t}^{G}\right)^2} \tag{47}$$

where S_m is the rated power of the DC/AC converter in m-th MG, $Q_{m,t}^{G}$ is the reactive power that the DC/AC converter can output to the distribution network, $P_{m,t}^{G}$ is the active power output by DC/AC converter.

4.8. Particle Swarm Algorithm

4.8.1. Procedure of PSO

The steps of particle swarm optimization (PSO) are shown in Figure 5 [31].

4.8.2. Coding

In the inner-loop, the coding about the economic dispatch of RESs, EVs, and DC/AC converters can be represented by a real-valued matrix. k is the index of the particle of the inner-loop. M is the number of MG. T is the dispatching cycle.

$$I_{\mathrm{MG}}^{k} = \begin{bmatrix} I_{\mathrm{MG1}} \\ I_{\mathrm{MG2}} \\ \vdots \\ I_{\mathrm{MGm}} \\ \vdots \\ I_{\mathrm{MGM}} \end{bmatrix} \tag{48}$$

$$I_{\mathrm{MGm}} = \begin{bmatrix} P_{m,1}^{PV} & P_{m,2}^{PV} & \cdots & P_{m,t}^{PV} & \cdots & P_{m,T}^{PV} \\ P_{m,1}^{WT} & P_{m,2}^{WT} & \cdots & P_{m,t}^{WT} & \cdots & P_{m,T}^{WT} \\ P_{m,1}^{G} & P_{m,2}^{G} & \cdots & P_{m,t}^{G} & \cdots & P_{m,T}^{G} \\ P_{m,1}^{EV} & P_{m,2}^{EV} & \cdots & P_{m,t}^{EV} & \cdots & P_{m,T}^{EV} \end{bmatrix} \tag{49}$$

$P_{m,t}^{EV}$, $P_{m,t}^{PV}$, $P_{m,t}^{WT}$, and $P_{m,t}^{G}$ are the power outputs of EVs, PVs, WTs, and DC/AC converters in the m-th MG in the t-th hour, respectively.

In the outer-loop, the coding about the reactive power output by DC/AC converters can be represented by a real-valued matrix. s is the index of the particle of the outer-loop.

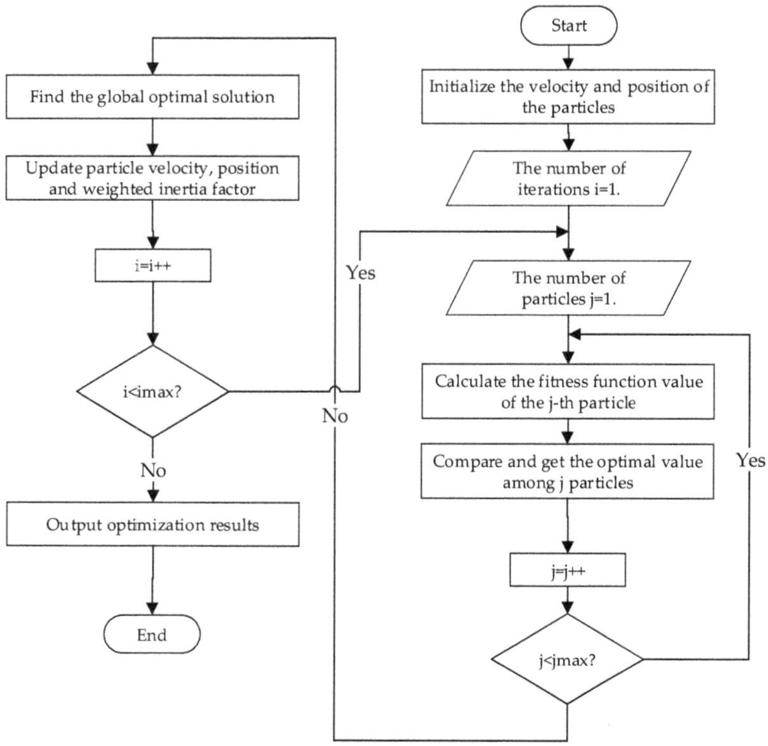

Figure 5. Process of the PSO algorithm.

$$O_{\mathrm{MG}}^{s} = \begin{bmatrix} O_{\mathrm{MG1}} \\ O_{\mathrm{MG2}} \\ \vdots \\ O_{\mathrm{MGm}} \\ \vdots \\ O_{\mathrm{MGM}} \end{bmatrix} \tag{50}$$

$$O_{MGm}^{s} = \begin{bmatrix} Q_{m,1}^{G} & Q_{m,2}^{G} & \cdots & Q_{m,t}^{G} & \cdots & Q_{m,T}^{G} \end{bmatrix} \tag{51}$$

$Q_{m,t}^{G}$ is the reactive power output of DC/AC converters in the m-th MG at t-th hour.

However, the dispatch range of the outer-loop variable also changes when the variable of the DC/AC converters changes in the inner-loop. Therefore, a dynamic range adjustment algorithm is added to the outer-loop model.

$$\left|Q_{m,t}^{G}\right|^{\max} = \sqrt{S_m^2 - \left(P_{m,t}^{G}\right)^2} \tag{52}$$

The inner-loop and outer-loop cooperate to generate the optimal optimization results.

5. Case Study and Discussion

5.1. Case Description

There are 30 EVs concentrated in OBMG/RMG for the charging/discharging service [31]. The dispatching cycle is 24 h. This paper sets up four cases to analyze the optimization model. By using NWPs from Wuhan City, Hubei Province, China in June 2020, a day's renewable power generation in summer is predicted as the input of the model.

5.1.1. Case 1

In this case, the EVs do not participate in the energy dispatch of the MG. Once they reach the MG, the EVs will be charged until the batteries are fully charged. MMGS does not optimize the reactive power output by DC/AC converters.

5.1.2. Case 2

In this case, after EVs are connected to the MG, they participate in the energy management system of each MG. Once they reach the MG, the energy in the EV battery will be dispatched by the MG's energy management system until they leave. When the EVs leave the MG at the end of the dispatching, the energy of EVs should be fully charged. This case takes advantage of the across-time energy transmission of EVs in each independent MG, and the optimization of the reactive power output of DC/AC converters is not considered.

5.1.3. Case 3

In this case, only the inner-loop economic dispatch model is used to minimize the total cost of MMGS by optimizing the active power output of RESs, EVs, and DC/AC converters. The ATSET of EV between RMGs and OBMGs is used. However, the reactive power output of DC/AC converters is also not optimized. Case 3 can be used as a reference.

5.1.4. Case 4

In this case, cooperative multi-objective optimization combines the inner-loop economic dispatch model and the outer-loop reactive power optimization model. The ATSET of EV between RMG and OBMG is considered. By optimizing the active power output of RES, EVs, and DC/AC converters, the total daily operating cost of MMGS is reduced. By optimizing the reactive power output of DC/AC converters, the loss of the distribution network is reduced, and the total economic cost of MMGS is reduced synergistically.

5.2. Simulation System Construction

5.2.1. System Introduction

The modified IEEE 33-node system is used to prove the model, whose structure is shown in Figure 6, and its parameters can be obtained from [32]. According to the principle of distribution [32], OBMG and RMG are set at node 19 and node 20, respectively.

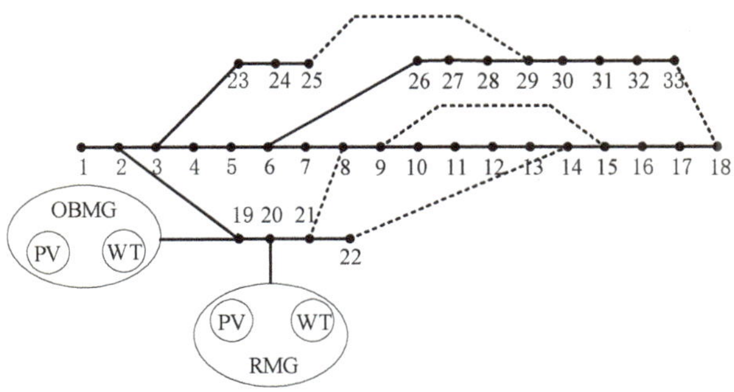

Figure 6. Topological diagram of the modified IEEE 33-node system.

5.2.2. Parameters of RESs

According to the principle of renewable energy consumption [27], the RESs installed in each MG and the power generation cost are given in Table 1. The optimization time interval is 1 h, and the optimization cycle is 1 day (24 h).

Table 1. Installed RESs and the power generation cost.

MG Type	RES Type	Installed Capacity/kW	Power Generation Cost/¥·kWh^{-1}
OBMG	PV1	800	0.24
	WT1	800	0.38
RMG	PV2	400	0.24
	WT2	400	0.38

The daily wind speed, radiation intensity, temperature, and load data are adopted in this area. The renewable energy output and load curves of each MG come from [27].

5.2.3. Parameters of DC/AC Converter

Considering the performance of the DC/AC converters of MMGS, S_m = 1000 kW, the power limit is set as [33]:

$$0 \leq \left| P_{m,t}^G \right| \leq 1000\text{kW} \tag{53}$$

$$0 \leq \left| Q_{m,t}^G \right| \leq 1000\text{kVar} \tag{54}$$

5.2.4. Parameters of EVs

Take a BYD E6 electric vehicle as an example, whose parameters are from [34]. The battery capacity is 80 kWh, and the upper limit of charging and discharging power of EVCDI is 7 kW. The charging and discharging efficiency are all 90% [34]. An EV consumes an average of 8% of electricity per way between RMG and OBMG [31]. The additional battery charging/discharging cycle cost of EV is CNY 50 each time [35]. The minimum power of the battery of EV is not less than 20% [36]. Considering the needs of users, the upper and lower limits for the battery are 100% and 35% [27].

5.2.5. Other Parameters

The time-of-use (TOU) energy prices in RMG/OBMG from [31] are shown in Figure 7. The carbon emissions factor e_c is 86.47 g/kWh [29], and the carbon cost factor k_c is 0.21 CNY/kg [37]. The loss cost coefficient of the distribution network k_{il} is 0.74 CNY/kWh [38].

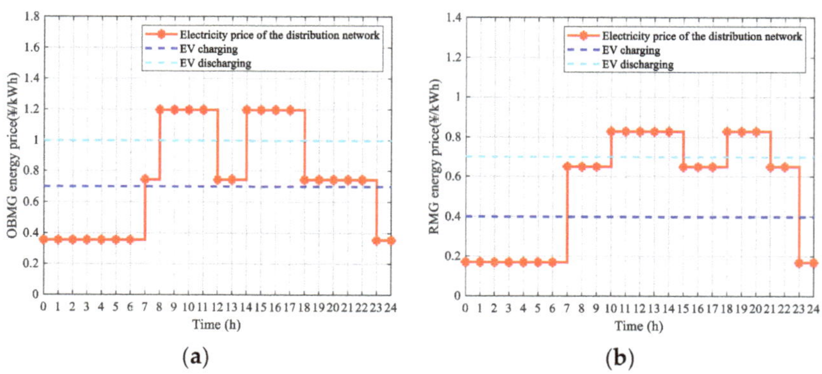

Figure 7. (**a**) Prices of energy exchanging in OBMG; (**b**) prices of energy exchanging in RMG.

5.3. Simulation Results

5.3.1. Inner-Loop Optimization Results

1. Case 1

In this case, when EVs are connected to the MG, they are charged immediately. In case 1, the 24 h curve of RESs, EVs, load, and DC/AC converter active power output in OBMG/RMG is shown in Figure 8. EVs are charged as soon as they reach RMG/OBMG.

The active power curve of the DC/AC converter represents the active power output curve of the MG to the distribution network. When it is below the X-axis, it means that the MG sells electric energy to the distribution network. When it is above the X-axis, it means that the MG purchases electric energy from the distribution network. The power curve of EVs has a similar definition. In Figure 8, RMG will allow EVs to be charged at maximum power from 19:00–20:00, and when RESs are insufficient, MEMS will purchase electricity from the distribution network. OBMG is also charging EVs at 9:00–10:00. The total daily operating cost of RMG is CNY 2776.3, and the total daily operating cost of OBMG is CNY 5732.6. Therefore, the total daily operating cost of MMGS is CNY 8508.9.

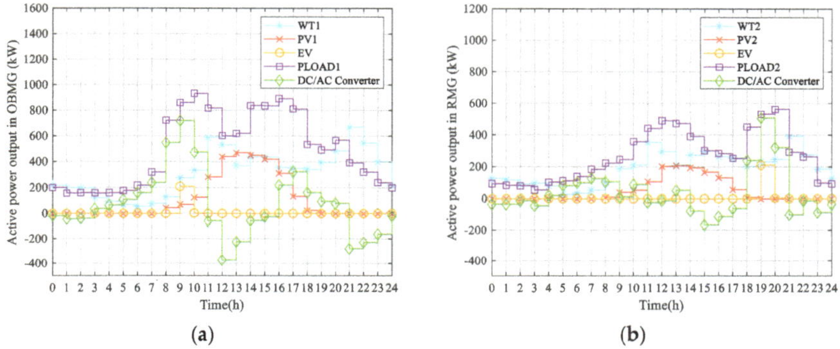

Figure 8. (**a**) Power output of RESs, EVs, and the DC/AC converter in OBMG; (**b**) power output of RESs, EVs, and the DC/AC converter in RMG.

2. Case 2

In this case, since EVs can participate in the energy dispatching of independent MGs, their across-time energy transmission is used. When the total generated power of the RESs in the MGs is greater than the load, the MEMS will sell the remaining energy to the distribution network or charge the EVs according to the energy prices. When the total power generation of RES is less than the load, the MEMS will purchase electricity from the distribution network or EVs according to the energy prices. In Figure 9, the active power output of RES, EV and DC/AC converters in OBMG and MG are optimized. In OBMG, due to the high energy prices of the distribution network and EVs from 9:00 to 12:00, MEMS choose to let EVs release electric energy. OBMG lowers costs by selling energy to the distribution network. When energy prices are low between 12:00 and 15:00, MEMS fully charges EVs. In RMG, MEMS chooses to charge EVs at 23:00 when energy prices are low. This is to avoid additional battery charge–discharge cycle costs due to discharge, so EVs are only charged. The across-time energy transmission of the EV in the independent MG is fully utilized. Through optimization model calculation, the total daily operating cost of RMG is CNY 2644.1, and the total daily operating cost of OBMG is CNY 5642.1. Therefore, the total daily operating cost of MMGS is CNY 8286.2. Compared with Case 1, the across-time energy transmission of EVs can reduce the overall operating cost of MMGS.

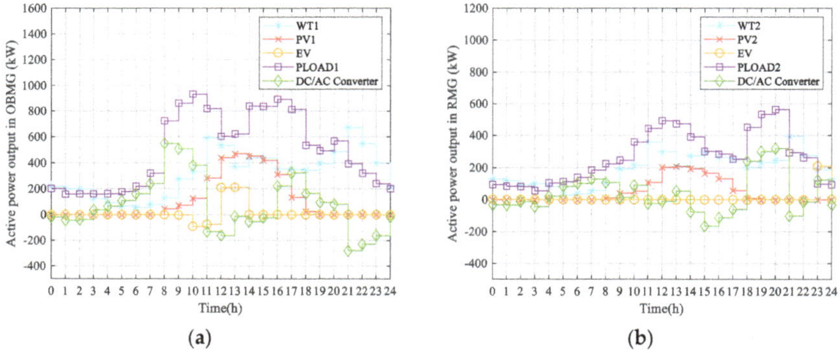

Figure 9. (**a**) Power output of RESs, EVs, and the DC/AC converter in OBMG; (**b**) power output of RESs, EVs, and the DC/AC converter in RMG.

3. Case 3

In this case, EVs can transfer energy among multiple MGs, and EVs participate in MMGS energy dispatching. For OBMG, the energy price of the distribution network and the discharging price of EVs are both high, and the difference between the energy price of the distribution network and the discharging price of EV is much higher than that of RMG. Therefore, MMGS's energy management system will discharge almost all EVs as much as possible when EVs are connected to OBMG, and earn more profits. For RMG, its advantage is that the charging price is lower, so MMGEMS will try its best to allow almost all EVs to be charged during the low energy price of RMG to reduce the charging cost of EVs. In Figure 10, EVs are discharged as much as possible in OBMG and then charged as much as possible in RMG. After optimization model calculation, the total daily operating cost of RMG is CNY 2391.8, and the total daily operating cost of OBMG is CNY 5404.1. Therefore, the total daily operating cost of MMGS is CNY 7795.9. However, compared with case 1 case 2, by using the across-time-and-space energy transmission of EVs, the total daily operating cost of the MMGS is the lowest in this case.

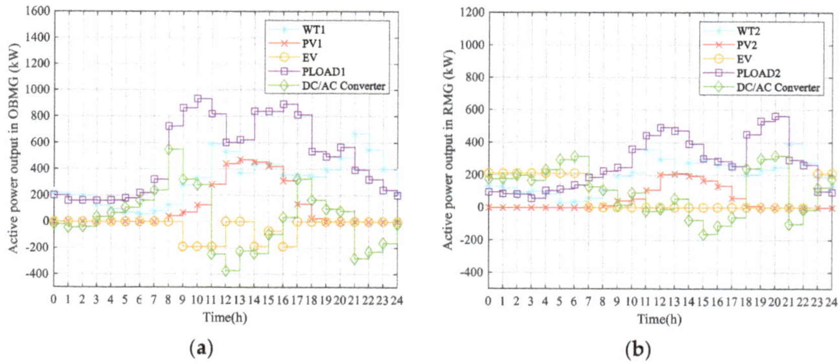

Figure 10. (**a**) Power output of RESs, EVs, and the DC/AC converter in OBMG; (**b**) power output of RESs, EVs, and the DC/AC converter in RMG.

Table 2 is the comparison of the results of the inner-loop economic dispatch in the three cases. Since case 4 and case 3 use the same inner-loop economic dispatch model, their inner-loop output conditions are the same. Here, the effectiveness of the inner-loop economic dispatch model is mainly discussed, so there is no need to show the results in case 4.

In case 1, EVs do not participate in the energy dispatching of the MG, and MMGS has the highest total operating cost. In case 2, the across-time energy transmission of EVs in the independent MG is used to reduce the cost. In case 3 and case 4, the across-time-and-space energy transmission of EVs is considered to further reduce the total daily operating cost of MMGS, which achieves the lowest daily operating cost C_{OTC}.

Table 2. Comparison of operating cost in three cases.

Case	MGs	C_{ex} Energy Transaction Cost/CNY	C_{ci} Carbon Emissions Cost/CNY	C_{OTC} Total Operating Cost/CNY
Case 1	RMG	2760.0	16.3	2776.3
	OBMG	5700.8	31.8	5732.6
	MMGS	8460.8	48.1	8508.9
Case 2	RMG	2627.8	16.3	2644.1
	OBMG	5609.6	32.5	5642.1
	MMGS	8237.4	48.8	8286.2
Case 3	RMG	2348.9	42.9	2391.8
	OBMG	5394.6	9.5	5404.1
	MMGS	7743.5	52.4	7795.9

Figure 11 is the remaining capacity curve of EVs, which proves that EVs meet the power constraint in the four cases. It is also verified that the charging and discharging behaviors analyses of EVs in the three cases are correct.

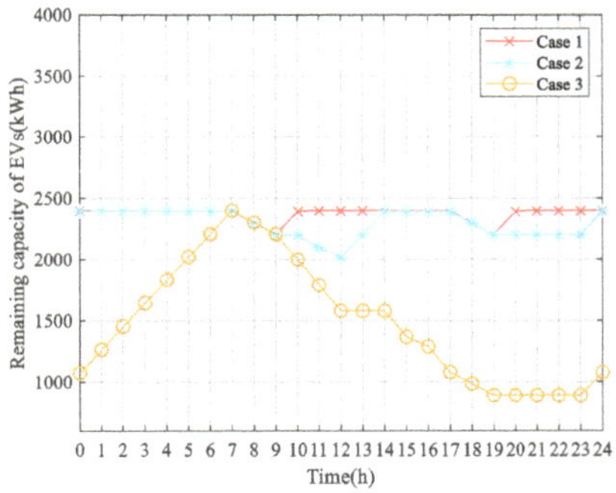

Figure 11. Remaining capacity of EVs in three cases.

Table 3 is the cost of EVs' users. Among the three cases, the user cost of case 3 is the lowest. In case 1, EVs do not participate in dispatching, and the cost of users is the highest. In case 2, the cost of users is reduced by the across-time energy transmission. In case 3, the inner-loop economic dispatch is adopted, which makes full use of the across-time-and-space energy transmission of EVs. Additionally, the cost of users is further reduced. Combining with the lowest daily operating cost of MMGS, the inner-loop economic dispatch model using ATSET of EVs achieved a win–win situation for MMGS and EVs' users.

Table 3. The cost of EVs' users in three cases in one day.

Case	Case 1	Case 2	Case 3
The Cost of EVs' Users/CNY	211.2	173.4	−410.8

5.3.2. Outer-Loop Optimization Results

In case 1, case 2, and case 3, the reactive power output of the DC/AC converters is not optimized. In case 4, the outer-loop reactive power optimization model is used to optimize the reactive power output of the DC/AC converters. The optimized reactive power output of the DC/AC converters of RMG and OBMG in case 4 is shown in Figure 12. The converters will absorb or output a certain amount of reactive power to the distribution network at every moment, which is used to optimize the operating network loss of the distribution network, thereby reducing the energy loss cost C_{WTC} and total carbon emissions E_C of MMGS and the distribution network, and cooperating with the inner-loop model to reduce the total economic cost C_{ETC} of MMGS. The comparison of the results under the four cases is shown in Table 4.

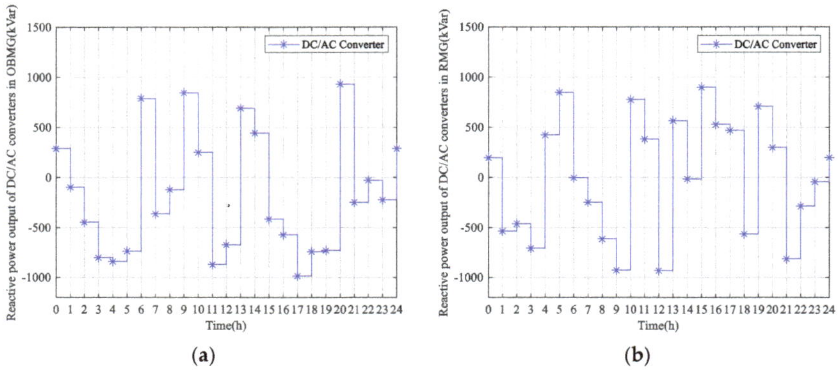

Figure 12. (**a**) Reactive power output by the DC/AC converter of OBMG in case 4; (**b**) reactive power output by the DC/AC converter of RMG in case 4.

Table 4. Comparison of network loss in the four cases.

Case	Case 1	Case 2	Case 3	Case 4
W_S^G Total Network Loss/kW	14,889.2	14,872.5	14,876.0	13,987.2
W_S^I Increased Network Loss/kW	101.3	84.6	88.1	−800.7

By analyzing the distribution network loss under the above different cases, it can be concluded that the reactive power output of the DC/AC converters to the distribution network will affect the distribution network loss. When MMGS is not integrated into the distribution network to work, the original baseline loss W_S^B of the distribution network is 14,787.9 kW. The distribution network loss under the first three cases is all greater than W_S^B, while the distribution network loss under case 4 is less than W_S^B and lower than the first three cases. Case 3 and case 4 are a set of comparisons. Under the common premise of using the inner-loop optimization model, case 4 that uses reactive power optimization has lower network loss. Figure 13 is the increased network loss diagram for each period of the distribution network which further proves that intelligently optimizing the reactive power output of DC/AC converters through the outer-loop model can effectively reduce the daily network loss of the distribution network.

Table 5 is the network loss cost and energy loss cost in four cases. Among the four cases, the network loss cost C_{il} and the carbon emissions cost C_{co} derived from the optimization of the outer-loop model are the lowest, which proves that the outer-loop optimization model plays a role in the cooperative optimization of the economic cost of MMGS.

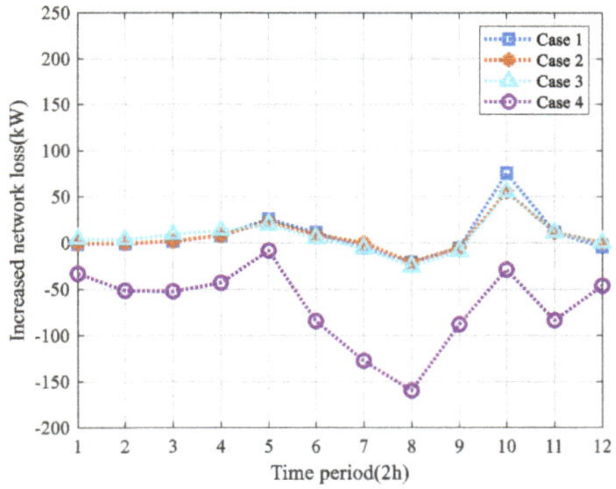

Figure 13. Increased network loss in four cases.

Table 5. Network loss cost and energy loss cost in the four cases.

Case	C_{il} Network Loss Cost/CNY	C_{co} Carbon Emissions Cost/CNY	C_{WTC} Energy Loss Cost/CNY
Case 1	75.0	1.8	76.8
Case 2	62.6	1.5	64.1
Case 3	65.2	1.6	66.8
Case 4	−592.5	−14.5	−607.0

5.3.3. Cooperative Multi-Objective Optimization Results

It can be concluded from Table 6 that, under the cooperative multi-objective optimization model, the total daily economic cost C_{ETC} of MMGS is the lowest. The cost of case 4 adopting the cooperative multi-objective model is 16.3% lower than that for case 1, 13.9% lower than that for case 2, 8.6% lower than that for case 3 which only uses the economic dispatch model of the inner-loop without optimizing reactive power output of DC/AC converters. It is proved that the cooperative multi-objective optimization model improves the economy of MMGS.

Table 6. The final economic cost of MMGS in four cases.

Case	Case 1	Case 2	Case 3	Case 4
C_{ETC} Total Economic Cost of MMGS/CNY	8585.7	8350.3	7862.7	7188.9

It can be concluded by analyzing the carbon emissions data in Table 7 that the total carbon emissions of the MMGS and distribution network with cooperative multi-objective optimization are the lowest among the four cases, which is 24.0% lower than that for case 1, 24.6% lower than that for case 2, and 29.8% lower than that for case 3, which does not optimize the reactive power. The economic cost of MMGS, the network loss of the distribution network, and the total carbon emission of MMGS and the distribution network were all optimized, which fully proves that the cooperative multi-objective optimization achieved the effect.

Table 7. Total carbon emissions in the four cases.

Case	Case 1	Case 2	Case 3	Case 4
E_C Total Carbon Emissions/kg	237.6	239.5	257.1	180.5

5.3.4. Further Verification

To further verify the correctness and validity of the model, the weather type of a certain day in the winter of December 2020 in Wuhan City, Hubei Province, China was used as the input of the model. The results of the economic dispatch of the inner-loop model are shown in Table 8, and the results of the network loss optimization of the outer-loop model are shown in Table 9. The network loss cost and energy loss cost in four cases are shown in Table 10. The final economic cost of MMGS in the four cases on another day is shown in Table 11.

Table 8. Comparison of operating cost in three cases on another day.

Case	MGs	C_{ex} Energy Transaction Cost/CNY	C_{ci} Carbon Emissions Cost/CNY	C_{OTC} Total Operating Cost/CNY
Case 1	RMG	2968.9	23.4	2992.3
	OBMG	6446.4	48.6	6495.0
	MMGS	9415.3	72.0	9487.3
Case 2	RMG	2836.7	23.4	2860.1
	OBMG	6304.5	44.6	6349.1
	MMGS	9141.2	68.0	9209.2
Case 3	RMG	2574.4	48.5	2622.9
	OBMG	6154.2	27.6	6181.8
	MMGS	8728.6	76.1	8804.7

Table 9. Comparison of network loss in the four cases on another day.

Case	Case 1	Case 2	Case 3	Case 4
W_S^G Total Network Loss/kW	14,933.1	14,915.8	14,919.4	14,110.8
W_S^I Increased Network Loss/kW	145.2	127.9	131.5	−677.1

Table 10. Network loss cost and energy loss cost on four cases on another day.

Case	C_{il} Network Loss Cost/CNY	C_{co} Carbon Emissions Cost/CNY	C_{WTC} Energy Loss Cost/CNY
Case 1	107.4	2.6	110.0
Case 2	94.6	2.3	96.9
Case 3	97.3	2.4	99.7
Case 4	−501.1	−12.3	−513.4

Table 11. The final economic cost of MMGS in the four cases on another day.

Case	Case 1	Case 2	Case 3	Case 4
C_{ETC} Total Economic Cost of MMGS/CNY	9597.3	9306.1	8904.4	8291.3

It can be concluded from the above tables that the multi-objective optimization of the model is still achieved after using the weather data of one day in winter. The optimal total economic cost of MMGS C_{ETC} and the lowest distribution network loss W_S^G are obtained, which further proves the correctness and effectiveness of the model.

6. Conclusions

A cooperative multi-objective optimization strategy for MMGS containing EVs and RESs is proposed, including dynamic economic dispatch and optimization of reactive power output by DC/AC converters. Dynamic economic dispatch optimizes the active power output of RESs, EVs, and DC/AC converters in MMGS to obtain the optimal daily operating cost of MMGS. Reactive power optimization reduces the daily operating network loss of the distribution network by optimizing the reactive power output of the DC/AC converters to the distribution network. By comparing the results of the four cases, the following conclusions are drawn:

1. According to the simulation results, the economic dispatch model of the inner-loop in the cooperative multi-objective optimization can reduce the operating cost of MMGS, which makes full use of the ATSET of EVs. Additionally, the optimization of the output reactive power output of the DC/AC converters of the outer-loop can reduce network loss cost and carbon emissions cost of the distribution network. The two cooperate to realize the improvement of the economy of MMGS and the efficient operation of the distribution network.
2. The cooperative multi-objective optimization model not only realizes the optimization of the economic cost of MMGS and the network loss of the distribution network, but also reduces the total carbon emissions of MMGS and the distribution network, which greatly responds to the calls for national carbon neutrality and carbon peak.

Author Contributions: Conceptualization, Z.X. and C.C.; methodology, Z.X.; software, Z.X.; validation, Z.X., M.D. and J.Z.; formal analysis, Z.X.; investigation, D.H.; resources, H.C.; data curation, Z.X.; writing—original draft preparation, Z.X.; writing—review and editing, Z.X. and C.C.; visualization, Z.X.; supervision, C.C.; project administration, C.C.; funding acquisition, C.C. All authors have read and agreed to the published version of the manuscript.

Funding: This research was funded by the National Natural Science Foundation of China, grant number 51977086.

Institutional Review Board Statement: Not applicable.

Informed Consent Statement: Not applicable.

Data Availability Statement: Not applicable.

Conflicts of Interest: The authors declare no conflict of interest.

Nomenclature

B_{ij}	Susceptance of branch ij in the distribution network.
C_{ETC}	Economic total cost of MMGS.
C_{OTC}	Operating total cost of MMGS from the inner-loop model.
C_{WTC}	Energy loss cost of the MMGS from the outer-loop model.
C_{il}	Network loss cost.
C_{co}	Carbon emissions cost.
C_{OCm}	Operating cost of the m-th MG from the inner-loop model.
C_{exm}	Energy transaction cost in the m-th MG.
C_{cim}	Carbon emissions cost in the m-th MG.
C_{resm}	Cost of RESs of the m-th MG.
C_{PVm}	Cost of PVs in the m-th MG.
C_{WTm}	Cost of WTs in the m-th MG.
$C_{m,t}^{PV}$	PV power generation cost in the m-th MG in the t-th hour.
$C_{m,t}^{WT}$	WT power generation cost in the m-th MG in the t-th hour.
C_{evm}	Cost of energy exchange between the m-th MG and EVs.
$C_{m,t}^{CEV}$	Charging price of EVs in m-th MG in the t-th hour.
$C_{m,t}^{DEV}$	Discharging price of EVs in m-th MG in the t-th hour.
C_{gm}	Energy exchange cost between the m-th MG and the distribution network.
$C_{m,t}^{G}$	Electricity price that m-th MG purchases/sells to the distribution network in the t-th hour.
C_{cym}	Additional cycle cost of EV batteries in m-th MG.
C_{cyn}^{EV}	Additional battery charging/discharging cycle cost of n-th EV.

E_C	Total carbon emissions of MMGS and the distribution network.
E_{Cim}	Carbon emissions generated by m-th MG in the inner-loop model.
E_{CO}	Carbon emissions in the outer-loop model.
e_c	Carbon emissions factor.
f	The main objective function of the cooperative optimization model.
f_1	The objective functions of the inner-loop model.
f_2	The objective functions of the outer-loop model.
G_{ij}	Conductance of branch ij in the distribution network.
$I_{ij}^{\max}$	Upper limit of branch ij current carrying capacity in the distribution network.
i,j	Nodes of the distribution network.
k_{il}	Loss cost coefficient.
k_c	Carbon cost factor.
k_{cy}	Number of additional charging/discharging cycles.
k_i	The state variable of the i-th branch switch.
M	Number of MGs in the MMGS.
N	Number of EVs in m-th MG.
N_{br}	Number of branches in the distribution network.
$P_{m,n,t}^{EV}$	Exchanging power in the t-th hour of the n-th EV in the m-th MG.
$P_{m,t}^{EV}$	Exchanging power in the t-th hour of the EVs in the m-th MG.
$P_{m,n,t}^{EVCDIs}$	Power of the EVCDIs of the n-th EV in the m-th MG in the t-th hour.
$P_{m,t}^{PV}$	Power output of PVs in the m-th MG in the t-th hour.
$P_{m,t}^{WT}$	power output of WTs in the m-th MG in the t-th hour.
$P_{m,t}^{G}$	Active power output between the m-th MG and the distribution network in the t-th hour through the DC/AC converters.
$P_{m,n,R}^{EVCDIs}$	Rated power of the EVCDI serving the n-th EV in the m-th MG.
$P_{m,t}^{L}$	Total load of the m-th MG in the t-th hour.
$P_{ij,t}$	Active power of branch ij in the t-th hour.
$P_{ij,t}^{0}$	Initially active power of branch ij when connected without MG in the t-th hour.
$P_{i,t}^{0}$	Initial input active power of node i in the t-th hour.
$P_{Li,t}$	Active load of node i in the t-th hour.
$P_{ij,\max}$	Maximum active power of the branch ij.
$Q_{ij,t}$	Reactive power of branch ij in the t-th hour.
$Q_{ij,t}^{0}$	Initially reactive power of branch ij when connected without MMGS in the t-th hour.
$Q_{m,t}^{G}$	Reactive power output between the m-th MG and the distribution network in the t-th hour through the DC/AC converters.
$Q_{i,t}^{0}$	Initial input reactive power of node i in the t-th hour.
$Q_{Li,t}$	Reactive load of node i in the t-th hour.
$Q_{ij,\max}$	Maximum reactive power of the branch ij.
R_{ij}	The resistance of branch ij.
$SOC_{EVm,n,t}$	Remaining power capacity of the n-th EV in the m-th MG in the t-th hour.
$SOC_{EVm,n,\min}$	Minima capacity, respectively, of the n-th EV in the m-th MG.
$SOC_{EVm,n,\max}$	Maximum capacity, respectively, of the n-th EV in the m-th MG.
S_m	Rated power of the DC/AC converter in the m-th MG.
T	Scheduling cycle, one day, 24 h.
$V_{ij,t}$	Voltage of branch ij in the t-th hour.
$V_i^{\min}$	Lower limits of the node i voltage amplitude.
$V_i^{\max}$	Upper limits of the node i voltage amplitude.
W_S^B	Original baseline network loss.
W_S^G	Total daily operating network loss of the distribution network.
W_S^1	Daily operating increased network loss of the distribution network.
σ	Self-discharge coefficient of EV's battery.
Δt	Length of the time slot set for the optimization.
η_{DEV}	Efficiency for EV discharging.
η_{CEV}	Efficiency for EV charging.
δ_{ij}	Phase angle difference of branch ij.

References

1. Chan, C.; Zhou, G.Y.; Zhang, D. Intelligent energy ecosystem based on carbon neutrality. In Proceedings of the 2017 IEEE Conference on Energy Internet and Energy System Integration (EI2), Beijing, China, 26–28 November 2017; pp. 1–5.
2. Wang, X.; Barnett, A. The evolving value of photovoltaic module efficiency. *Appl. Sci.* **2019**, *9*, 1227. [CrossRef]
3. Li, H.; Ye, Y.; Lin, L. Low-Carbon Economic Bi-Level Optimal Dispatching of an Integrated Power and Natural Gas Energy System Considering Carbon Trading. *Appl. Sci.* **2021**, *11*, 6968. [CrossRef]
4. Izquierdo-Monge, O.; Peña-Carro, P.; Villafafila-Robles, R.; Duque-Perez, O.; Zorita-Lamadrid, A.; Hernandez-Callejo, L. Conversion of a Network Section with Loads, Storage Systems and Renewable Generation Sources into a Smart Microgrid. *Appl. Sci.* **2021**, *11*, 5012. [CrossRef]
5. Saponara, S.; Saletti, R.; Mihet-Popa, L. Hybrid micro-grids exploiting renewables sources, battery energy storages, and bi-directional converters. *Appl. Sci.* **2019**, *9*, 4973. [CrossRef]
6. Van Der Meer, D.; Mouli, G.R.C.; Mouli, G.M.-E.; Elizondo, L.R.; Bauer, P. Energy management system with PV power forecast to optimally charge EVs at the workplace. *IEEE Trans. Ind. Inform.* **2018**, *14*, 311–320. [CrossRef]
7. Puma-Benavides, D.S.; Izquierdo-Reyes, J.; Calderon-Najera, J.d.D.; Ramirez-Mendoza, R.A. A Systematic Review of Technologies, Control Methods, and Optimization for Extended-Range Electric Vehicles. *Appl. Sci.* **2021**, *11*, 7095. [CrossRef]
8. Ko, K.S.; Han, S.; Sung, D.K. Performance-based settlement of frequency regulation for electric vehicle aggregators. *IEEE Trans. Smart Grid* **2018**, *9*, 866–875. [CrossRef]
9. Alam, M.J.E.; Muttaqi, K.M.; Sutanto, D. Effective utilization of available PEV battery capacity for mitigation of solar PV impact and grid support with integrated V2G functionality. *IEEE Trans. Smart Grid* **2016**, *7*, 1562–1571. [CrossRef]
10. Kisacikoglu, M.C.; Erden, F.; Erdogan, N. Distributed control of PEV charging based on energy demand forecast. *IEEE Trans. Ind. Inform.* **2018**, *14*, 332–341. [CrossRef]
11. Yang, H.; Pan, H.; Luo, F.; Qiu, J.; Deng, Y.; Lai, M.; Dong, Z.Y. Operational planning of electric vehicles for balancing wind power and load fluctuations in a microgrid. *IEEE Trans. Sustain. Energy* **2017**, *8*, 592–604. [CrossRef]
12. Panwar, L.K.; Konda, S.R.; Verma, A.; Panigrahi, B.K.; Kumar, R. Operation window constrained strategic energy management of microgrid with electric vehicle and distributed resources. *IET Gener. Transm. Distrib.* **2017**, *11*, 615–626. [CrossRef]
13. Masrur, M.A.; Skowronska, A.G.; Hancock, J.; Kolhoff, S.W.; McGrew, D.Z.; Vandiver, J.C.; Gatherer, J. Military-based vehicle-to-grid and vehicle-to-vehicle microgrid—System architecture and implementation. *IEEE Trans. Transp. Electrif.* **2017**, *4*, 157–171. [CrossRef]
14. Zou, H.; Mao, S.; Wang, Y.; Zhang, F.; Chen, X.; Cheng, L. A survey of energy management in interconnected multi-microgrids. *IEEE Access* **2019**, *7*, 72158–72169. [CrossRef]
15. Nikmehr, N.; Ravadanegh, S.N. Optimal power dispatch of multi-microgrids at future smart distribution grids. *IEEE Trans. Smart Grid* **2015**, *6*, 1648–1657. [CrossRef]
16. Kaloudas, C.G.; Ochoa, L.F.; Marshall, B.; Majithia, S.; Fletcher, I. Assessing the future trends of reactive power demand of distribution networks. *IEEE Trans. Power Syst.* **2017**, *32*, 4278–4288. [CrossRef]
17. Ganguly, S. Multi-objective planning for reactive power compensation of radial distribution networks with unified power quality conditioner allocation using particle swarm optimization. *IEEE Trans. Power Syst.* **2014**, *29*, 1801–1810. [CrossRef]
18. Syed, I.; Khadkikar, V.; Zeineldin, H.H. Loss reduction in radial distribution networks using a solid-state transformer. *IEEE Trans. Ind. Appl.* **2018**, *54*, 5474–5482. [CrossRef]
19. Bolognani, S.; Zampieri, S. A distributed control strategy for reactive power compensation in smart microgrids. *IEEE Trans. Autom. Control* **2013**, *58*, 2818–2833. [CrossRef]
20. Arefifar, S.A.; Mohamed, Y.A.-R.I. Probabilistic optimal reactive power planning in distribution systems with renewable resources in grid-connected and islanded modes. *IEEE Trans. Ind. Electron.* **2014**, *61*, 5830–5839. [CrossRef]
21. Cho, Y.-W.; Kwon, J.-M.; Kwon, B.-H. Single power-conversion AC–DC converter with high power factor and high efficiency. *IEEE Trans. Power Electron.* **2014**, *29*, 4797–4806. [CrossRef]
22. Sayed, M.A.; Takeshita, T.; Kitagawa, W. Advanced PWM switching technique for accurate unity power factor of bidirectional three-phase grid-tied DC–AC converters. *IEEE Trans. Ind. Appl.* **2019**, *55*, 7614–7627. [CrossRef]
23. Chen, C.; Duan, S.; Cai, T.; Liu, B. Online 24-h solar power forecasting based on weather type classification using artificial neural network. *Sol. Energy* **2011**, *85*, 2856–2870. [CrossRef]
24. Aien, M.; Fotuhi-Firuzabad, M.; Rashidinejad, M. Probabilistic optimal power flow in correlated hybrid wind–photovoltaic power systems. *IEEE Trans. Smart Grid* **2014**, *5*, 130–138. [CrossRef]
25. Quan, H.; Srinivasan, D.; Khosravi, A. Short-term load and wind power forecasting using neural network-based prediction intervals. *IEEE Trans. Neural Netw. Learn. Syst.* **2014**, *25*, 303–315. [CrossRef]
26. Luo, Z.; Song, Y.; Hu, Z.; Xu, Z.; Yang, X.; Zhan, K. Forecasting charging load of plug-in electric vehicles in China. In Proceedings of the 2011 IEEE Power and Energy Society General Meeting, Detroit, MI, USA, 24–28 July 2011; pp. 1–8.
27. Chen, C.; Xiao, L.; Duan, S.; Chen, J. Cooperative optimization of electric vehicles in microgrids considering across-time-and-space energy transmission. *IEEE Trans. Ind. Electron.* **2019**, *66*, 1532–1542. [CrossRef]
28. Chen, C.; Duan, S. Optimal integration of plug-in hybrid electric vehicles in microgrids. *IEEE Trans. Ind. Inform.* **2014**, *10*, 1917–1926. [CrossRef]

29. Li, Y.; Zou, Y.; Tan, Y.; Cao, Y.; Liu, X.; Shahidehpour, M.; Tian, S.; Bu, F. Optimal stochastic operation of integrated low-carbon electric power, natural gas, and heat delivery system. *IEEE Trans. Sustain. Energy* **2018**, *9*, 273–283. [CrossRef]
30. Zhang, T.; Yu, L. Optimal allocation of DSTATCOM considering the uncertainty of photovoltaic systems. *IEEJ Trans. Electr. Electron. Eng.* **2020**, *15*, 355–363. [CrossRef]
31. Chen, C.; Chen, J.; Wang, Y.; Duan, S.; Cai, T.; Jia, S. A price optimization method for microgrid economic operation considering across-time-and-space energy transmission of electric vehicles. *IEEE Trans. Ind. Inform.* **2020**, *16*, 1873–1884. [CrossRef]
32. Saha, S.; Banerjee, S.; Maity, D.; Chanda, C. Optimal sizing and location determination of distributed generation in distribution networks. In Proceedings of the 2015 International Conference on Energy, Power and Environment: Towards Sustainable Growth (ICEPE), Shillong, India, 12–13 June 2015; pp. 1–5.
33. Yang, J.; He, Z.; Ke, J.; Xie, M. A new hybrid multilevel DC–AC converter with reduced energy storage requirement and power losses for HVDC applications. *IEEE Trans. Power Electron.* **2018**, *34*, 2082–2096. [CrossRef]
34. Chen, J.; Chen, C.; Duan, S. Cooperative optimization of electric vehicles and renewable energy resources in a regional multi-microgrid system. *Appl. Sci.* **2019**, *9*, 2267. [CrossRef]
35. Bocca, A.; Baek, D. Optimal Life-Cycle Costs of Batteries for Different Electric Cars. In Proceedings of the 2020 AEIT International Conference of Electrical and Electronic Technologies for Automotive (AEIT AUTOMOTIVE), Turin, Italy, 18–20 November 2020; pp. 1–6.
36. Yang, H.; Zhou, W.; Lu, L.; Fang, Z. Optimal sizing method for stand-alone hybrid solar–wind system with LPSP technology by using genetic algorithm. *Sol. Energy* **2008**, *82*, 354–367. [CrossRef]
37. Xu, K.; Wang, S.; Zhao, J.; Dou, X.; Zhang, P. Research on Microgrid Dispatch Considering the Charge and Discharge Strategy of Electric Vehicles. In Proceedings of the 2020 12th IEEE PES Asia-Pacific Power and Energy Engineering Conference (APPEEC), Nanjing, China, 20–23 September 2020; pp. 1–5.
38. Li, Z.; Lu, Q.; Fu, Y. State-splitting multi-objective dynamic programming algorithm for dynamic reconfiguration of active distribution network. *Proc. CSEE* **2019**, *39*, 5025–5036.

applied sciences

MDPI

Article

Performance Analysis of a Stand-Alone PV/WT/Biomass/Bat System in Alrashda Village in Egypt

Hoda Abd El-Sattar [1], Salah Kamel [1], Hamdy Sultan [2], Marcos Tostado-Véliz [3,*], Ali M. Eltamaly [4,5,6] and Francisco Jurado [3]

1 Department of Electrical Engineering, Faculty of Engineering, Aswan University, Aswan 81542, Egypt; eng_ha20@yahoo.com (H.A.E.-S.); skamel@aswu.edu.eg (S.K.)
2 Department of Electrical Engineering, Faculty of Engineering, Minia University, Minia 61111, Egypt; hamdy.soltan@mu.edu.eg
3 Electrical Engineering Department, University of Jaen, EPS, 23700 Linares, Spain; fjurado@ujaen.es
4 Saudi Electricity Company Chair in Power System Reliability and Security, King Saud University, Riyadh 11421, Saudi Arabia; eltamaly@ksu.edu.sa
5 Sustainable Energy Technologies Center, King Saud University, Riyadh 11421, Saudi Arabia
6 Electrical Engineering Department, Mansoura University, Mansoura 35516, Egypt
* Correspondence: mtostado@ujaen.es

Abstract: This paper presents an analysis and optimization of an isolated hybrid renewable power system to operate in the Alrashda village in the Dakhla Oasis, which is situated in the New Valley Governorate in Egypt. The proposed hybrid system is designed to integrate a biomass system with a photovoltaic (PV), wind turbine (WT) and battery storage system (Bat). Four different cases are proposed and compared for analyzing and optimizing. The first case is a configuration of PV and WT with a biomass system and battery bank. The second case is the integration of PV with a biomass system and battery bank. The third case is WT integrated with biomass and a battery bank, and the fourth case is a conventional PV, WT, and battery bank as the main storage unit. The optimization is designed to reduce component oversizing and ensure the dependable control of power supplies with the objective function of reducing the levelized cost of energy and loss of power supply probability. Four optimization algorithms, namely Heap-based optimizer (HBO), Franklin's and Coulomb's algorithm (CFA), the Sooty Tern Optimization Algorithm (STOA), and Grey Wolf Optimizer (GWO) are utilized and compared with each other to ensure that all load demand is met at the lowest energy cost (COE) for the proposed hybrid system. The obtained results revealed that the HBO has achieved the best optimal solution for the suggested hybrid system for case one and two, with the minimum COE 0.121171 and 0.1311804 $/kWh, respectively, and with net present cost (NPC) of $3,559,143 and $3,853,160, respectively. Conversely, STOA has achieved the best optimal solution for case three and four, with a COE of 0.105673 and 0.332497 $/kWh, and an NPC of $3,103,938 and $9,766,441, respectively.

Keywords: PV; wind turbine; biomass system; heap-based optimizer; Franklin's and Coulomb's algorithm; sooty tern optimization; energy cost

Citation: El-Sattar, H.A.; Kamel, S.; Sultan, H.; Tostado-Véliz, M.; Eltamaly, A.M.; Jurado, F. Performance Analysis of a Stand-Alone PV/WT/Biomass/Bat System in Alrashda Village in Egypt. *Appl. Sci.* **2021**, *11*, 10191. https://doi.org/10.3390/app112110191

Academic Editor: Amjad Anvari-Moghaddam

Received: 23 September 2021
Accepted: 20 October 2021
Published: 30 October 2021

Publisher's Note: MDPI stays neutral with regard to jurisdictional claims in published maps and institutional affiliations.

Copyright: © 2021 by the authors. Licensee MDPI, Basel, Switzerland. This article is an open access article distributed under the terms and conditions of the Creative Commons Attribution (CC BY) license (https://creativecommons.org/licenses/by/4.0/).

1. Introduction

The world's need is increasing every day to reduce dependence on the use of fossil fuels, so finding means, solutions, and alternatives for how to produce the required energy has become of paramount importance. Thus, the push to develop and produce renewable energy globally increases every year, and many countries have managed to develop renewable energy projects based on solar and wind energy on a large scale. This progress is essential to the plan to replace renewable energy sources that depend on fossil fuels and establish a solid foundation for a sustainable society [1].

Off-grid power generation is a viable option for supplying electricity to small communities in developing countries that do not have enough money to spend on a continuous connection to the public electric grid, and places that are very remote and cannot be easily connected to the grid due to their distance from basic infrastructure. In such circumstances, the utilization of renewable energies can help these places develop more quickly [2]. The most common methods of generating renewable energy are solar and wind energy solutions. However, it often depends on the area to decide which resources will be used to get the best results. This could include hydropower and/or biomass energy as additional means of producing renewable energy. Hybrid Renewable Energy System stations are generally characterized as a combination of two or more various power sources to supply the electrical power required for the loads, and can be a mixture of either traditional and renewable sources, or only renewable sources [3].

An off-grid power generation system causes reliability issues because of an unavailability of electricity backup from the utility grid. Moreover, solar and wind energy's variable nature causes non-linear and erratic energy production, which leads to a power mismatch where the load requirements of the consumer are not satisfied by the capacity production [4]. To overcome this, a hybrid renewable energy system is used with an energy backup unit to meet consumer demand. Where the energy storage system consists of fuel cells (FCs), batteries (Bats), etc., thereby the wind and solar energy complimentary characteristics are integrated with the energy storage system backup unit to make the system credible and sustainable [5].

Several researchers have introduced popular software-based, classical, and metaheuristic techniques for the unit sizing of hybrid renewable systems. One of the most known of these software programs used for the optimization process is the Multiple Energy Sources Hybrid Optimization Model (HOMER). The authors in [6] utilized the HOMER simulation to study the performance of six different configurations of hybrid systems based on a photovoltaic (PV)/wind turbine (WT)/FC/Bat model. This research paper aims to look at the energy production potential and creation of hydrogen using solar and wind power resources in various regions throughout Saudi Arabia, including Dhahran, Riyadh, Jeddah, Abha, and Yanbu. The results revealed that integrating PV/WT/Bat storage bank is the optimal option for achieving the lowest energy cost (COE) with 0.609 $/kWh in the Yanbu area. Ref. [7] investigated a design of a hybrid stand-alone renewable energy model for the Azad National Institute of Technology, Bhopal in the Indian state of Madhya Pradesh using 5 kW PV, 5 kW biomass gasifier generator and a 5 kW fuel cell. The HOMER program was employed for obtaining the optimized results, where the COE of the proposed power system has been found to be 15.064 Rs/kWh and total net present cost (TNPC) Rs. 5189003. Authors in [8] introduced a techno-economic analysis and optimum analysis planning of different configurations of a hybrid renewable energy system based on PV/WT/ diesel generator (DG)/Bat, and converter to meet up with the electric load requirements for a rural area in Dongola, Sudan. This was achieved by studying various layouts of the suggested hybrid system to explore the optimal solution for the lowest NPC and greenhouse gas emissions using the HOMER program. The results evidenced that the construction of the PV/WT/DG/Bat converter unit achieved the best performances for both the TNPC with 24.16 M$ and COE with 0.387 $/kWh.

Ajlan et al. [9] examined the feasibility of introducing a micro-grid hybrid system using five alternative energy scenarios (DG-only, PV/DG, WT/DG, PV/WT and PV/WT/DG) for a rural community in the Shafar village, Hajjah province, Yemen. From an environmental and economic standpoint, the results obtained from the HOMER software showed that PV/WT/DG scenario was the optimal hybrid system in CO_2 emission reduction with 70%, system cost reduction with 45%, and high system reliability. Dufo-López et al. [10] formulated a new multi-objective evolutionary algorithm (MOEA) to identify the best feasible way of a stand-alone hybrid power system based on PV/WT/DG/Bat/converter to satisfy the required load in the Tindouf area, Algeria. The main objective functions of this suggested system are to reduce the NPC and maximize both Human Development Index (HDI), as well as job creation (JC).

Antonio et al. [11] evaluated an optimal configuration analysis using HOMER software for an off-grid hybrid system based on PV/BG/hydrokinetic turbines/Bats bank located in Southern Ecuador. Mehran et al. [12] applied the multi-objective crow search algorithm for optimum sizing and the techno-economic analysis of a hybrid system consisting of PV/DG/FCs and batteries. Suresh et al. [13] developed the multi-objective improved genetic algorithm to find the optimal sizing of an off-grid hybrid model for rural areas by considering the minimization of the COE. This proposed system was based on PV/WT/DG/Bat components. Kharrich et al. [14] discussed improving a hybrid system consisting of PV/WT/DG/Bat in the Dakhla area in Morocco by considering the minimization of the NPC. This optimization problem is based on using a novel Equilibrium Optimizer (EO), and the obtained results of this optimizer were compared with the results obtained from the use of the Harris Hawks optimizer (HHO), Artificial Electric Field optimizer (AEFO) Algorithm, GWO Algorithm, and Sooty Tern Optimization Algorithm (STOA).

Ramli et al. [15] developed a multi-objective self-adaptive differential evolution (MOSaDE) technique for the optimal scheduling of a microgrid system composed of PV/WT/DG/Bat for Yanbu, Saudi Arabia. This optimization technique has been used to analyze the COE, LPSP, and the Renewable Factor (RF) simultaneously. Ashraf et al. [16] presented the PV/WT/DG hybrid system as the optimal configuration for providing the required loads with least minimum COE, the total emissions generated, and maximum LPSP in the Gobi Desert in China. The optimized design of the proposed hybrid system is based on a new Elephant Herding Optimization (EHO) algorithm. Diab et al. [17] formed an optimal grid system to reduce the energy cost while satisfying the operational constraints by using a Modified Farmland Fertility Algorithm (MFFA), while the hybrid system is a combination of PV, WT, and FC units as a case study for Ataka region in Egypt.

Geleta et al. [18] proposed and analyzed an optimized sizing of PV/WT/Bat bank hybrid system as the optimal configuration for supplying the needed load with the least COE. The GWO algorithm is the proposed technique used for solving the optimization problem. Shakti and Subhash [19] studied an optimized sizing of an off-grid PV/biomass system compared to grid-connected PV/biomass system. The assessment of various viewpoints of multiple technical and economic performance were made using two optimization techniques, the Artificial Bee Colony (ABC) optimization technique and HOMER software. The results showed that the grid-connected model outperformed the off grid model in terms of cost. Bukar et al. [20], determined the optimal hybrid energy system composed of PV/WT/DG/Bat that would fulfill the load required to reliably supply residential housing in Yobe State, Nigeria, based on reducing the COE and LPSP. Optimization of the suggested hybrid power system was done using the grasshopper optimization algorithm (GOA) and the obtained results were compared with the results obtained from CS, PSO algorithms.

Heydari and Askarzadeh [21], evaluated an approach for optimal sizing of an off grid hybrid system based on PV/biomass in Bardsir, Iran, with objectives of minimizing NPC and the LPSP. This research is focused on utilizing the harmony search (HS) optimization algorithm on modeling the optimal hybrid system. Sarkar et al. [22] analyzed the operational behavior of an optimized hybrid micro-grid consists of PV/WT/biomass/Bat storage unit using the HOMER program to supply the required load of the investigated area in India with least COE, and to ensure zero LPSP. Li et al. [23] addressed the issue of techno-economic optimal design of stand-alone PV/WT/Biomass/Bat hybrid model utilizing HOMER program for a town in West China.

Ghosh et al. [24] discussed the optimal sizing and cost reduction solution for a micro-grid hybrid system that both includes PV and biomass. The dragonfly algorithm has been applied to simulate and perform this optimization analysis and the results have been compared with the obtained results from the ABC method. Eteiba et al. [25] evaluated the effect of four optimization techniques (Flower Pollination Algorithm (FPA), the HS, ABC, and the Fire-fly Algorithm (FA)) to determine the optimal sizing of an off-grid hybrid PV/biomass/Bat storage system while utilizing the minimization of NPC as the fitness function for the suggested optimization methods. Sawle et al. [26] presented different optimization strategies based on GA, BFPSO, PSO and Teaching-Learning-Based Optimization (TLBO) to construct an optimal PV/WT/Biomass/Bat hybrid system with different objectives which are COE, LPSP, RF, Particular matter (PM), HDI, JC, and GHG. According to the results, the TLBO technique is an effective tool for dealing with all problem objectives and providing the best solution. Alshammari and Asumadu [27] discussed the optimization of an off-grid hybrid system consisting of PV/WT/biomass/Bat units to supply customers' electrical demands in a cost-effective, efficient, and reliable manner. To determine the optimal solution, two optimization methods were used (HS and PSO techniques). The major objectives of this work are as follows:

- The paper contains the study of four scenarios of a stand-alone hybrid system utilizing real-time meteorological data for a remote area located in the New Valley Governorate of Egypt called Alrashda village in Dakhla Oasis. The first system scenario is PV/WT/Biomass/Bat, the second is PV/Biomass/Bat, the third is WT/Biomass/Bat, and the fourth one is PV/WT/Bat.
- Studying a new optimization algorithm, which is the Heap-based optimizer (HBO) technique, while make a comparison with a three recent types of optimization methods namely, Franklin's and Coulomb's algorithm (CFA), the Sooty Tern Optimization Algorithm (STOA), and Grey Wolf Optimizer (GWO).
- The study includes exploiting the capabilities of the proposed algorithms to optimize and minimize COE with increasing the reliability and efficiency of the suggested hybrid systems, and performs different sensitivity analyses on an optimal design to predict the upcoming system implementation.

The suggested work is structured as follows: Section 2 explains the modeling of the suggested system units. Section 3 discusses the description of the studied area. Section 4 discusses the formulation of the optimization problem. Section 5 discusses a brief explanation of the optimization methodology of HBO, CFA, GWO, and STOA. Section 6 presents the results of the optimal sizing for the stand-alone hybrid power system. Finally, the conclusions are provided in Section 7.

2. Modeling of the Proposed System

The stand-alone hybrid system considered in this paper consists of PV/WT/Biomass/Bat units. The layout of the suggested hybrid model is illustrated in Figure 1. This section explains in detail the description of the major system units and the optimization methodology of the suggested hybrid model.

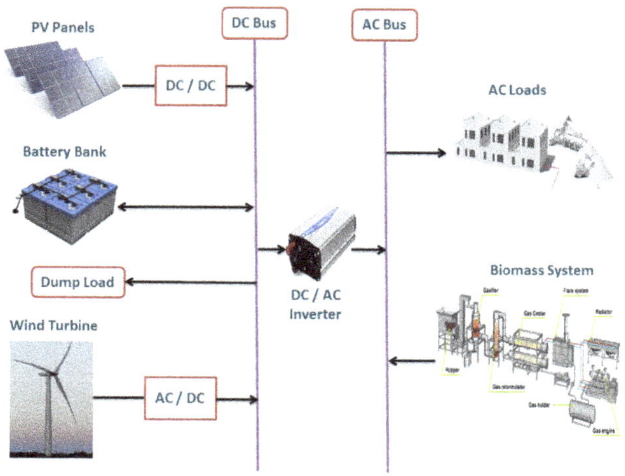

Figure 1. The architecture of the proposed stand-alone hybrid system.

2.1. Photovoltaic System (PV)

A simplified model based into account ambient temperature and solar irradiation is employed in this study to compute the energy generated by the PV panels $PV_p(t)$, as indicated in the equation below [28,29].

$$PV_p(t) = N_{PV} \times PV^P_{rat}\left(\frac{R_{int}(t)}{R_{STC}}\right)\left[1 + \gamma_T\left((R_{int}(t)\left(\frac{T_{nor} - 20}{0.8}\right) + T_{amb}(t)) - T_{STC}\right)\right] \cdot \eta_w \, \eta_{PV} \quad (1)$$

where, PV^P_{rat} indicates the rated power of the PV panel at standard test condition (STC) [kW], $R_{int}(t)$ is the intensity of solar radiation at time (t), R_{STC} denote the intensity of solar radiation at standard conditions [1000 W/m^2], N_{PV} is the number of PV units, γ_T is the PV module temperature coefficient [%/°C], η_w is the wiring efficiency, η_{PV} is the PV module efficiency, T_{nor} is the cell temperature under at normal operating conditions, $T_{amb}(t)$ denote the ambient temperature (°C), T_{STC} denote the cell temperature under standard operating conditions (°C). The technical specifications of the PV panel modeling are shown in Table 1.

Table 1. The main parameters of the selected photovoltaic model [30].

Parameter	Value	Unit
Model type		PV-MLT260HC
PV panel cost (C_{PV})	14,854	\$/m^2
γ_T	0.0037	-
η_{PV}	15	%
T_{STC}	25	°C
PV^P_{rat}	1	kW
Length	1625	mm
Width	1019	mm
Thickness	46	mm
lifetime of PV system (PV_S)	20	year
PV replacement cost (C^{PV}_{rep})	13,885	\$

2.2. Wind Turbine

Every month, NASA supplies data on wind speed, which has been utilized as input data for this study (NASA, 2020). The following mathematical formulas are employed to calculate the wind turbine output $WT_P(t)$ based on a comprehensive literature review [2,31].

$$WT_p(t) = \left\{ \begin{array}{cc} 0, & V(t) < V^{in}_{cut} \text{ or } V(t) > V^{off}_{cut} \\ N_{WT} \times WT^P_{rat} \times {}_{WT}\left(\frac{V^2(t) - V^2_{cut}}{V^2_{rat} - V^2_{cut}}\right), & V^{in}_{cut} < V(t) < V_{rat} \\ N_{WT} \times WT^P_{rat} \times \eta_{WT}, & V_{rat} < V(t) < V^{off}_{cut} \end{array} \right\} \quad (2)$$

In which, $V(t)$, V^{in}_{cut}, V^{off}_{cut}, and V_{rat} are WT speed at time t, WT speeds cut-in, WT speeds cut-off wind speed, and rated speed respectively. N_{WT} denotes the number of WTs modules, η_{WT} is the WT efficiency, and WT^P_{rat} is the rated power of the WT (kW).

Wind speed increases with height above ground level, and the wind turbine hub's height has also a major impact on wind speed, which affects power generation, according to the below power law equation [13]:

$$\left(\frac{V_n}{V_{ref}}\right) = \left(\frac{H_n}{H_{ref}}\right)^{\varepsilon_{wt}} \quad (3)$$

where, V_n represents the WT speed (m/s) at the new height H_n (m), V_{ref} is the WT speed (m/s) at the original turbine hub height H_{ref} (m), and ε_{wt} denotes the WT friction coefficient.

According to the International Electro technical Committee (IEC), the value of the coefficient of friction in the case of normal wind conditions is 0.20 and in the case of intensive wind conditions is 0.11. The technical specifications of the selected WT modeling are presented in Table 2.

Table 2. The main parameters of the selected wind turbine model [30].

Parameter	Value	Unit
Model type		Fuhrländer FL 30
WT^{P}_{rat}	30	kW
wind turbine height	50	m
η_{WT}	80	%
V^{in}_{cut}	2.5	m/s
V_{rat}	12	m/s
V^{off}_{cut}	25	m/s
lifetime of WT (WT_S)	20	year
wind turbine cost (C_{WT})	3200	$/kW

2.3. Biomass System

Biomass comprises of the stored chemical energy from solar energy, so biomass can be used for heating by direct burning or transformed through many operations into liquid fuels and renewable gases [32,33]. One of the major aspects in determining the type of technology used to generate biomass energy is the type of biomass to be used and the type of fuel to be produced from the conversion process [34].

In this work, biomass gasification is the conversion process used which is a pyrolysis process in which the raw materials of biomass are heated in closed and pressurized vessels, the output gaseous fuel by this process is usually called the producer gas [35].

In this study, sugarcane bagasse was used as a raw material for biomass to feed a small-scale downdraft gasifier, as the cane crop is one of the agricultural crops available in the New Valley city. The biomass generator was utilized as the primary generator to satisfy the electrical load requirement beside the PV and WT systems, the technical characteristics of the biomass system are illustrated in Table 3. The hourly generated power from the biomass system $BG_P(t)$ can be expressed according to the following mathematical formula [25,27];

$$BG_P(t) = FS_{rat}(t) \times HHV_{fs} \times \eta_{gas} \times \omega \tag{4}$$

where, $FS_{rat}(t)$ is the biomass raw material rate per hour (kg/h), HHV_{fs} indicates the higher heat value of the biomass raw material, η_{gas} denotes the efficiency of the gasifier reactor (75%), and ω represents a factor for converting units from kJ to kWh (27.78×10^{-5}).

The load ratio of the considered biomass generator is set to operate at no less than 30% (Gen_{min} = 30%) of its rated capacity to avoid running at much lower demands, while its maximum load is 80% (Gen_{max} = 80%) of its rated capacity. The generator output power (Gen_{out}) can be described according to the following constraints [25,27,36];

$$Gen_{out} = \begin{cases} 0 \; BG_P < Gen_{min} \\ BG_P \; Gen_{max} > Gen_{max} \\ BG_P \; Gen_{max} < BG_P < Gen_{min} \end{cases} \tag{5}$$

Table 3. The main parameters characteristics of the biomass system.

Parameter	Value	Unit
η_{gas}	75	%
Generator rated	50	kW
Capital Cost [37]	23,700	$/kW
Lifespan [37]	15,000	h
Replacement cost [37]	15,000	$/unit
Yearly O&M cost [37]	0.05	$/ h

Based on the previous mathematical expressions, $F^{BG}_{con}(t)$ is the average fuel consumption per hour, and E_{Bio} (kWh) is the annual energy output which can be computed as following;

$$E_{Bio} = \sum_{t=1}^{8760} N_g \times Gen_{out} \times t \quad (6)$$

where, N_g is the number of generators.

2.4. Battery Bank Model

The battery bank serving as a backup system of storing energy in the event that the renewable sources are unable to deliver the needed power. The hourly total power generated by the PV, WTs, and biomass system $P_{re}(t)$ is obtained based on the below equation [19,25];

$$P_{re} = P_{PV} + WT_P + BG_P/\eta_{inv} \quad (7)$$

The technical specifications of the battery bank model are illustrated in Table 4. The following equations explain the energy production and consumption of the battery system from time t–1 to time t [30,38];

During the charging phase Bat_{CH},

$$Bat_{CH}(t) = (P_{re}(t) - (P_L(t)/\eta_{inv})) \times \Delta t \times \eta_{CH} \quad (8)$$

$$SOC_{Bat}(t) = SOC_{Bat}(t-1) \times (1-\sigma) + Bat_{CH}(t) \quad (9)$$

During discharging phase Bat_{DIS},

$$Bat_{DIS}(t) = ((P_L(t)/\eta_{inv}) - P_{re}(t)) \times \Delta t \times \eta_{DIS} \quad (10)$$

$$SOC_{Bat}(t) = SOC_{Bat}(t-1) \times (1-\sigma) - Bat_{DIS}(t) \quad (11)$$

Table 4. The main parameters characteristics of the batter bank.

Parameter	Value	Unit
Model		RS lead acid battery
Nominal battery voltage [39]	5	V
Nominal battery capacity [39]	360	Ah
Storage capacity of battery [7]	4.8	kWh
η_{CH} [30]	90	%
η_{DIS} [30]	85	%
σ [30]	0.005	-
Battery cost (C_{Bat}) [30]	3880	$
Battery lifetime (Bat_S) [30]	25	year

In which, η_{CH} and η_{DIS} indicate the battery charging and discharging efficiencies, respectively, σ is self-discharge rate, and SOC_{Bat} is the battery state of charge. η_{inv} denotes the inverter efficiency.

2.5. Bi-Directional Converter Model

A bidirectional transducer is adopted to maintain power flow between DC and AC components. There are two kinds of power conversion devices in a power system, the inverter which converts DC current to AC current and the rectifier which converts AC current to DC current. The technical characteristics of the inverter model are presented in Table 5. The hourly input power of the inverter $P_{inv}(t)$ can be expressed as below [16];

$$P_{inv}(t) = P_L(t)/\eta_{inv} \quad (12)$$

In which, η_{inv} represents the inverter efficiency.

Table 5. The main parameters characteristics of the inverter.

Parameter	Value	Unit
η_{inv}	95	%
Max. power	1	kW
inverter lifetime (Inv_S)	10	year
Inverter cost (C_{inv})	711	$/kW
Inverter replacement cost (C_{rep}^{inv})	650	$/kW
η_{inv}	95	%

3. Description of the Studied Area

The considered area for this study is Alrashda village, which is located 10 km northwest of Mut town, the administrative center of the Dakhla Oasis in the New Valley Governorate in Egypt, at 28.938° east longitude, 25.576° north latitude, and an altitude of 243 m. The reason of choosing this village because of its comparatively high solar, wind, and biomass energy potential. The proposed mathematical model is used for designing a small scale stand-alone hybrid system to feed a range of loads which are represented in residential loads, where the peak loads are occurred during the summer and in the evening period from 19:00 to 23:00 p.m. In Figure 2, the profile of the proposed loads during a year is depicted, which shows that the average residential load of the village has reached about 260 kW, with a maximum load of 410 kW. Figures 3–5 illustrate the plot of hourly data of the solar radiation, temperature, and wind speed which are obtained from the NASA Surface Meteorology and Solar Energy website for 20 years for the selected area. Figure 3 presents the short-wave solar irradiance of the studied area during a year, where the yearly radiation rate is between 2.45 kWh/m^2/day to 10.94 kWh/m^2/day, with the average yearly radiation on this site's horizontal surface is around 6.89 kWh/m^2/day, while the yearly ambient temperature of the selected site is indicated in Figure 4, which showed that the maximum ambient temperature can be reached, is 40°. Figure 5 illustrates the annual wind speed for the selected location with a maximum wind speed of about 13.9 m/s and an average in the range from 8.71 m/s to 9.89 m/s. As previously mentioned, the biomass feedstock used in this study was the sugarcane bagasse. The sugar cane crop is considered one of the strategic crops in Egypt, where the harvest period begins during January of each year and extends until May. The amount of biomass feedstock available at the selected site was assumed to have a variable values over the year, the monthly biomass consumption rate is presented in Figure 6, with an average of one ton/day.

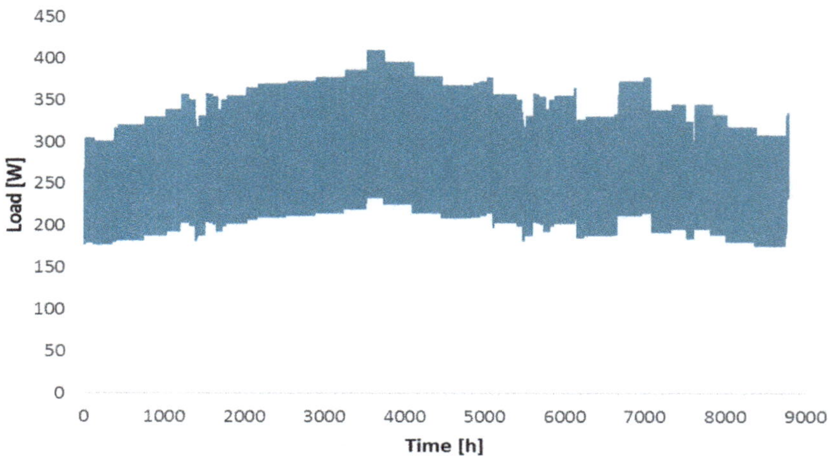

Figure 2. Load profile of the studied area.

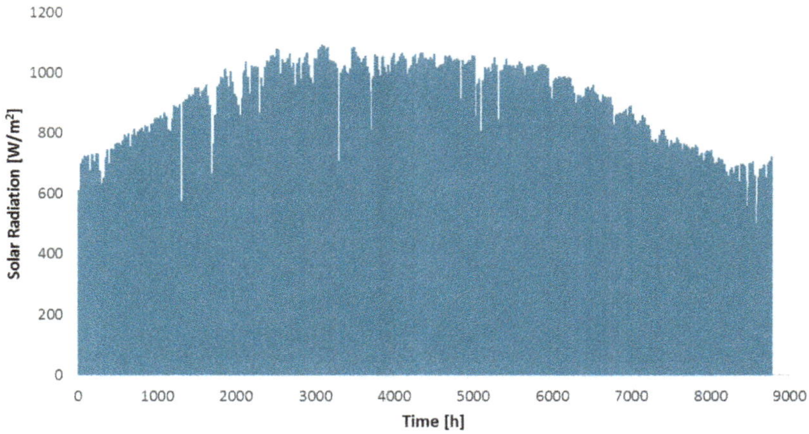

Figure 3. The annual short-wave solar irradiance of the studied area.

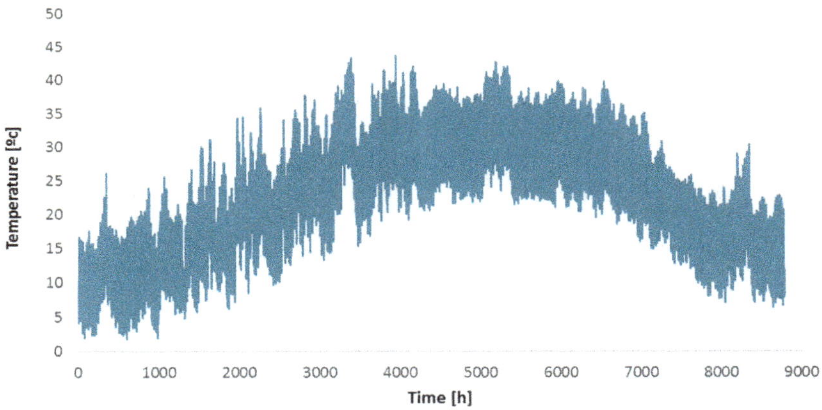

Figure 4. The yearly ambient temperature of the studied area.

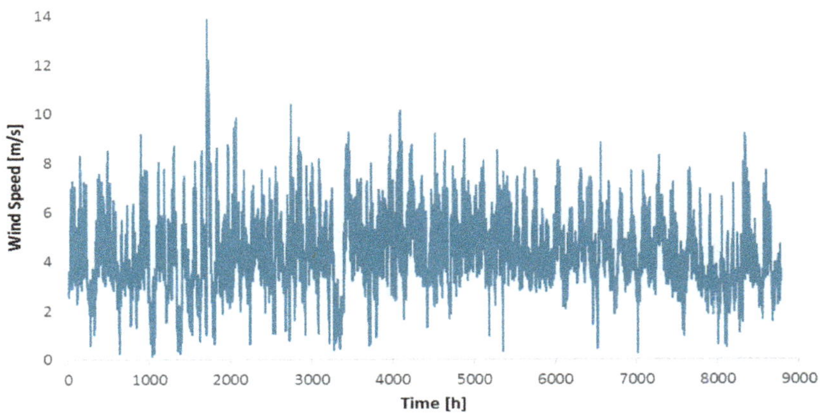

Figure 5. The yearly wind speed of the studied area.

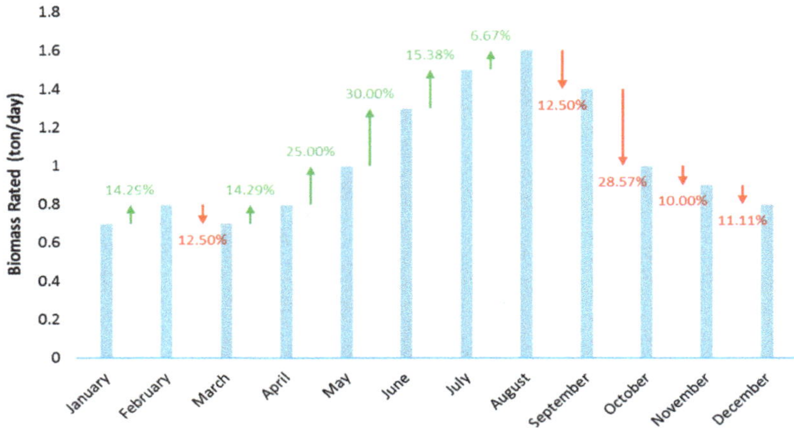

Figure 6. The biomass consumption of the studied area.

4. Optimization Problem

The primary aim of this work is to indicate the capacity to optimize the suggested stand-alone hybrid power system in order to provide a guaranteed supply of power at the lowest feasible cost. In this section, the economic and cost analysis, the main objective function, the optimization constraints, and system management strategy are discussed.

4.1. Economic and Cost Analysis

The COE for a specific system is an economic evaluation of the system's costs and of the associated cost in its lifespan. The COE is a function of the NPC, it actually helps to select the lowest energy prices from different feasible hybrid configurations, which means the least overall investment cost in a renewable power system plant, after fulfilling the energy dependability limitations. While the NPC represents the current value of the capital investment and operating costs over the lifespan. The NPC and the COE in ($/kWh) can be computed as follows [3,30]:

$$\mathrm{COE} = \frac{\mathrm{NPC}}{\sum_{1}^{8760} \mathrm{P_L}} \mathrm{C_{RF}} \tag{13}$$

$$\mathrm{NPC} = \mathrm{C_{Ann}^{T}} / \mathrm{C_{RF}} \tag{14}$$

where, C_{Ann}^{T} is the total annual cost of the proposed hybrid system, and C_{RF} is the capital recovery factor, which is a ratio for the current cash value calculation and it can be estimated over a lifespan of years (S = 25 years) and an interest rate (I_r = 6%). C_{RF} and C_{Ann}^{T} are modeled as:

$$\mathrm{RF}\ (\mathrm{I_r}\ ,\ \mathrm{S}) = \frac{\mathrm{I_r} \times (\mathrm{I_r} + 1)^{\mathrm{S}}}{(\mathrm{I_r} + 1)^{\mathrm{S}} - 1} \tag{15}$$

$$\mathrm{C_{Ann}^{T}} = \sum \mathrm{C_{Ann}^{u}} = \mathrm{C_{Ann}^{PV}} + \mathrm{C_{Ann}^{WT}} + \mathrm{C_{Ann}^{BG}} + \mathrm{C_{Ann}^{Bat}} + \mathrm{C_{A}^{inv}} \tag{16}$$

$$\mathrm{C_{A}^{u}} = \mathrm{C_{Ann\_Cap}^{u}} + \mathrm{C_{OM}^{u}} + \mathrm{C_{Ann\_Rep}^{u}} + \mathrm{C_{Ann\_fuel}} \tag{17}$$

where, C_{Ann}^{u} is the annual cost of each unit, $C_{Ann\_Cap}^{u}$ is the total annualized cost of each unit, C_{OM}^{u} is the operation and maintenance cost of each unit, $C_{Ann\_Rep}^{u}$ is the replacement cost for each unit, and C_{Ann_fuel} is the annual fuel cost of the biomass unite which is computed by applying the following formula [25,27]:

$$\mathrm{C_{Ann\_fuel}} = \mathrm{C_{Bio}} \times \mathrm{Bio_T} \tag{18}$$

$$\mathrm{Bio_T} = \sum_{1}^{8760} \mathrm{FS_{rat}}(t) \tag{19}$$

where, C_{Bio} is biomass fuel cost, and Bio_T is the total feedstock consumption of the generator (kg/year).

4.2. Objective Function

The aim is to create the optimal combination of units for the hybrid renewable energy system to achieve maximum energy supply. To achieve this aim, the COE is minimized, high power supply reliability is maintained, the LPSP is minimized, excess power (P_{EXC}) absorption is reduced dummy load (P_{dum}) to reduce the total system costs. To calculate this objective functions, the following formulas are applied:

$$\text{Min F(X)} = \text{Min} \left(\varphi_1 \text{ COE} + \varphi_2 \text{LSPS} + \varphi_3\right) \tag{20}$$

$$X = [N_{PV} \cdot N_{WT} \cdot N_g \cdot N_{Bat}] \tag{21}$$

$$\text{LPSP} = \sum_{1}^{8760} \frac{\text{LPS (t)}}{P_L \text{ (t)}} \tag{22}$$

$$\text{LPS (t)} = P_L \text{ (t)} - ((\text{t}) + \text{SOC}_{Bat}(t-1) - \text{SOC}_{min}) * \eta_{inv} \tag{23}$$

$$P_{EXC} = \sum_{1}^{8760} \frac{P_{dum} \text{ (t)}}{P_L \text{ (t)}} \tag{24}$$

where, φ is the weight factor value of each objective function, X represents the control variables of the optimization problem that must be optimized using the studied optimization algorithms, and LPS(t) is the loss of power supply at any time.

4.3. Constraints

The optimization procedure is based on the following limitations and on the upper and lower limit of the following decision variables;

$$1 \le \begin{bmatrix} N_{PV} \\ N_{WT} \\ N_g \\ N_{Bat} \end{bmatrix} \le \begin{bmatrix} N_{PV}^{max} \\ N_{WT}^{max} \\ N_g^{max} \\ N_{Bat}^{max} \end{bmatrix} \tag{25}$$

$$\text{LSPS} \le \text{LSPS}_{max} \tag{26}$$

where, N_{PV}^{max} is the maximum number of PV, and N_{WT}^{max} represent the maximum number of WTs units, based on the maximum load and rated power of PV/wind unit, which set to be 410 (410 kW/1 kW) and 13 (410 kW/30 kW), respectively. N_g^{max} is the maximum number of generator units which set to be 8 (410 kW/50 kW), N_{Bat}^{max} is the maximum number of batteries which is set to be 1000.

4.4. System Management Strategy

The methodology provided in this work aims to optimize the combination of PV, WT, biomass generators as the main power sources, and batteries which work to keep the energy supply continuous to the loads and enhancing the power supply, which reduces the costs of LPSP and P_{EXC}. The flowchart explaining the operational strategy of the proposed hybrid system is presented in Figure 7, while the operating management methodology can be stated according to the following steps:

- Initially, the charge (SOC) state of the battery bank remains unchanged when P_{re} meets the charge requirement ($P_{re}(t) = P_{inv}(t)$), and the loss in power supplies is zero (LPS (t) = 0) during this time interval.
- When the system's generation of P_{re} exceeds the load demand ($P_{re}(t) > P_{inv}(t)$), and the battery system's SOC is less than the maximum permissible charging limit for that interval, then the battery is charged with the surplus power ($P_{Sur}(t)$) until it reaches its highest limit, $P_{Sur}(t)$ is computed according to the following formula:

$$P_{Sur}(t) = P_{re}(t) - P_{inv}(t) \tag{27}$$

- When the maximum charge limit of the battery is reached, the storage system charging status remains unmodified and identical to the previous charge state ($SOC_{Bat}(t) = SOC_{Bat}(t-1)$), while the surplus energy remaining is treated as waste energy (P_W) that can be discharged into the dummy load.

$$P_W(t) = P_{Sur}(t) - (SOC_{max} - SOC_{Bat}(t-1)) \quad (28)$$

The energy stored in the battery bank shall be used to satisfy the load demand if the P_{re} generated from the proposed system cannot meet the load need and if the battery storage system charge is higher than a minimum permissible limit $P_{re}(t) < P_{inv}(t)$ and $SOC_{Bat}(t-1) \times (1-\sigma) > SOC_{min}$).

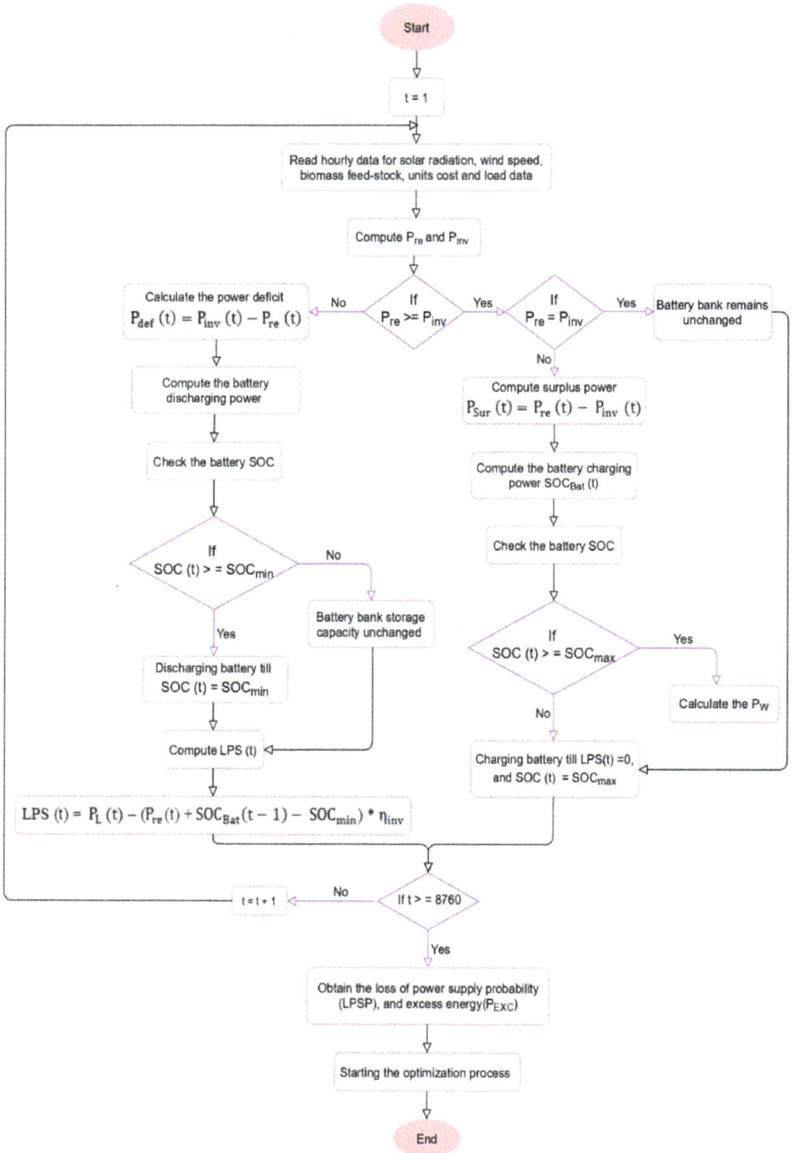

Figure 7. Flowchart of the operating management methodology.

5. Optimization Techniques

To find the solution of optimal sizing problem, four optimization algorithms with the highest efficiency have been utilized.

5.1. HBO

A new meta-heuristic optimization technique based on human-behavior called Heap-based optimizer (HBO) has been created by Qamar Askari in [40]. The HBO technique is based on the hierarchy of corporate rank (CRH) and the interplay of individuals in this hierarchy, HBO's mathematical model is based on three phases: interplay between employees and their direct manager, interplay amongst coworkers, and employee self-contribution. Unlike numerous of previous meta-heuristics algorithms, the relative fitness of search agents is used to organize them in a hierarchy, and the notion of minimum or maximum heap is used to allow interplay between them while maintaining their relative difference. Furthermore, a factor termed Gamma (γ) is established to help the algorithm avoid premature convergence without compromising the exploitation capability by allowing it to escape local optima. The pseudo-code of the HBO technique is indicated in Algorithm 1.

To prove the efficiency and performance of the HBO technique, it has been tested and compared with seven well-known algorithms and 97 diverse test functions involving 29 CEC-BC-2017 functions. The exploitative and explorative behavior of HBO has assessed from the obtained results of using 24 unimodal and 44 multimodal functions. Experiments and the Friedman mean rank test reveal that HBO outperforms and takes first place.

Algorithm 1. HBO [40]

```
Initialize general parameters, N, D, T_max, and (L_i, U_i)
Generate a random population P of N search agents
Building the heap
Algorithm 1; Heapify_Up (i)
Input: i (the index of the node we are trying to heapify)
Assuming that the rest of the nodes fulfill the heap property
while i ≠ root and heap[i].key < heap[parent (i)].key do
        swap(heap[i], heap[parent (i)])
        i ← parent (i)
end
Algorithm 2; Build_Heap (P, N)
Input: P, N
for i ← 1 to N do
    heap[i].value ← i
    heap[i].key ← f (X_i)
    Heapify_Up (i)
end
Update Search agents positions repeatedly
Algorithm 3; HBO_Main_Body
for t ← 1 to T_max do
    Compute γ, p1, p2
    for I ← N down to 2 do
        i ← heap[I].value
        bi ← heap[parent (I)].value
        ci ← heap[colleague(I)].value
        B⃗ ← x⃗_c1
        S⃗ ← x⃗_b1
        for k ← 1 to D do
            p ← rand ()
            x^k_temp ← update x^k_i(t)
        end
        if  f(x⃗_temp) < f(x⃗_i(t)) then
                x⃗_i(t + 1) ← x⃗_i(t)
            end
            Heapify_Up (I)
    end
end
return x_heap[1].value
```

5.2. CFA

A meta-heuristic optimization algorithm called Franklin's and Coulomb's algorithm (CFA) has been created by Ghasemi et al. [41], is based on the theories of the Coulomb's and Franklin's law. For optimal outcomes, the CFA employs two separate theories. First is the Coulomb's Law, which is based on the attraction and repulsion of electrons. This phenomenon governs the interaction of two independent point charges separated by a certain distance. Second is the Franklin's Law, which is based on that every item has an equal quantity of positive and negative charges, according to this law. CFA's mathematical model is based on four steps: Initialization, Attraction/repulsion, Probabilistic ionization, and Probabilistic contact. The pseudo-code of the CFA technique is indicated in Algorithm 2.

Algorithm 2. CFA [41]

Generate the initial point charges (population)
Initialize parameters
Initial fitness evaluation of whole population
While criteria not satisfied **do**
    for (i = 1: N) **do**
        The attraction/repulsion phase is applied for point charges of objects.
        Evaluate the fitness values.
        The probabilistic ionization phase is applied for elementary charges (control variables) of point charge and fitness evaluation.
        The probabilistic contact phase is applied for objects.
        Selection of the best solution.
    end for
end while

5.3. STOA

A novel bio-inspired optimization algorithm called Sooty Tern Optimization Algorithm (STOA) has been created by Dhiman and Kaur [42] to address the constraints of the industrial issues. The movement and attacking habits of the sea bird sooty tern in nature are the key motivations for modeling the STOA technique. STOA was validated using 44 benchmark test functions and compared it with nine well-known optimization techniques in terms of performance. The results of CEC 2005 and CEC 2015 standard test functions prove that the STOA is capable of addressing difficult and high dimensionality bound constrained actual situations. The pseudo-code of the STOA technique is indicated in Algorithm 3.

Algorithm 3. STOA [42]

Initialize the population $X_i^P = (X_1^P, X_2^P, X_3^P, \ldots\ldots, X_N^P$ within the limits $X_i^{min} \leq X_i^P \leq X_i^{max}$
Initialize parameters D_A and C_B
Evaluate the fitness of whole population
Best search agent $\rightarrow X_{best}$
While (it < Max_{it})
    for (i = 1: N) **do**
        Update the position of the current search agent
    end for
    Initialize parameters D_A and C_B
    Evaluate the fitness of whole population
    Update X_{best}
    it = it +1
end while
return X_{best}

5.4. GWO

The GWO has been proposed by Mirjalili et al. [43], it's a heuristic optimization technique created to find a candidate solution from a large solution space without requiring any explicit input parameters. Such qualities are ideal for dealing with nonlinear issues, such as controller parameter tweaking. Grey wolves' natural behavior and social structure in seeking prey served as inspiration for GWO. There is a hierarchical framework that governs each wolf pack. The alpha wolf, who heads the entire group, is the most formidable. In the absence of the alpha wolf, the second strongest wolf, called as the beta wolf, assumes leadership. The weaker wolves are the delta and omega wolves. The pseudo-code of the GWO technique is indicated in Algorithm 4.

Algorithm 4. GWO [43]

1. Initialize the grey wolf population $X_1^P = (X_1^P, X_2^P, X_3^P, \ldots\ldots, X_N^P$ within the limits $X_i^{min} \le X_i^P \le X_i^{max}$
2. Initialize parameters a, A, and C
3. Evaluate the fitness of whole population
4. X_α = the best search agent
5. X_β = the second best search agent
6. X_δ = the third best search agent
7. **While** (it < Max_{it})
8.   **for** (i = 1: N) **do**
9.   Update the position of the current search agent
10.   **end for**
11.   Update a, A, and C
12.   Evaluate the fitness of whole population
13.   Update X_α, X_β, and X_δ
14. **end while**
15. **return** X_α

6. Results and Discussion

In this work, a novel HBO technique is suggested to determine the optimal sizing of four alternatives off-grid hybrid system scenarios based on PV, WT, biomass, and battery units. These four scenarios of the hybrid system are namely PV/WT/biomass/Bat, PV/biomass/Bat, WT/biomass/Bat, and PV/WT/Bat. In order to validate the effectiveness of this HBO as a way to provide optimal reliability and least cost, the results achieved by the suggested algorithms are compared with other recent optimization techniques CFA, GWO and STOA. The control parameters used in the optimization process for each algorithm are listed in Appendix A.

Figure 8 presents the graphic form of the final values of the target function over the 50 executes for the four analyzed configurations scenarios utilizing the optimization techniques namely, HBO, CFA, GWO, and STOA. It can be noted that, the fitness values for the suggested HBO method in the four system cases were within a limited range, which demonstrated the stability of the suggested technique over the other techniques. Therefore, parametric and nonparametric metric values are superior using the HBO method compared to the rest of the optimization techniques.

Figure 9 displays the best optimal solution convergence curve for each scenario utilizing HBO, CFA, GWO, and STOA. For Case (1), the best solution achieved by using HBO technique which is 0.0643767 after 27 iterations, followed by CFA technique with 0.06437783 after 44 iterations. For Case (2), the best solution achieved by using HBO technique which is 0.0703404 after 49 iterations, followed by best solution achieved by CFA technique with 0.07034462 after 32 iterations. For Case (3), the best solution achieved by using HBO technique with 0.0705909 after 41 iterations, followed by best solution achieved by CFA technique with 0.0651240320 after 39 iterations. Finally for Case (4), the best solution achieved by using HBO technique with 0.151991724 after 41 iterations, followed by best solution achieved by CFA technique with 0.152001799 after 58 iterations. It can be noticed that the HBO method provides a good convergence characteristic over the other optimization algorithms CFA, GWO, and STOA in all suggested cases.

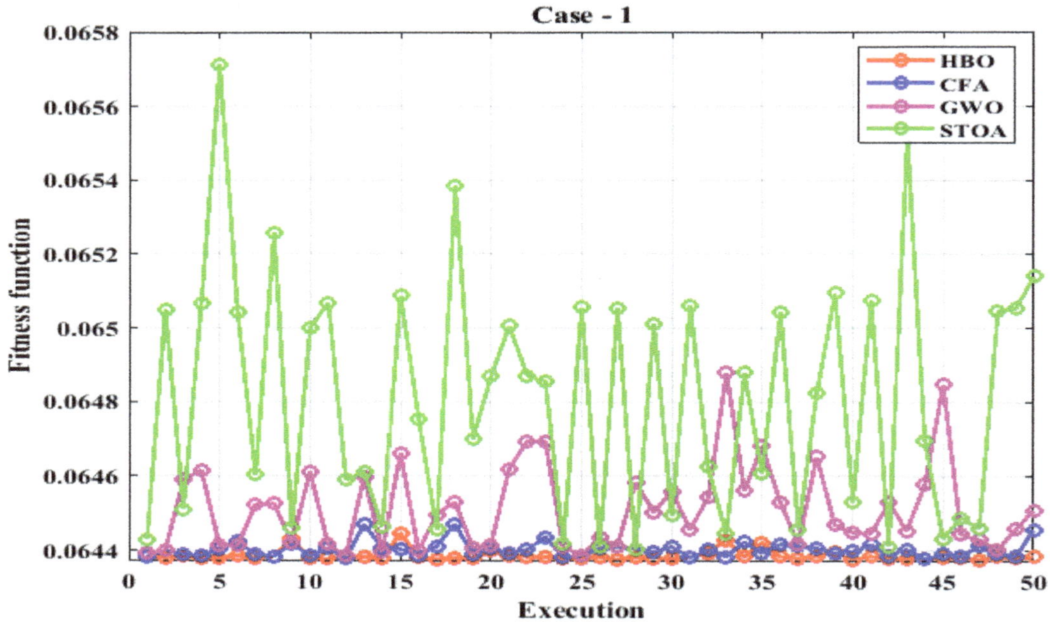

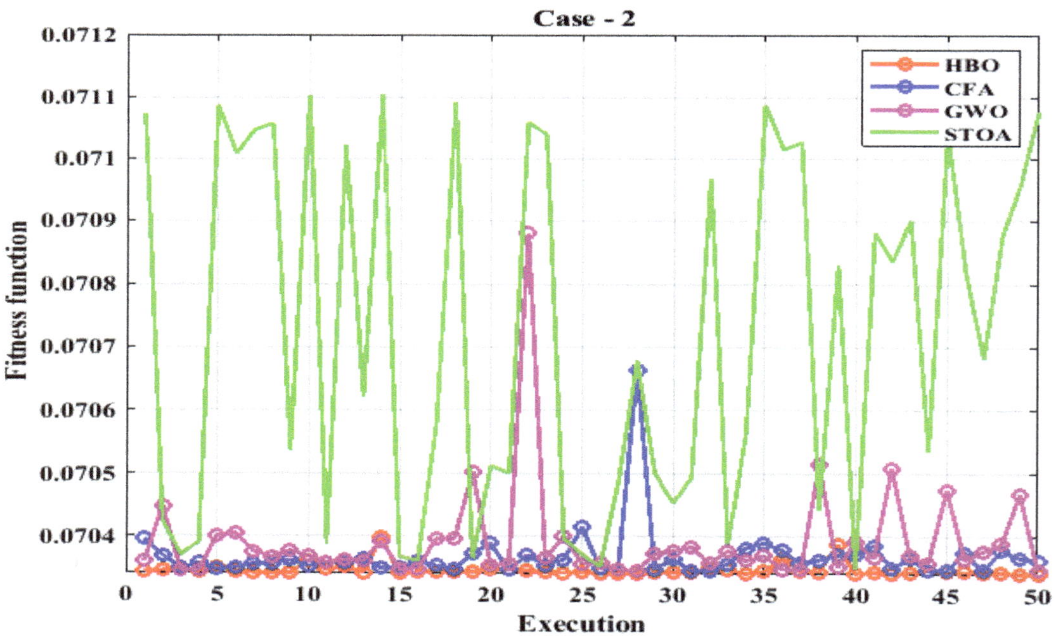

Figure 8. *Cont.*

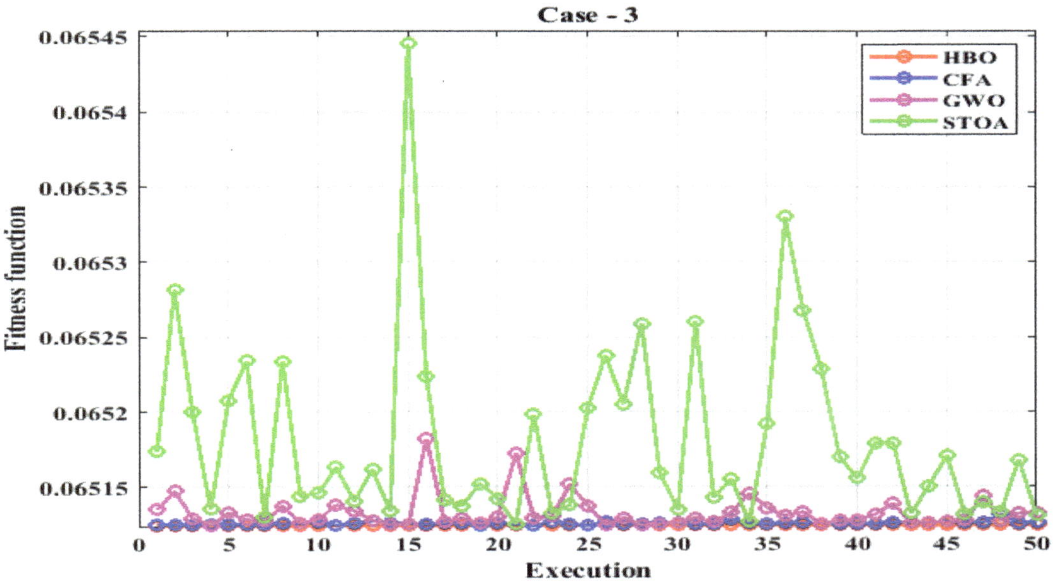

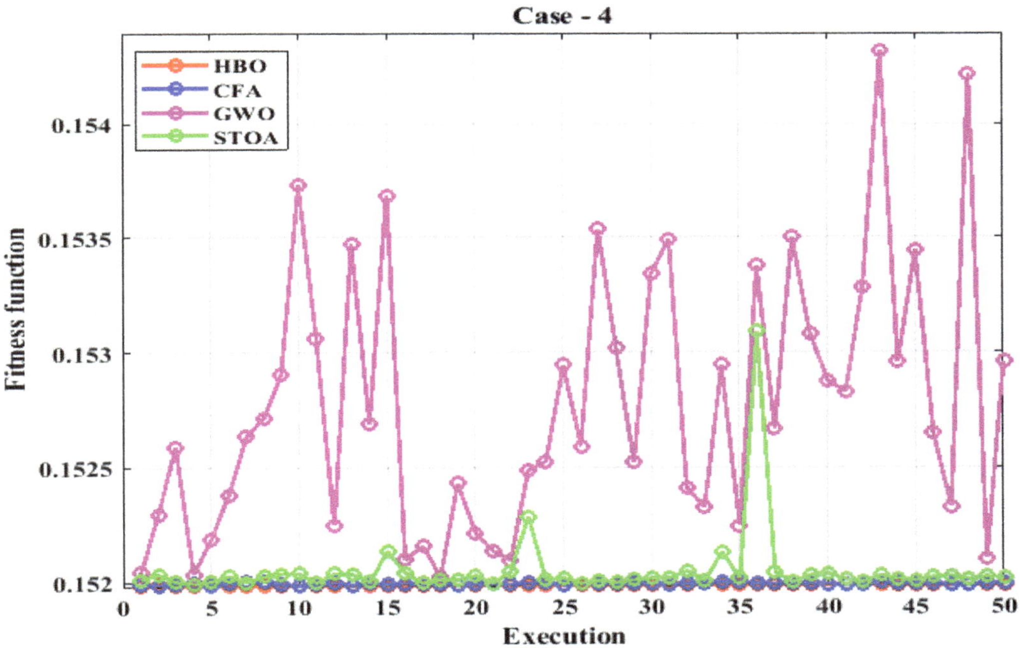

Figure 8. End values of the fitness function for 50 executions using HBO, CFA, GWO, and STOA methods: case-1: PV/WT/Biomass/Bat system, case-2: PV/Biomass/Bat system, case-3: WT/Biomass/Bat system, case-4: PV/WT/Bat system.

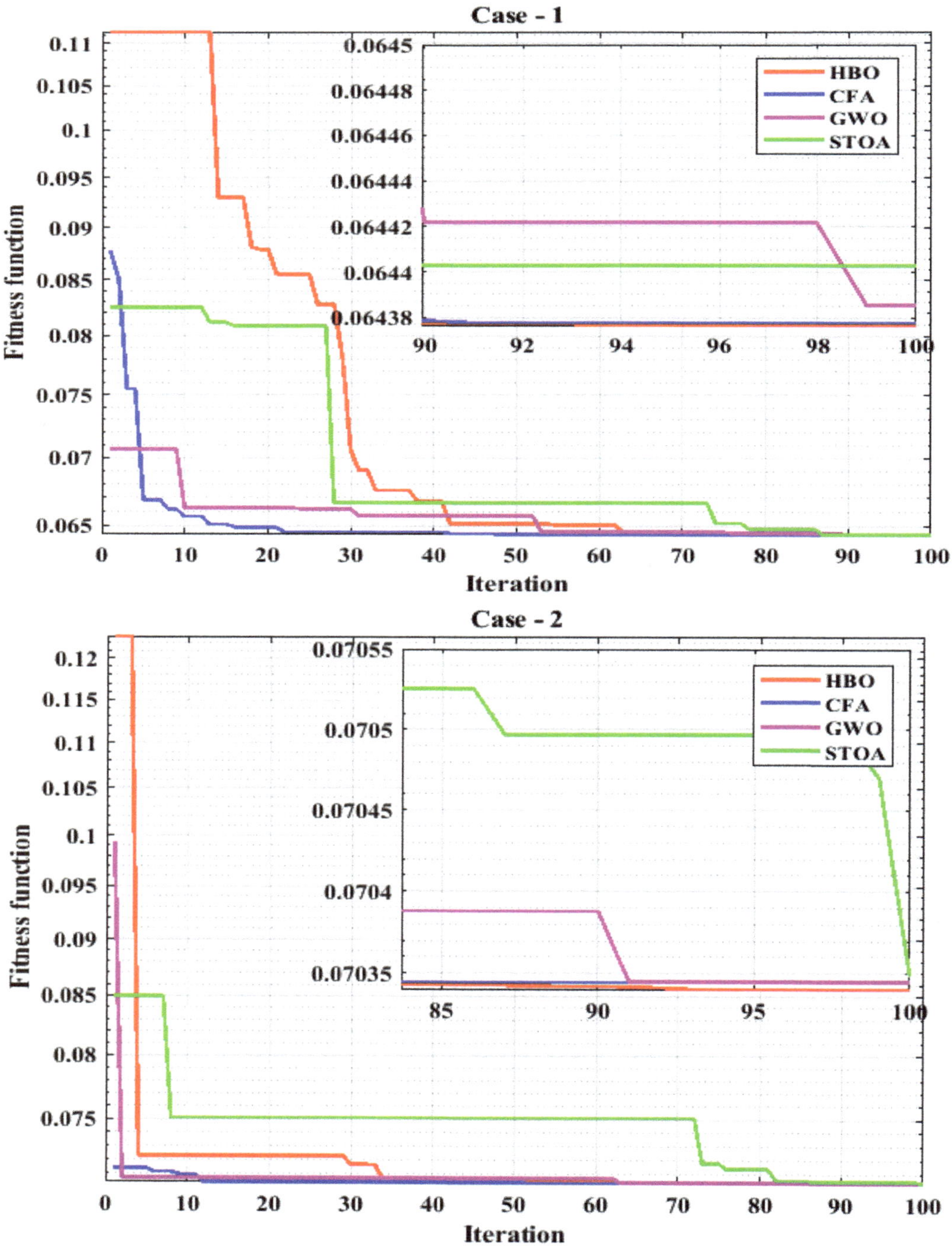

Figure 9. *Cont.*

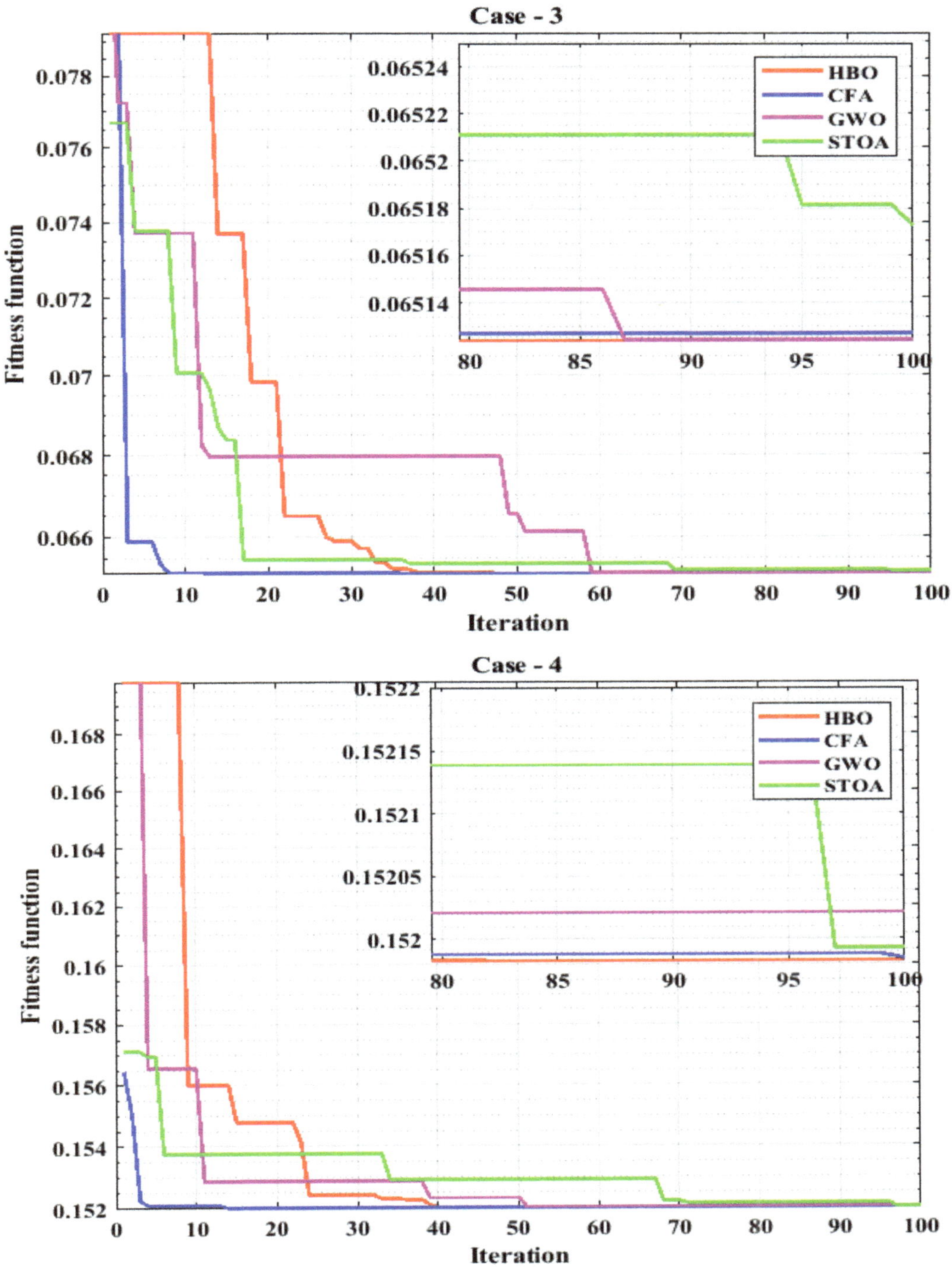

Figure 9. The Convergence curves for 100 iterations using HBO, CFA, GWO, and STOA methods: case-1: PV/WT/Biomass/Bat system, case-2: PV/Biomass/Bat system, case-3: WT/Biomass/Bat system, case-4: PV/WT/Bat system.

Tables 6–9 illustrate the results of the optimization properties for the four system scenarios proposed, which is based on many factors including the best value of the objective function, the decision variables (N_{PV}, N_{WT}, N_g and N_{Bat}), the COE, LPSP, and NPC of the suggested optimization algorithms (HBO, CFA, GWO and STOA).

Table 6. The optimization properties for the proposed hybrid system based on using HBO, CFA, GWO and STOA for an isolated PV/WT/Biomass/Bat.

	Best	It. Num.	N_{PV}	N_{WT}	N_g	N_{Bat}	COE	LPSP	NPC
HBO	0.0643767	27	15	1	2	400	0.121171	0.026789	3,559,143
CFA	0.06437783	44	15	1	2	401	0.1213225	0.0267184	3,563,603
GWO	0.0643857	12	17	1	2	406	0.1236759	0.0262159	3,632,730
STOA	0.0644027	28	15	1	2	441	0.1291354	0.0096015	3,793,092

Table 7. The optimization properties for the proposed hybrid system based on using HBO, CFA, GWO and STOA for an isolated PV/Biomass/Bat.

	Best	It. Num.	N_{PV}	N_g	N_{Bat}	COE	LPSP	NPC
HBO	0.0703404	49	17	2	447	0.1311804	0.0298557	3,853,160
CFA	0.07034462	32	17	2	447	0.1315446	0.0291707	3,863,857
GWO	0.07034495	28	16	2	450	0.1316478	0.0287446	3,866,887
STOA	0.0703484	40	21	2	468	0.1384959	0.0163783	4,068,036

Table 8. The optimization properties for the proposed hybrid system based on using HBO, CFA, GWO and STOA for an isolated WT/Biomass/Bat.

	Best	It. Num.	N_{WT}	N_g	N_{Bat}	COE	LPSP	NPC
HBO	0.0651238467	41	1	2	413	0.11216872966	0.0300781042	3,294,729.69
CFA	0.0651240320	39	1	2	403	0.11097771588	0.0306442748	3,259,746.07
GWO	0.0651241750	87	1	2	412	0.11213661772	0.0301290735	3,293,786.47
STOA	0.0651252636	99	1	2	375	0.10567322935	0.0504315519	3,103,937.501

Table 9. The optimization properties for the proposed hybrid system based on using HBO, CFA, GWO and STOA for an isolated PV/WT/Bat.

	Best	It. Num.	N_{PV}	N_{WT}	N_{Bat}	COE	LPSP	NPC
HBO	0.1519917239	41	182	94	999	0.3471381051	0.05921588	10,196,480.121
CFA	0.1520017987	58	182	94	1000	0.3470716848	0.059466541	10,194,529.161
GWO	0.1543193268	51	181	98	996	0.3469565942	0.05992429	10,191,148.605
STOA	0.1530886015	97	170	88	983	0.3324974563	0.089288773	9,766,440.657

In Table 6, for the PV, WT, Biomass, and Bat system, the results indicate that the HBO has the best configuration by using 15 PV panels, 1 WTs, 2 biomass generators, and 400 batteries, achieving the least COE, and NPC with 0.121171$/kWh and $ 3,559,143, respectively. In Table 7, for the second system case based on PV, Biomass, and Bat, the results prove that the HBO has the best configuration by using 17 PV panels, 2 biomass generators, and 447 batteries, achieving the least COE, and NPC with 0.1311804$/kWh and $ 3,853,160, respectively.

While Table 8, for the WT, Biomass, and Bat system, the results prove that the STOA has the best configuration by using 1 WT, 2 biomass generators, and 375 batteries, achieving the least COE, and NPC with 0.1056732 $/kWh and $ 3,103,938, respectively. In Table 9, for the fourth system case based on PV, WT, and Bat, the results illustrate that the STOA has the best configuration by using 170 PV panels, 88 WTs, and 983 batteries, achieving the least COE, and NPC with 0.3324975$/kWh and $ 9,766,441, respectively.

By comparing the COE and NPC of the four suggested cases, it finds that Case-3 achieved the lowest COE and NPC, followed by the Case-1. Although the third scenario which based on WT/biomass/Bat units produces the minimum value of COE and NPC, but it is not the optimal and efficient system for use. As the design of this case is based on batteries and biomass generators only, which have the highest yearly sharing of the capital cost, operating and maintenance cost. While the first scenario which is consists of PV/WT/biomass/Bat units considered an appropriate solution with minimal investment cost for the suggested case study area.

Parametric and non-parametric statistical measurements were performed for a more accurate comparison between the four optimization methods (HBO, CFA, GWO and STOA) on the basis of the acquired values of the objective function across a hundred individual runs for all analyzed cases. Parametric measurements comprise the lower value (Min.), maximum value (Max.) and mean of the target function, whereas the non-parametric measurements contain the median, relative error (RA), mean absolute error (MAE), standard deviation (SD), and efficiency. The efficiency here referred to the ratio of the lower value to the mean value of the goal function. For all four system scenarios, the results for statistical metrics for HBO, CFA, GWO, and STOA are shown in Table 10. On the basis of the results obtained, the proposed HBO in each case proved the best sensitivity and stability results compared to other optimization methods.

Table 10. The statistical performance of the studied optimization algorithms for the four system cases.

	HBO	CFA	GWO	STOA
		Case 1		
Max.	0.0644471	0.0644714	0.0648796	0.0657149
Min.	0.0643767	0.0643778	0.0643857	0.0644027
Mean	0.064386	0.064403	0.064513	0.064812
Median	0.064382	0.064401	0.064484	0.064838
SD	0.001476	0.002092	0.011667	0.032557
RE	0.007403	0.019536	0.099052	0.317511
MAE	0.0000095	0.000025	0.000128	0.000409
RMSE	0.000017	0.000033	0.000172	0.000521
Efficiency	99.9852	99.9609	99.8026	99.3714
		Case 2		
Max.	0.0703979	0.0706643	0.0708796	0.0711048
Min.	0.0703404	0.0703446	0.07034495	0.0703484
Mean	0.070347	0.070368	0.070392	0.070722
Median	0.070344	0.070358	0.070368	0.070681
SD	0.001061	0.004539	0.008255	0.028633
RE	0.004321	0.016689	0.033444	0.265769
MAE	0.000006	0.000023	0.000047	0.000374
RMSE	0.000012	0.000051	0.000094	0.000469
Efficiency	99.9914	99.9667	99.9333	99.4729
		Case 3		
Max.	0.0651243	0.06513126	0.0651819	0.06544516
Min.	0.065123847	0.065124032	0.06512418	0.06512526
Mean	0.065123988	0.065125588	0.06513273	0.06518123
Median	0.065123916	0.065125436	0.06512924	0.06516043
SD	0.000014838	0.000112835	0.00111241	0.00612358
RE	0.0001089	0.001194718	0.00657080	0.04296589
MAE	0.000000142	0.000001556	0.00000856	0.00005596
RMSE	0.000000204	0.000001915	0.00001395	0.00008250
Efficiency	99.9998	99.9976	99.9869	99.91423

Table 10. *Cont.*

	HBO	CFA	GWO	STOA
		Case 4		
Max.	0.151983478	0.15198479	0.1520216	0.151993102
Min.	0.151991724	0.152001799	0.154319327	0.153088601
Mean	0.151985107	0.151991986	0.152775666	0.152047201
Median	0.151984263	0.151991439	0.152656294	0.152014582
SD	0.000194530	0.000389174	0.057991385	0.015732186
RE	0.000535826	0.002367153	0.248002849	0.01779644
MAE	0.000001629	0.000007195	0.000754036	0.000054099
RMSE	0.000002522	0.000008162	0.000947705	0.000164869
Efficiency	99.99893	99.99527	99.50784	99.96452

Figure 10, illustrate the sensitivity analysis of studying the impact of the variation of the decision parameters on the stand-alone system objective functions, (a) COE, (b) NPC, (c) LPSP, (d) EXP. Where "0" on the x-axis refers to the nominal values of the sensitivity factors.

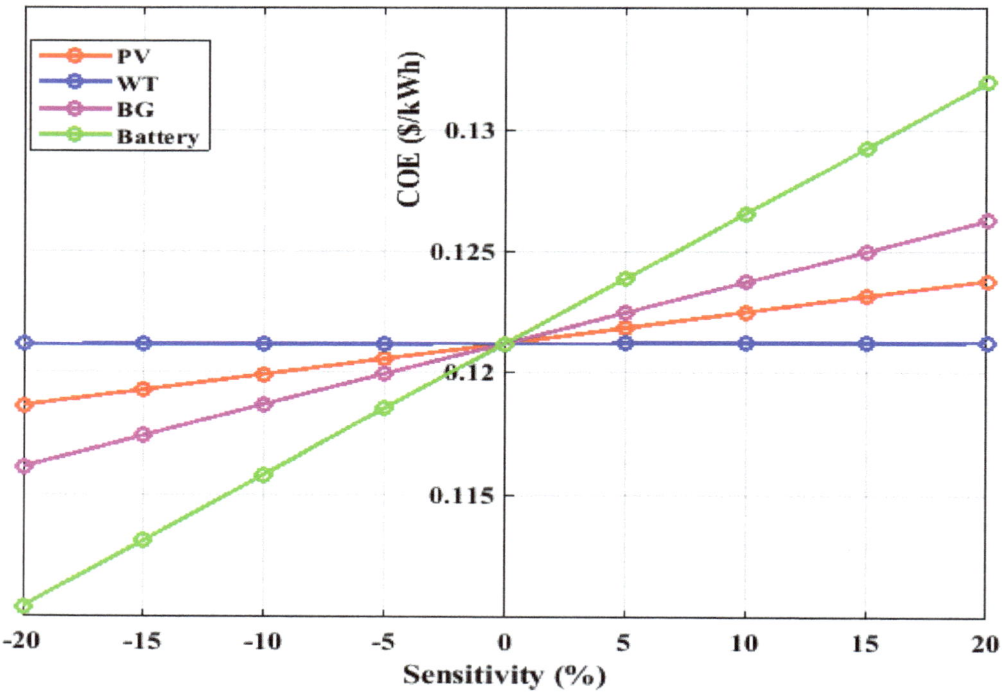

(A)

Figure 10. *Cont.*

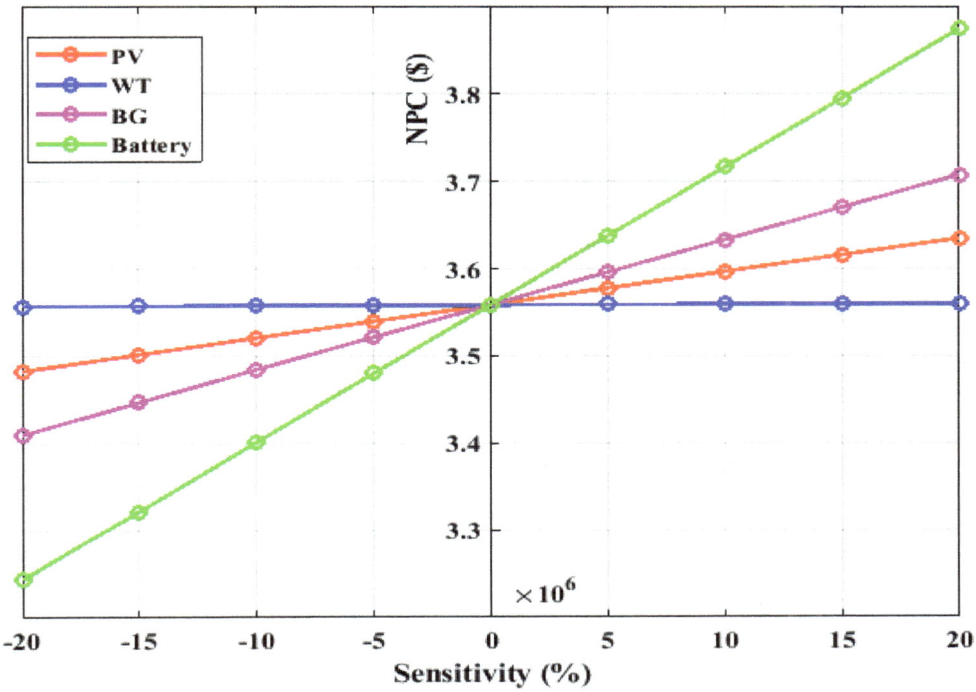

(B)

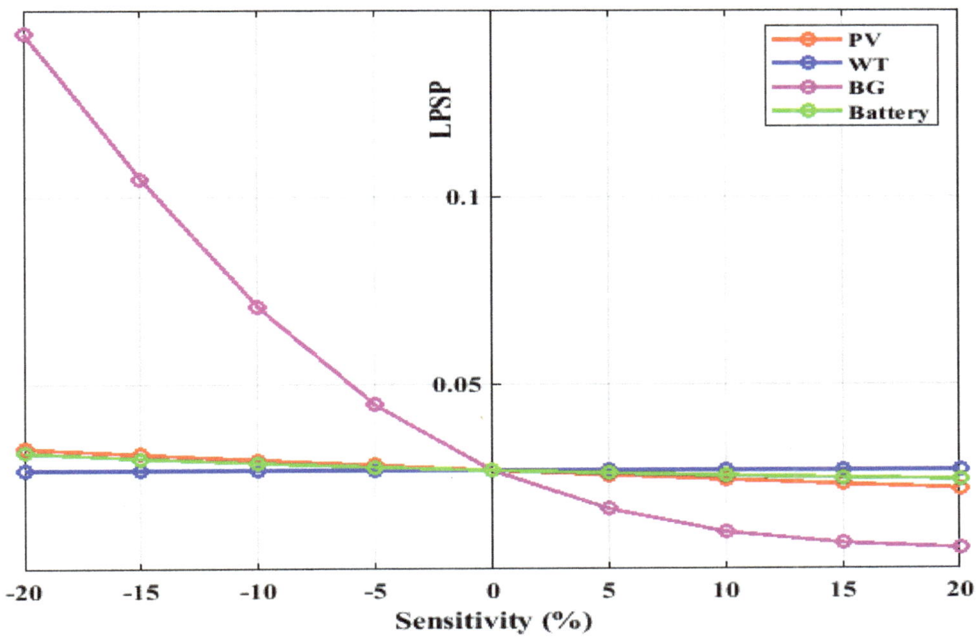

(C)

Figure 10. *Cont.*

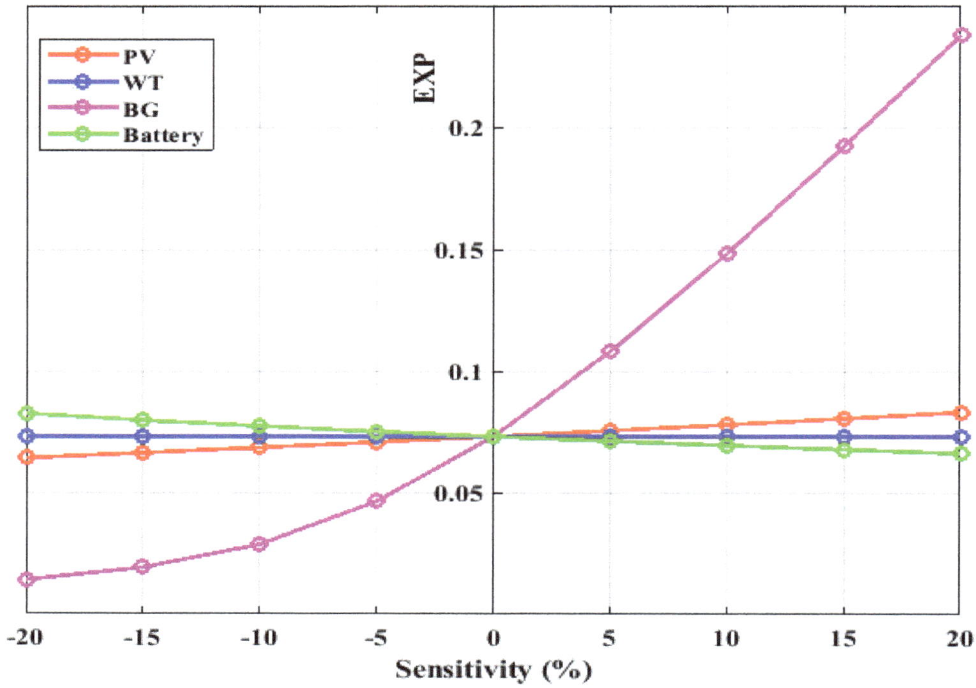

(D)

Figure 10. Sensitivity analysis of studying the influence of sizing parameters variation on the stand-alone system variables, (a) COE, (b) NPC, (c) LPSP, (d) EXP. (**A**) The influence of sizing change on the energy cost. (**B**) The influence of sizing change on the NPC. (**C**) The influence of sizing variation on the LPSP. (**D**) The influence of sizing variation on an excess of energy.

Figure 10A,B illustrate the effect on the COE and the NPC. As it can be noted that, at lower values of the specified parameters, both COE and NPC drop when the number of each PV panels, biomass generators, and batteries decreased. While, at a higher parameter values, the COE and NPC raise with increasing the number of each PV panels, biomass generators, and batteries. For the number of wind turbines, it can be noted that both the COE and NPC are nearly constant with the variation of the wind turbines number.

Figure 10C,D indicates that the chosen parameters has an effect on the system parameters, especially the number of the biomass generators.

Table 11, illustrate the yearly expenses breakdown of the hybrid system units and in turns show the system's NPC. The reader can notice that, for all suggested system cases the battery storage system has the highest yearly sharing of the capital cost compared to other system units. While the Biomass system has the highest operating and maintenance cost compared with other generating units in the suggested hybrid power system.

Table 11. The contribution of the yearly cost of the system regarding the type of cost which includes the yearly interest of capital cost, operating and maintenance costs, and annual replacement cost.

	PV			WT			Biomass			Battery			Inverter
	Cap ($)	O&M ($)	Rep ($)	Cap ($)	O&M ($)	Rep ($)	Cap ($)	O&M ($)	Fuel ($)	Cap ($)	O&M ($)	Rep ($)	Cap ($)
Case 1													
HBO	21,881.639	168.965	11,730.398	329.481	100	213.333	2359.588	19,385.299	36,345.833	122392.681	2016.223	39,645.509	26676
CFA	21,335.236	164.74584	11,437.480	329.480	100	213.333	2333.178	19,168.3297	35,939.033	121501.132	2001.536	39,645.510	26676
GWO	14,779.440	114.123	7923.022	512.442	155.530	331.797	2432.913	19,987.708	37,475.301	135376.241	2230.106	39,645.510	26676
STOA	17,705.137	136.7150	9491.442	1011.176	306.899	654.719	2348.039	19,290.416	36,167.936	125554.249	2068.305	39,645.510	26676
Case 2													
HBO	21,436.	165.530	11,491.98				2621.525	21,537.259	40,380.580	135,236.988	2227.812	39,645.510	26676
CFA	21,786.985	168.234	11,679.655				2622.910	21,548.632	40,401.905	135,494.976	2232.062	39,645.510	26676
GWO	21,159.690	163.390	11,343.373				2633.295	21,633.953	40,561.875	136,429.340	2247.454	39,645.510	26676
STOA	26,555.090	205.052	14,235.760				2700.724	22,187.919	41,600.514	142,081.934	2340.572	39,645.510	26676
Case 3													
HBO				329.481	100	213.333	2579.229	21,189.775	39,729.077	125,210	2062.647	39,645.510	26676
CFA				329.480	100	213.333	2577.439	21,175.068	39,701.502	122,561.892	2019.010	39,645.510	26676
GWO				329.480	100	213.333	2578.854	21,186.690	39,723.293	125,147.341	2061.601	39,645.510	26676
STOA				341.761	103.727	221.284	2448.266	20,113.842	37,711.79	113,676.027	1872.630	39,645.510	26676
Case 4													
HBO	235,285.415	1816.820	126,132.763	30,769.429	9338.761	19,922.690				303,057.402	4992.384	39,645.510	26676
CFA	235,117.876	1815.526	126,042.948	30,739.268	9329.607	19,903.162				303,219.603	4995.056	39,645.510	26676
GWO	233,972.041	1806.678	125,428.684	32,047.435	9726.645	20,750.177				302,188.858	4978.077	39,645.510	26676
STOA	219,195.935	1692.581	117,507.449	28,791.202	8738.353	18,641.821				298,195.455	4912.292	39,645.510	26676

7. Conclusions

This study offered a new meta-heuristic optimization method, called the Heap-based optimizer (HBO) technique, for design four different scenarios of a stand-alone hybrid power model based on PV, WT, Biomass system, and battery storage unit. The major objective function is minimizing the COE with increasing the reliability LPSP, which fulfils the required load of an isolated society in Alrashda village, Dakhla Oasis in the New Valley Governorate, Egypt. The operation of the suggested off-grid system relies on the meteorological data of wind speed, radiation and temperature per hour which obtained from NASA Surface Meteorology and Solar Energy website for 20 years for the selected region. The result obtained by the proposed methodology HBO which is based on utilizing MATLAB software was compared with other optimization methods CFA, GWO, STOA techniques.

The simulation results clearly showed that the suggested optimization method HBO is an effective method in identifying the optimal capacities of the generating and energy storage units and ensured good execution in the different scenarios of proposed the hybrid system. The results from the analyses presented in this study show the following:

- ✓ The obtained results indicated that the HBO technique showed the optimal convergence between the investigated algorithms in reaching the best solution.
- ✓ The HBO method has achieved the best optimal solution for Case-1 scenario. This case is a combination of PV/WT/biomass/Bat units, the best optimal solution has achieved after 27 iterations with the minimal COE of 0.121171 $/kWh, NPC of $ 3,559,143, and LPSP of 0.026789, followed by the results obtained from the CFA technique with the minimal COE of 0.1213225 $/kWh, NPC of $ 3,563,603, and LPSP of 0.0267184 after 44 iterations, followed by the results obtained from GWO, and STOA methods
- ✓ The HBO technique has achieved the best optimal solution for Case-2 scenario. This case is consisted of PV/biomass/Bat units, the best optimal solution has achieved after 49 iterations with the minimal COE of 0.1311804 $/kWh, NPC of $3,853,160, and LPSP of 0.0298557, followed by the results obtained from the CFA method with the minimal COE of 0.1315446 $/kWh, NPC of $3,863,857, and LPSP of 0.0291707 after 32 iterations, followed by the results obtained from GWO, and STOA techniques.
- ✓ Based on the results of parametric and non-parametric statistical measurements performed for a more accurate comparison of the four optimization methods, the proposed HBO in all studied cases showed the best results compared to other optimization methods.
- ✓ STOA has achieved the best optimal solution for Case-3, and Case-4 with COE of 0.105673 and 0.332497 $/kWh, and NPC of $3,103,938 and $9,766,441, respectively.
- ✓ By comparing the NPC of the four suggested cases, it finds that Case-3 achieved the lowest NPC, followed by the Case-1. Although the third scenario which based on WT/biomass/Bat units is the least NPC, but it is not the optimal and efficient system for use. As the design of this case is based on batteries and biomass generators only, which have the highest yearly sharing of the capital cost, operating and maintenance cost.
- ✓ The hybrid power system which is consists of PV/WT/biomass/Bat units would be an appropriate solution with minimal investment cost for rural communities, small industries, isolated wells and isolated farming areas where grid access is too costly or even impossible.

Author Contributions: Conceptualization, H.A.E.-S., S.K. and H.S.; Data curation, H.A.E.-S., S.K., A.M.E. and F.J.; Formal analysis, S.K., H.S., A.M.E. and F.J.; Investigation, H.A.E.-S., S.K., H.S. and F.J.; Methodology, H.A.E.-S., S.K. and H.S.; Project administration, M.T.-V.; Resources, S.K., H.S., M.T.-V., A.M.E. and F.J.; Software, S.K., H.S. and M.T.-V.; Supervision, S.K., M.T.-V. and F.J.; Validation, S.K., H.S., M.T.-V. and A.M.E.; Visualization, M.T.-V. and F.J.; Writing—original draft, H.A.E.-S., S.K. and H.S.; Writing—review & editing, M.T.-V., A.M.E. and F.J. All authors have read and agreed to the published version of the manuscript.

Funding: The authors thank the support of the National Research and Development Agency of Chile (ANID), ANID/Fondap/15110019.

Institutional Review Board Statement: Not applicable.

Informed Consent Statement: Not applicable.

Data Availability Statement: Not applicable.

Conflicts of Interest: The authors declare no conflict of interest.

Appendix A

- HBO: Search agents number (N_S) = 20, Maximum number of iterations (Max_{it}) = 100, and dimension size (D_S) = 4.
- CFA: N_S = 20, T = 50, and dimension size = 4.
- GWO: N_S = 20, Max_{it} = 100, D_S = 4, and a linearly decreased from 2 to 0.
- STOA: N_S = 20, Max_{it} = 100, =4, and a linearly decreased from 2 to 0.

References

1. Ellabban, O.; Abu-Rub, H.; Blaabjerg, F. Renewable energy resources: Current status, future prospects and their enabling technology. *Renew. Sustain. Energy Rev.* **2014**, *39*, 748–764. [CrossRef]
2. Sultan, H.M.; Menesy, A.S.; Kamel, S.; Korashy, A.; Almohaimeed, S.A.; Abdel-Akher, M. An improved artificial ecosystem optimization algorithm for optimal configuration of a hybrid PV/WT/FC energy system. *Alex. Eng. J.* **2021**, *60*, 1001–1025. [CrossRef]
3. Abd El-Sattar, H.; Sultan, H.M.; Kamel, S.; Menesy, A.S.; Rahmann, C. Optimal Design of Hybrid Stand-alone Microgrids Using Tunicate Swarm Algorithm. In Proceedings of the 2021 IEEE International Conference on Automation/XXIV Congress of the Chilean Association of Automatic Control (ICA-ACCA), Valparaíso, Chile, 22–26 March 2021; pp. 1–6.
4. Zhao, H.; Wu, Q.; Hu, S.; Xu, H.; Rasmussen, C.N. Review of energy storage system for wind power integration support. *Appl. Energy* **2015**, *137*, 545–553. [CrossRef]
5. Khan, A.; Javaid, N. Optimal sizing of a stand-alone photovoltaic, wind turbine and fuel cell systems. *Comput. Electr. Eng.* **2020**, *85*, 106682. [CrossRef]
6. Al-Sharafi, A.; Sahin, A.Z.; Ayar, T.; Yilbas, B.S. Techno-economic analysis and optimization of solar and wind energy systems for power generation and hydrogen production in Saudi Arabia. *Renew. Sustain. Energy Rev.* **2017**, *69*, 33–49. [CrossRef]
7. Singh, A.; Baredar, P. Techno-economic assessment of a solar PV, fuel cell, and biomass gasifier hybrid energy system. *Energy Reports* **2016**, *2*, 254–260. [CrossRef]
8. Elkadeem, M.R.; Wang, S.; Sharshir, S.W.; Atia, E.G. Feasibility analysis and techno-economic design of grid-isolated hybrid renewable energy system for electrification of agriculture and irrigation area: A case study in Dongola, Sudan. *Energy Convers. Manag.* **2019**, *196*, 1453–1478. [CrossRef]
9. Ajlan, A.; Tan, C.W.; Abdilahi, A.M. Assessment of environmental and economic perspectives for renewable-based hybrid power system in Yemen. *Renew. Sustain. Energy Rev.* **2017**, *75*, 559–570. [CrossRef]
10. Dufo-López, R.; Cristóbal-Monreal, I.R.; Yusta, J.M. Optimisation of PV-wind-diesel-battery stand-alone systems to minimise cost and maximise human development index and job creation. *Renew. Energy* **2016**, *94*, 280–293. [CrossRef]
11. Cano, A.; Arévalo, P.; Jurado, F. Energy analysis and techno-economic assessment of a hybrid PV/HKT/BAT system using biomass gasifier: Cuenca-Ecuador case study. *Energy* **2020**, *202*, 117727. [CrossRef]
12. Jamshidi, M.; Askarzadeh, A. Techno-economic analysis and size optimization of an off-grid hybrid photovoltaic, fuel cell and diesel generator system. *Sustain. Cities Soc.* **2019**, *44*, 310–320. [CrossRef]
13. Suresh, M.; Meenakumari, R. An improved genetic algorithm-based optimal sizing of solar photovoltaic/wind turbine generator/diesel generator/battery connected hybrid energy systems for standalone applications. *Int. J. Ambient Energy* **2021**, *42*, 1136–1143. [CrossRef]
14. Kharrich, M.; Kamel, S.; Abdeen, M.; Mohammed, O.H.; Akherraz, M.; Khurshaid, T.; Rhee, S.B. Developed approach based on equilibrium optimizer for optimal design of hybrid PV/Wind/Diesel/Battery Microgrid in Dakhla, Morocco. *IEEE Access* **2021**, *9*, 13655–13670. [CrossRef]
15. Ramli, M.A.M.; Bouchekara, H.R.E.H.; Alghamdi, A.S. Optimal sizing of PV/wind/diesel hybrid microgrid system using multi-objective self-adaptive differential evolution algorithm. *Renew. Energy* **2018**, *121*, 400–411. [CrossRef]
16. Ashraf, M.A.; Liu, Z.; Alizadeh, A.; Nojavan, S.; Jermsittiparsert, K.; Zhang, D. Designing an optimized configuration for a hybrid PV/Diesel/Battery Energy System based on metaheuristics: A case study on Gobi Desert. *J. Clean. Prod.* **2020**, *270*, 122467. [CrossRef]
17. Zaki Diab, A.A.; El-Ajmi, S.I.; Sultan, H.M.; Hassan, Y.B. Modified farmland fertility optimization algorithm for optimal design of a grid-connected hybrid renewable energy system with fuel cell storage: Case study of Ataka, Egypt. *Int. J. Adv. Comput. Sci. Appl.* **2019**, *10*, 119–132.

18. Geleta, D.K.; Manshahia, M.S.; Vasant, P.; Banik, A. Grey wolf optimizer for optimal sizing of hybrid wind and solar renewable energy system. *Comput. Intell.* **2020**, 1–30. [CrossRef]
19. Singh, S.; Kaushik, S.C. Optimal sizing of grid integrated hybrid PV-biomass energy system using artificial bee colony algorithm. *IET Renew. Power Gener.* **2016**, *10*, 642–650. [CrossRef]
20. Bukar, A.L.; Tan, C.W.; Lau, K.Y. Optimal sizing of an autonomous photovoltaic/wind/battery/diesel generator microgrid using grasshopper optimization algorithm. *Sol. Energy* **2019**, *188*, 685–696. [CrossRef]
21. Heydari, A.; Askarzadeh, A. Optimization of a biomass-based photovoltaic power plant for an off-grid application subject to loss of power supply probability concept. *Appl. Energy* **2016**, *165*, 601–611. [CrossRef]
22. Sarkar, T.; Bhattacharjee, A.; Samanta, H.; Bhattacharya, K.; Saha, H. Optimal design and implementation of solar PV-wind-biogas-VRFB storage integrated smart hybrid microgrid for ensuring zero loss of power supply probability. *Energy Convers. Manag.* **2019**, *191*, 102–118. [CrossRef]
23. Li, J.; Liu, P.; Li, Z. Optimal design and techno-economic analysis of a solar-wind-biomass off-grid hybrid power system for remote rural electrification: A case study of west China. *Energy* **2020**, *208*, 118387. [CrossRef]
24. Ghosh, S.; Karar, V. Assimilation of optimal sized hybrid photovoltaic-biomass system by dragonfly algorithm with grid. *Energies* **2018**, *11*, 1892. [CrossRef]
25. Eteiba, M.B.; Barakat, S.; Samy, M.M.; Wahba, W.I. Optimization of an off-grid PV/Biomass hybrid system with different battery technologies. *Sustain. Cities Soc.* **2018**, *40*, 713–727. [CrossRef]
26. Sawle, Y.; Gupta, S.C.; Bohre, A.K. Socio-techno-economic design of hybrid renewable energy system using optimization techniques. *Renew. Energy* **2018**, *119*, 459–472. [CrossRef]
27. Alshammari, N.; Asumadu, J. Optimum unit sizing of hybrid renewable energy system utilizing harmony search, Jaya and particle swarm optimization algorithms. *Sustain. Cities Soc.* **2020**, *60*, 102255. [CrossRef]
28. Sultan, H.M.; Kuznetsov, O.N.; Menesy, A.S.; Kamel, S. Optimal Configuration of a Grid-Connected Hybrid PV/Wind/Hydro-Pumped Storage Power System Based on a Novel Optimization Algorithm. In Proceedings of the 2020 International Youth Conference on Radio Electronics, Electrical and Power Engineering (REEPE), Moscow, Russia, 12–14 March 2020. [CrossRef]
29. Jung, W.; Jeong, J.; Kim, J.; Chang, D. Optimization of hybrid off-grid system consisting of renewables and Li-ion batteries. *J. Power Sources* **2020**, *451*, 227754. [CrossRef]
30. Diab, A.A.Z.; Sultan, H.M.; Mohamed, I.S.; Kuznetsov, O.N.; Do, T.D. Application of Different Optimization Algorithms for Optimal Sizing of PV/Wind/Diesel/Battery Storage Stand-Alone Hybrid Microgrid. *IEEE Access* **2019**, *7*, 119223–119245. [CrossRef]
31. Sawle, Y.; Gupta, S.C.; Bohre, A.K. Review of hybrid renewable energy systems with comparative analysis of off-grid hybrid system. *Renew. Sustain. Energy Rev.* **2018**, *81*, 2217–2235. [CrossRef]
32. El-Sattar, H.A.; Kamel, S.; Tawfik, M.A.; Vera, D.; Jurado, F. Modeling and Simulation of Corn Stover Gasifier and Micro-turbine for Power Generation. *Waste Biomass Valorization* **2018**, *10*, 1–14. [CrossRef]
33. El-Sattar, H.A.; Kamel, S.; Tawfik, M.A.; Vera, D. Modeling of a Downdraft Gasifier Combined with Externally Fired Gas Turbine using rice straw for generating electricity in Egypt. In Proceedings of the 2016 Eighteenth International Middle East Power Systems Conference (MEPCON), Cairo, Egypt, 27–29 December 2016; pp. 747–752.
34. El-Sattar, H.A.; Kamel, S.; Vera, D.; Jurado, F. Tri-generation biomass system based on externally fired gas turbine, organic rankine cycle and absorption chiller. *J. Clean. Prod.* **2020**, *260*, 121068. [CrossRef]
35. El-Sattar, H.A.; Kamel, S.; Jurado, F. Fixed bed gasification of corn stover biomass fuel: Egypt as a case study. *Biofuels Bioprod. Biorefining* **2020**, *14*, 7–19. [CrossRef]
36. Samy, M.M.; Barakat, S.H. biomass and fuel cells Multi-objective optimization of hybrid renewable energy system based on biomass and fuel cells. *Int. J. Energy Res.* **2021**, *45*, 8214–8230. [CrossRef]
37. Alturki, F.A.; Awwad, E.M. Sizing and Cost Minimization of Standalone Hybrid WT/PV/Biomass/Pump-Hydro Storage-Based Energy Systems. *Energies* **2021**, *14*, 489. [CrossRef]
38. Diab, A.A.Z.; Sultan, H.M.; Kuznetsov, O.N. Optimal sizing of hybrid solar/wind/hydroelectric pumped storage energy system in Egypt based on different meta-heuristic techniques. *Environ. Sci. Pollut. Res.* **2020**, *27*, 32318–32340. [CrossRef] [PubMed]
39. Hossain, M.S.; Jahid, A.; Islam, K.Z.; Rahman, M.F. Solar PV and Biomass Resources-Based Sustainable Energy Supply for Off-Grid Cellular Base Stations. *IEEE Access* **2020**, *8*, 53817–53840. [CrossRef]
40. Askari, Q.; Saeed, M.; Younas, I. Heap-based optimizer inspired by corporate rank hierarchy for global optimization. *Expert Syst. Appl.* **2020**, *161*, 113702. [CrossRef]
41. Ghasemi, M.; Ghavidel, S.; Aghaei, J.; Akbari, E.; Li, L. CFA optimizer: A new and powerful algorithm inspired by Franklin's and Coulomb's laws theory for solving the economic load dispatch problems. *Int. Trans. Electr. Energy Syst.* **2018**, *28*, 1–30. [CrossRef]
42. Dhiman, G.; Kumar, V. Seagull optimization algorithm: Theory and its applications for large-scale industrial engineering problems. *Knowl.-Based Syst.* **2019**, *165*, 169–196. [CrossRef]
43. Mirjalili, S.; Mirjalili, S.M.; Lewis, A. Grey Wolf Optimizer. *Adv. Eng. Softw.* **2014**, *69*, 46–61. [CrossRef]

applied sciences

MDPI

Article

Design and Development of a Management System for Energy Microgrids Using Linear Programming

Mateo Espitia-Ibarra [1,*], Pablo Maya-Duque [1] and Álvaro Jaramillo-Duque [2]

1 Research Group of Analytics for Decision Making (ALIADO), Industrial Engineering Department, Universidad de Antioquia, Calle 67 No. 53-108, Medellín 050010, Colombia; pablo.maya@udea.edu.co
2 Research Group in Efficient Energy Management (GIMEL), Electrical Engineering Department, Universidad de Antioquia, Calle 67 No. 53-108, Medellín 050010, Colombia; alvaro.jaramillod@udea.edu.co
* Correspondence: mateo.espitia@udea.edu.co; Tel.: +57-310-376-7962

Abstract: Energy is a fundamental tool for human development and this paper presents an approach that seeks to improve its use in Colombian off-grid communities. This approach is based on microgrid concepts where generation, storage, and consumption units interact with each other, and these interactions are presented through a linear programming model. In this approach, specific strategies are implemented according to the Colombian context, where some isolated communities already have diesel-based solutions for energy access, and the type of element that is studied, finding that the proposed optimization model is capable of adequately managing the loads of the microgrid,, thus improving the way in which the generated energy is stored and used through said horizon. Finally, different characteristics of the model are evaluated against multiple indicators and it is concluded that there may be much more specific strategies that improve its operation.

Keywords: microgrids management; renewable energy resources; isolated microgrids; optimization techniques

Citation: Espitia-Ibarra, M.; Maya-Duque, P.; Jaramillo-Duque, Á. Design and Development of a Management System for Energy Microgrids Using Linear Programming. *Appl. Sci.* **2022**, *12*, 3980. https://doi.org/10.3390/app12083980

Academic Editors: Marcos Tostado-Véliz, Paul Arévalo, Salah Kamel, Ragab A. El-Sehiemy, Tomonobu Senjyu and Davide Astiaso Garcia

Received: 3 March 2022
Accepted: 3 April 2022
Published: 14 April 2022

Publisher's Note: MDPI stays neutral with regard to jurisdictional claims in published maps and institutional affiliations.

Copyright: © 2022 by the authors. Licensee MDPI, Basel, Switzerland. This article is an open access article distributed under the terms and conditions of the Creative Commons Attribution (CC BY) license (https://creativecommons.org/licenses/by/4.0/).

1. Introduction

Electricity is an essential commodity for economic growth and the well-being of developing communities. However, almost 15% of the world's population does not have access to it [1]. The use of energy is absolutely necessary in the production, distribution, and consumption activities of human society [2] and the supply of electrical energy is one of the primary needs in modern societies [3]. However, meeting this need over all the areas in a country is a challenge for governments and companies in the energy and electricity sector. Therefore, electricity and its management have been one of the most important topics in scientific and industrial research for the past few decades [4].

In recent years, discussion has also been focused around sustainable energy and its environmental obligations, specifically on greenhouse gas emissions such as carbon dioxide (CO_2) [5], resulting from companies in the energy and electricity sector. Energy diversification arises as a solution to the emission of gases and is focused on guaranteeing the security of the energy supply through the incorporation of non-conventional renewable energy sources (NCRES). This diversification aims to have a hedge against energy supply problems, such as those caused by intense summers (climate change) or by problems resulting from a shortage of hydrocarbons.

Consequently, new approaches based on renewable technologies could provide electricity to communities in off-grid areas. Those technologies are usually less polluting and easy to maintain and operate [6]. "Microgrids" are one of the most typical approaches in scientific and industrial research that considers renewable energy sources to optimize the use of, and access to, electricity. They are defined as a group of interconnected loads and distributed energy resources, such as generators and energy storage devices, with defined electrical boundaries that form a local electric power system at distribution voltage levels, acting as a single, controllable entity that is able to operate in either grid-connected or island mode [7]. A microgrid serves as a system that integrates the actions of all users connected to it and uses advanced information, control, and communication technologies to save energy, reduce costs, and increase reliability and transparency [8]. Microgrids are emerging as an alternative solution to the problem of accessibility to service isolated areas where the electric power system cannot reach, or even as a sustainable alternative that allows communities to autonomously manage part (or even all) of their energy requirements.

This research studies the methods and tools used to make optimal use of renewable energy sources in microgrids within the context of isolated communities in Columbia in order to develop a new approach that could optimize the generation and use of electricity in off-grid areas where the current energy solution is the use of fossil sources such as diesel to supply energy demand.

Consequently, this research presents a novel approach within a Colombian context for the use of renewable resources to produce energy and its storage capacity within a microgrid. This approach seeks to meet the energy demand from isolated communities at the lowest possible cost, implementing strategies that ensure good load distribution and the minimum amount of unattended demand.

2. Materials and Methods

2.1. Microgrid Control System

One important component of microgrids is the energy source. There are different sources for the generation of energy that can be classified according to their predominance, called conventional and non-conventional sources, or according to the type of resources used, that is, renewable and non-renewable sources [9]. Conventional sources refers to the most widely used energy sources worldwide, among which the energy supplied by means of hydroelectric plants or the burning of fossil fuels, such as diesel, coal, and gas, stand out. Less predominate sources in the market are called unconventional sources, where the primary source is produced by natural resources or it is an emerging energy source under development. In the same way, renewable energy sources are those whose potential is abundant and include solar energy, hydraulic energy, wind energy, biomass energy, and geothermal energy [10]. Conversely, there are other sources of energy that are not renewable, that is, they are found in a limited quantity in the world and the rate of their consumption is higher than their regeneration time; some of these non-renewable energy sources are fossil fuels such as coal, oil, and natural gas, as well as uranium. A traditional microgrid system generally consists of a set of wind turbines, photo-voltaic panels, small hydro power, diesel micro-turbine engine, and battery storage.

The unpredictable nature of renewable energy sources such as solar and wind affects the performance and reliability of the microgrid, due to excess electricity generation or lack of generation, which is considered to be the main drawback for its adoption [11]. However, some approaches such as those presented by [12] argue that this problem can be solved by combining two or more power sources together with a backup unit to form a hybrid renewable energy system. In other words, the system is operated with a set of energy sources and storage devices to satisfy the demand, even when some of them are not available.

The microgrid control system is described in [13] as a four-levels system: the fourth level has a system that assigns the production of active and reactive power to each generator element of the grid according to the demand. The third level sets the voltage and frequency references in the nodes, while the second level of the control system corrects the voltage and frequency deviations in the network. Finally, the primary control level executes actions locally on the generation sources, keeping the voltage and power at the reference values.

Within the fourth level of the control system mentioned above, setting the active and, sometimes, reactive power from each generator unit at each time period is a core element of the strategy adopted to manage the power generated by the microgrid. This strategy would depend of the system goals, minimizing costs and/or maximizing coverage. In this study, the management strategy is a Unit Commitment (UC) that defines an optimization problem that seeks the optimal scheduling of generation units in a specific time horizon (hourly, daily, or weekly). The objective of the UC strategy described in [14] is to satisfy the demand at minimum cost, considering the on/off states of each generation unit, its ramps, reliability restrictions, system capacity, transmission, environmental impact, etc.

2.2. Literature Review

The UC has usually been formulated as a non-convex and nonlinear combinatorial optimization problem by some authors. However, in some cases, the UC strategy can be applied to models like the one described by [15], where a lineal cost function is considered and only management constraints are taken into account. This research adopts a linear definition of the cost function based on the amount of energy produced by each of the generation technologies.

Optimal power flow and UC has been also divided by [13] into two more specific problems: static and dynamic economic dispatch. The first one is a typical mode of power system planning and operation, which only studies the optimization scheme of a single time section rather than the connection between each time period. When modeling the static economic dispatch of the microgrid system, the objective function is usually to minimize the overall operating cost of the microgrid. Most of the constraints only consider the active power limits of the generating units in the microgrid and the power balance constraints within it, while ignoring the characteristics of battery storage units such as useful life, loading and unloading ramps, among other. On the other hand, dynamic economic dispatch is defined in [16] as one that considers the relationships between the values of the optimization variables in subsequent periods; for example, the battery level in a defined period affects the battery charge level in future periods depending on the state of charge or discharge that is defined through optimization.

The dynamic economic dispatch model of the microgrid takes into account factors such as the ramp restriction of the controllable sources and the operation restriction [17] of the energy storage units, etc., which is closer to the power system. The addition of energy storage units not only makes the microgrid more closely connected in time, but also makes the operation more economical and reliable. In [18], it is shown that dynamic economic scheduling with energy storage units can save about 37% of the operating cost compared to static economic scheduling without an energy storage unit.

One example of the implementation of a UC strategy to manage microgrids is presented in [15]. This research proposed a optimization model for planning an appropriate stand-alone, renewable-based electricity system for off-grid communities in Colombia. It used implicit stochastic optimization to make decisions regarding the sizing of renewable energy sources to meet energy demand during an average day. This research also considers the use of a unitary battery system for each zone in such a way that the energy generated from renewable sources during a certain period of time can be used later. This work concludes that the use of renewable energies must be adaptive according to the conditions associated with the environmental variables. In addition, the combination of these technologies provides a solution that is significantly cheaper for the community than typical diesel platforms because it is not necessary to buy or transport fuel.

Another important study related to power microgrid management is [19]. In this study, a multi-objective economic-emission dispatch problem of combined heat and power is developed. In their model, multiple energy sources are also considered: renewable and non-renewable, but its objective function involves unit operating costs, emission level, emission tax, and the cost of power purchase from the main external grid. Their case is also important since it is possible to study the chance of managing microgrids in non-isolated areas, whose access to energy from the main grid is partial and, therefore, energy from it can be accessed to fully or partially satisfy the demand in certain periods of time.

An important approach to mention is the one developed in [20], where the microgrid is approached from the basic consumer unit. In this case, it is considered that each client has an energy storage unit that can be used to store energy generated from renewable sources and that this is then used to reduce the total load on the general grid. In this case, the main objective is to reduce the load on the conventional network from the use of production and consumption forecasts. In this study, metaheuristics are used to solve the proposed model and the results and computation times for the methods used are compared.

In [21], a mathematical optimization approach is proposed for the optimal operation focused on the economic dispatch of a DC microgrid using renewable energy generators and energy storage systems through semi-defined programming. The proposed mathematical approach contemplates the operation of a DC microgrid over a period of time with variable energy purchase prices. This characteristic makes it a practical methodology to be applied in real-time operating conditions. Further, a nonlinear auto regressive exogenous model (NARX) is used to train an artificial neural network (ANN) to forecast solar radiation and wind speed for the integration and dispatch of renewable generation considering prediction periods of 0.5 h and a time horizon 24 h.

Finally, the work developed in [22] shows a more robust model in which a greater number of possible technologies (diesel, gas, fuel, solar, and wind power generators) can be used for power generation and are considered, as well as different costs associated with the power generation activity. Consideration of these multiple costs is then reflected in multiple objective functions. On the one hand, there is cost-effective operation, which is the minimization of the operation and the aging costs of the micro-grid components. On the other hand, there is the maximum islanding degree, which states either there is no physical connection with the macrogrid, or no power will be exchanged, or only a fixed and predefined power profile may be considered as exchange power; in this case, the formulation of the objective function is similar to the cost-effective operation objective function. Finally, there is eco-friendly operation, where the objective function is formulated as the minimization of the pollutant treatment costs. A genetic algorithm is used to solve instances and a case of study is presented.

2.3. Renewable Energy in Colombia

Colombia is one of the most privileged countries in Latin America thanks to its geographical location. It provides special characteristics like zones where wind speeds are twice the world average and there is also sunlight most days of the year [23]. As a consequence, a great research scenario is open for new ideas to propose and develop different methodologies to take advantage of these characteristics and then to focus on the optimal use of renewable resources, specifically, in off-grid zones.

In 2020, 29 off-grid zones supplied their own energy demand using renewable energies; however, it is a small number in front of 1798 off-grid localities that are distributed in almost 51% of national territory [24]. In most of the cases, off-grid energy demand is supplied with diesel-based generators as a single option and only 31.3% of them have electric service available 24 h a day. One of the reasons that could explain this limited use of renewable options to supply energy in those off-grid zones is that their installation cost used to be substantially high and made this solution financially impractical. However, their prices have come down and the renewable generator industry is more competitive than before [25].

In the same way, in recent years, the Colombian government has developed some laws, such as "Ley 1715" [26], that have encouraged the development of projects that promote the generation of electrical energy solutions in off-grid areas [27]. However, these laws do not present a clear framework on the benefits that the use of renewable generation sources can bring in specific contexts such as isolated areas. In this sense, and given the economic difficulties represented by the implementation of projects based on renewable energies in these contexts, it is necessary to develop alternatives that allow for the intelligent management of resources and that ensure their optimal operation.

The model we propose seeks to manage the energy microgrids considering uncertainty from different sources (environmental variables that affect the generation and demand of energy) at the same time as it implements specific strategies on some components of the microgrid, such as the battery system. This management is carried out through an optimization process by scenarios, which allows for adaptation to various situations whose conditions may affect the operation of the microgrid.

2.4. Proposed Optimization Approach

Mathematically, the problem is described as follows: Let $\mathcal{G}$ be a set of available generation technologies (Solar: S, Wind: W, Diesel: D), $\mathcal{I}$ be a set of electric generators, conventional and non-conventional, each of them with technical parameters and dimensions depending on the associated technology, and $\mathcal{T}$ be the set of time periods within the planning horizon. In addition, consider a battery system with the capacity to charge when the total power of the microgrid exceeds the demand load and also supports the microgrid by serving as another energy source.

The goal is to satisfy the expected load demand for each time period (d_t) into an energy microgrid, determining the functional capacity connected to the network of each generation technology in each time frame, all while minimizing the total cost of operation. Table A1 presents model variables.

According to the nature of the problem, the optimization model needs to calculate how much power needs to be produced by each generator at each time period ($t \in \mathcal{T}$). Therefore, based on the work of [15], Equations (1)–(3) describe, respectively, the solar, wind, and diesel generation. Let $f(i) \rightarrow G = \{S, W, D\}$ serve as the function that returns the type of generator technology $i \in I$. Thereby, power generation g_{it} depends on whether the generator is switched on or off.

$$f(i) = S \rightarrow g_{St} = \left(G_{S,test} * \frac{R_t}{R_{S,test}} \right) * x_{St} \tag{1}$$

$$f(i) = W \rightarrow g_{Wt} = \begin{cases} 0, & W_t < W_{min} \\ \frac{1}{2}(\rho_W A^W W_t^3 \eta_W) * x_{Wt}, & W_{min} \leq W_t < W_a \\ \frac{1}{2}(\rho_W A^W W_a^3 \eta_W) * x_{Wt}, & W_a \leq W_t \leq W_{max} \\ 0, & W_t > W_{max} \end{cases} \tag{2}$$

$$f(i) = D \rightarrow x_{Dt} * g^D_{min} \leq g_{Dt} \leq x_{Dt} * g^D_{max} \tag{3}$$

In the generation equations, the concept of installed capacity is used to measure how much energy can be generated from a specific source. In the case of renewable sources, the concept of peak power is used by the technology, which allows one to calculate how much electricity would be transformed according the amount of the primary source.

Considering the decisions and the associated generation functions, the mathematical model is defined as follows:

$$\min \sum_{i \in I} va\_op_i * g_{it} \tag{4}$$

The objective function (4) seeks to minimize the variable operational costs of the energy system. The variable operational cost, va_op_i, is associated with the maintenance of the equipment during its life-cycle and the fuel, if any is used by the generator.

$$\sum_{i \in I : f(i) \in \{S,W,D\}} g_{it} = el_t + eb_t + ew_t \tag{5}$$

$$d_t = el_t + g_{Bt} + g_{Ft} \tag{6}$$

The first group of constraints is related to the control of the use of the generated energy. Equation (5) splits the generated energy into the portion that seeks to satisfy the demand, the portion that is used to load the battery, and a potential portion that is discarded. Meanwhile, constraints in Equation (6) ensure that the demand is satisfied through the sum of the energy generated for that purpose, the energy coming from the battery, and the energy from a dummy generator representing the unmet demand.

$$b_t = b_{t-1} * (1-\sigma) + \left[(eb_t * \eta_{bat}) - \left(\frac{g_{Bt}}{\eta_{inv}} \right) \right] * \delta t \tag{7}$$

$$b_t \leq u_B \tag{8}$$

The second group of constraints is related to the state of charge of the battery. The constraint in Equation (7) calculates the state of charge of the battery system, at the end of the period t, taking into account the energy dissipation σ, and the charging and discharging efficiencies, η_{bat} and η_{inv}, respectively. Equation (8) sets the limit to the capacity of the battery u_B. The remaining sets of constraints implement strategies for managing the charging and discharging processes in order to enhance the battery health and long term performance

$$u_B * y_t \geq g_{Bt} \tag{9}$$

$$\epsilon * y_t \leq g_{Bt} \tag{10}$$

Constraints in Equations (9) and (10) set the value of the binary variable y_t that indicates if the battery has been discharged ($y_t = 1$) in period t.

$$Sc = \sum_{t \in T} z_t \tag{11}$$

$$z_t \geq y_t - y_{t-1} \tag{12}$$

The first strategy is modeled through constraints (11) and (12), where they count the number of times the battery system enters a discharge process.

$$eb_t \leq eb_{max} * (1 - y_t) \tag{13}$$

$$g_{Bt} \leq g_B^{max} \tag{14}$$

Constraints in Equations (13) and (14) implement the second strategy, which regulates the charging and discharging ramps. A third strategy to manage the battery ensures that the number of times that the battery reaches the deep discharge level (b_{min}), and the overcharge level (b_{max}), is less than L_{min} and L_{max}, respectively.

$$\frac{b_t}{u_B} + p_t^{min} \geq b_{min} \tag{15}$$

$$\frac{b_t}{u_B} - (1 - p_t^{min}) \leq b_{min} \tag{16}$$

$$\frac{b_t}{u_B} - p_t^{max} \leq b_{max} \tag{17}$$

$$\frac{b_t}{u_B} + (1 - p_t^{max}) \geq b_{max} \tag{18}$$

$$\sum_{t \in T} p_t^{min} \leq L_{min} \tag{19}$$

$$\sum_{t \in T} p_t^{max} \leq L_{max} \tag{20}$$

$$y_t, x_{it} \in \{0, 1\} \tag{21}$$

$$g_{it}, b_t, eb_t, el_t, ew_t, Ic_t, z_t \geq 0 \tag{22}$$

Hence, the equations from (15) to (18) determine when the discharge or overcharge levels are reached and constraints (19) and (20) set a limit to the number of times that it occurs. Finally, constraints (21) and (22) define the domain of the decision variables.

2.5. Computational Experiments

The model described in Section 2.4 is tested on different scenarios built on real operating conditions for non-interconnected zones (NIZ) in Colombia. We first described the case studies that were used and then the experiments designed to validate different scenarios upon them.

2.6. Instance Generation

This research considers energy microgrids in isolated communities of Colombia subsidized by the state. These areas are difficult to access and the energy solution currently implemented is 100% based on the use of diesel generators. This situation presents logistical challenges to ensure the correct and continuous operation of the microgrid.

Data was collected from three different regions in Colombia: San Andrés (SA), Providencia (P), and Puerto Nariño (PN). These locations already have diesel-based solutions to generate electricity and their installed capacities are presented in Table A3.

The parameters considered in the model are classified in technical, demand, and environmental parameters. Technical parameters are defined in Table A2. For batteries and generators, these parameters were drawn from existing bibliographic sources according to the used technology [28]. The default values for each parameter within this category were established by experts in the field. The parameters within the environmental category are solar radiation, wind speed, and temperature. The web tool RenewableNinja [29] was used to collect historical data for these parameters. This web tool provides a global meteorological database from the MERRA-2 system of the National Aeronautics and Space Administration [30] that gives timely (per hour) information. The downloaded data is from January to December 2019.

Finally, the demand parameter represented by the demand loads of the NIZ in Colombia were obtained and analyzed from the "Instituto de Planificación y Promoción de Soluciones Energéticas para Zonas No Interconectadas" (IPSE) that shared all NIZ's hourly energy supply reports from 2019. These data is supplied to the IPSE by the "Centro Nacional de Monitoreo" [24] that measures the actual electrical energy consumption of different isolated communities through telemetry systems. Data for every day of the year is available. However, when data was analyzed, it was observed that there was no significant variation between the same days of the week throughout the year. For this reason, an annual average is used as the actual hourly load for every day of the week. Therefore, 21 instances of the problem were built, seven for each of the three NIZ considered.

2.7. Design of Scenarios

The experiments carried out with the model seek to generate information to validate the model's sensitivity and to understand the impact that some selected parameters have on different management scenarios. For this, the following four research questions are formulated:

- How sensitive is the model to changes in demand? (Scenario 1)
- How sensitive is the model to changes in the availability of resources? (Scenario 2)
- What impact do the technical characteristics of the battery have? (Scenario 3)
- What is the impact of the penalty on the unserved demand? (Scenario 4)

For each of these questions, scenarios with the parameters of interest (called factors) will be identified and defined, and different levels will be experimented with.

Table 1 describes the four scenarios in terms of the parameters of interest in each case and the values that they will take. The experiments of each scenario correspond to a complete factorial design of the different levels for each factor.

Table 1. Sets of experimental conditions.

Component	Scenario 1	Scenario 2	Scenario 3	Scenario 4
Demand (d_t)	$[0.75 * d, 1 * d, 1.25 * d, 1.25 * rushhour]$	$[1 * d]$	$[1 * d]$	$[1 * d]$
Diesel capacity	$[0.8 * d_{max}]$	$[0.8 * d_{max}]$	$[0.8 * d_{max}]$	$[0.8 * d_{max}]$
Solar capacity	$[0.3 * d]$	$[0 * d, 0.05 * d, 0.1 * d, 0.15 * d, 0.20 * d, 0.25 * d, 0.3 * d]$	$[0.3 * d]$	$[0.3 * d]$
Wind capacity	$[0.3 * d]$	$[0 * d, 0.05 * d, 0.1 * d, 0.15 * d, 0.20 * d, 0.25 * d, 0.3 * d]$	$[0.3 * d]$	$[0.3 * d]$
Battery capacity	[500]	[500]	$[0.1 * d_{max}, 0.2 * d_{max}, 0.3 * d_{max}]$	[500]
L_{min}	[2]	[2]	[0, 1, 2]	[2]
L_{max}	[2]	[2]	[0, 1, 2]	[2]
eb_{max}	[100]	[100]	$[0.2 * ub, 0.35 * ub, 0.45 * ub]$	[100]
g_B^{max}	[100]	[100]	$[0.2 * ub, 0.3 * ub, 0.6 * ub]$	[100]
Unattended demand cost	$[1.6 * Diesel_{cost}]$	$[1.6 * Diesel_{cost}]$	$[1.6 * Diesel_{cost}]$	$[0.8 - 1.2 * Diesel_{cost}]$

The first scenario seeks to show the capacity of the model to properly manage the changes on demand under different conditions of availability of renewable resources. Therefore, it considers the actual demand, an increase of 25% and a decrease of 25% in the hourly demand, and a special case in which only the interval with the highest demand is increased by 25%. Additionally, the percentage of renewable energy generation varies from 0% to 30% with steps of 5%. The remaining factors are set to their default value according to the recommendations of experts in the field of study. The second scenario explores the model's sensitivity to changes in the availability of renewable energy resources. This is achieved by keeping demand and other parameters constant and making variations in the percentage of renewable energy generation of 5%, starting from 0% and reaching 30%. The third scenario evaluates the impact of the technical characteristics of the battery in meeting the demand and the performance of the microgrid. For this, the demand is left constant, a level of renewable resource capacity is selected where the use of batteries is evidenced, and different levels of the parameters U_B, eb_{max}, g_B^{max}, and L_{max} L_{min} are tested. Finally, in the fourth scenario, the effect of the penalty to the unmet demand is studied. In this case, different levels of penalty cost are tested for the unmet demand to check the behavior of the latter as said cost increases. These different test values are based on percentages according to the cost of the most expensive generation source, generally diesel generation. In this way, it is possible to know how much the cost of unattended energy should be raised so that the management model minimizes the unmet demand.

3. Results

According to the proposed experiments in each scenario, the analysis of the results is carried out in order to evaluate research hypotheses and review the sensitivities of the response variables regarding changes in the configuration and working conditions to which they are subjected to on a microgrid.

Figure 1 shows, for one of the experiments, the hourly balance of the load and each energy resource on the microgrid. This figure shows a trend, which is repeated for all scenarios, in the use of renewable generation resources as the operating base of the microgrid, ensuring that through this generation of renewable energy, renewable resources are used for 100% of the time that they are available, despite the need for a diesel source in all periods during the day, given that its total operating cost is higher than the other sources.

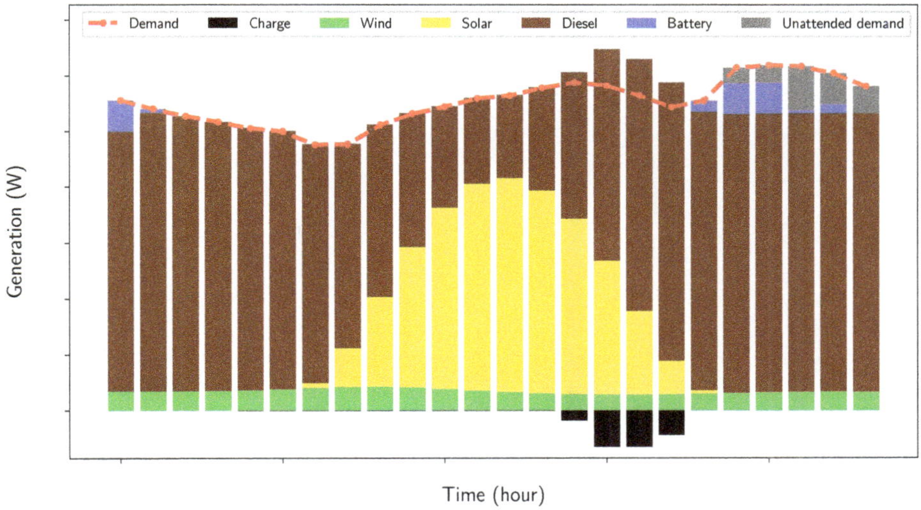

Figure 1. Microgrid load management over the time horizon (24 h).

Additionally, it shows that after the availability of renewable resources increases in a certain hour, charging processes start at the battery storage system. This process of storing energy from different generation sources starts just before of the disappearance of the generation from the solar source (which is the most available renewable generation source), so it can be understood that the model identifies a lack of energy generation at night time and tries to supply it by storing energy at the battery system to meet future demand.

Likewise, the model uses the energy produced and stored when the renewable sources decrease, as expected. In particular, it makes use of stored energy at times when solar generation is not available. In the same way, it uses renewable generation and allows for the modulation of attention to demand with the diesel generator, which in each period of time is available for regulation by the operator.

3.1. Sensitivity of the Model to Changes in Demand

To analyze how sensitive the management model is to different changes in the energy demand of the microgrid, Figure 2 shows the values of the objective function and the percentage of renewable energy used for different levels of demand as long as the installed capacity of renewable production is constant. It shows that, at different levels of demand, the management algorithm uses renewable energy resources in order to satisfy as much demand as possible through them.

Additionally, it is evident that since there is a relative increase in demand of 25% for each one of the three experimental levels, the decrease in the use of renewable sources to cover the demand is proportional. This shows that the percentage of demand coverage by renewable sources is determined by factors other than the level of demand, which are probably associated with the installed capacity of renewable sources.

For the special case in which within one hour period the demand is as its maximum peak during the day, the model handles demand coverage in a similar way as it does with medium load because it is capable of reconfiguring the battery system's usage processes to meet peak demand.

For its part, the operating cost of the microgrid by making use of renewable sources is lower compared to the use of diesel to fully satisfy the demand. This represents a saving in economic terms that is presented in the second part of Figure 2. In this case, a concordance is obtained in terms of demand coverage and savings generated by the use of renewable energies to satisfy different levels of demand from the same installed production capacity.

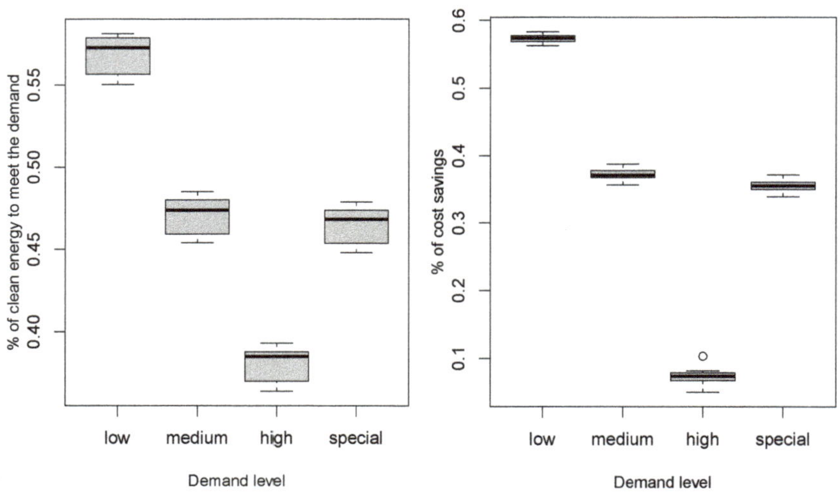

(a) (b)

Figure 2. Sensitivity of the model to changes in demand. (**a**) Percentage of renewable energy to meet the demand for each experimental demand level. (**b**) Cost savings comparison for each level of demand using the same installed capacity.

3.2. Sensitivity of the Model to the Availability of Renewable Resources

When the installed capacity of each renewable technology increases, measured as a percentage of the demand, the portion of the total demand satisfied by renewable generation sources increases.

Figure 3 shows that the demand coverage trend from renewable sources as they increase their availability is linear, marking more extensive use in terms of the solar source, as it uses approximately 100% of its installed capacity within the time horizon, which is not the case for the wind source. However, despite the inferior performance of the wind source compared to the solar source, its use in microgrids is justified due to its generation availability over the entire time horizon.

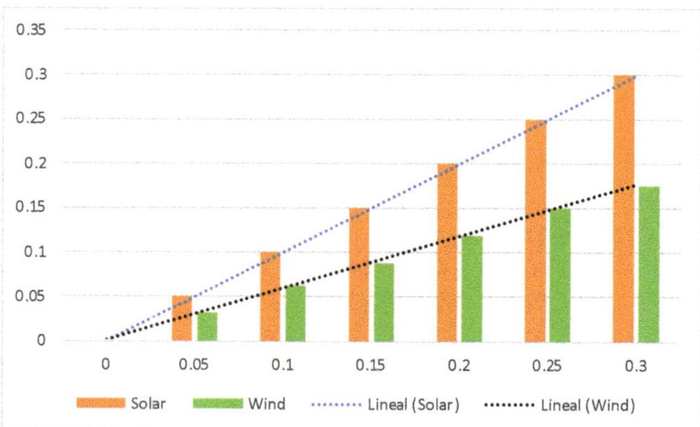

Figure 3. Trend of use for renewable sources.

In conjunction with the previous results, Figure 4 presents the operating result of the microgrid when the renewable capacity is established at 45% of the maximum demand. This result shows that, in order to better meet the demand from renewable sources, it will be necessary to install a greater wind generation capacity, since this generation source is available over the entire operating horizon of the microgrid. However, a similar effect can be achieved with a smaller additional solar capacity and a larger battery capacity to be able to supply the night demand from stored solar energy.

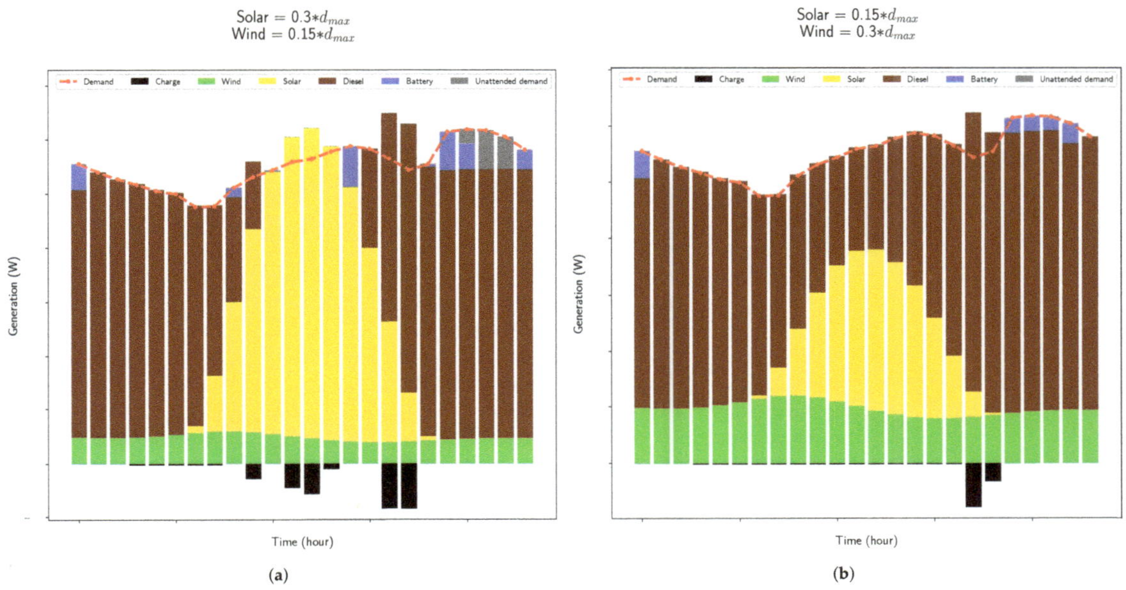

Figure 4. Use of renewable resources for a combined 45% of installed capacity. (**a**) Microgrid load management for 30% $* d_{max}$ and 15% $* d_{max}$ of solar and wind installed capacity. (**b**) Microgrid load management for 15% $* d_{max}$ and 30% $* d_{max}$ of solar and wind installed capacity.

3.3. Impact of the Characteristics of the Battery

The technical characteristics of the battery affect the performance of the microgrid and especially the use of the storage system significantly in situations when the installed capacity for renewable generation is extremely low or extremely high.

When the installed capacity of renewable sources is small and its generation is not significant, there is an overgeneration from diesel sources in order have enough energy to store in the battery system that will be used in periods of high demand. On the other hand, in situations with a large renewable capacity of renewable sources and its production is significant, the overgeneration is used to charge the battery system, in order to turn off any of the diesel generators in one or more periods to lower the total cost of covering the demand.

As shown in Table 2, as battery storage capacity increases, more intensive use is made of it. However, the percentage of variation for the use of the battery refers to how much percentage of the demand is covered from the use of the battery systems with respect to the immediately previous level on which it has been experienced,and it is evident that the use of this system to supply the demand does not increase in the same proportion as its capacity increases.

Battery capacity has a direct impact on the fraction of demand that is not met over the time horizon, since this is the first way to avoid unattended demand. On the one hand, greater storage capacity would mean a lower fraction of unattended demand at night when the solar (most available renewable source) is not in operation. This is reflected in the savings represented by the use of a larger installed energy storage capacity. In this case, the "Savings %" column contrasts the operating cost of the microgrid with each storage capacity, in front of the current option of covering the entire demand from diesel. However, a greater power capacity from renewable sources and a higher charge and discharge capacity per period in the ramp parameters of the battery system allow more charging and discharging cycles to be started over the optimization time horizon.

Table 2. Impact of battery capacity.

Battery Capacity	Demand Supply %	Variation %	Savings %
$0.1 * d_{max}$	0.90%	–	36.04
$0.2 * d_{max}$	1.54%	71.11%	36.51
$0.3 * d_{max}$	1.97%	27.92%	36.73

Regarding the battery charge and discharge ramp conditions (eb_{max}, G_B^{max}), it should be mentioned that there are significant variations in the operation of the microgrid for each of the experimental levels. As the charge and discharge value of the ramps increase together, the battery charge and discharge processes are carried out in a greater number of periods, thus allowing a more intensive use of the energy storage capacity and more immediately using the energy produced from renewable sources. This decreases the use of the diesel source, increasing the use of renewable sources.

Additionally, it should be noted that the deep discharge parameter L_{min} has no impact on the operation of the model; this can be attributed to the fact that due to the constant production of energy, the battery is not only used to supply the demand, thus allowing it to be recharged in several periods. However, due to the ability of the model to foresee low production in certain periods of time, the parameter of the number of periods in which overcharge is allowed, L_{max}, does significantly influence its operation. These annotations may affect the choice and configuration of the storage system, allowing it to increase the level at which the battery enters an overcharge state, keeping the installed capacity as low as possible and thus reducing purchase and installation costs.

3.4. Impact of the Penalty on Unmet Demand

Finally, for the analysis of changes in the price of unserved demand, it is necessary to study the percentage of unserved demand with respect to total demand in each of the scenarios previously proposed.

The model considers unattended demand because there are moments within the time horizon in which the generation units are not available for operation, or the installed capacity is not enough to fully meet the demand. In this unattended demand analysis, its cost is calculated based on what has been implemented in [31], multiplying the amount of unattended energy by the cost of the unattended load.

Figure 5 shows that the objective function presents significant increases for certain levels of unsatisfied demand penalty cost as long as these are less than the cost of the most expensive generation source. However, there is a limit to this value of unattended demand cost, beyond which this quantity reaches its virtual limit and the overall changes in the objective function are smaller.

It is important to highlight that when the value of unattended demand is equal to that of the most expensive generation source, the maximum demand is met without changing the load configuration in the microgrid, that is, the maximum possible demand is supplied without entailing cost overruns. This can be attributed to the fact that the model does not try to store energy and incur costs associated with this process to cover future demand when it is possible to just leave it unattended. In addition, it avoids the intensive use of the battery system due to the considerations of its self-discharge process.

It is important to note that since the unmet demand penalty strategy applies in the same way over the entire time horizon, the model seeks to supply demand from all generation and storage sources in the periods immediately following the low energy production. This illustrates that at night time, when the solar generation is not available, the stored energy is used early, preventing the stored energy from dissipating in a self-discharge processes, thus causing unattended demand at times when it is perhaps more important to supply it. For this situation, as with the ability of the model to better cover unattended demand, a segmentation of demand by type of demand and by hours could be considered, thus making the model identify the best times to use stored energy and availability of renewable and non-renewable resources. Thus, a new possible unattended demand penalty strategy should consider different costs according to the priority of the type of demand and the period of time in which it is presented.

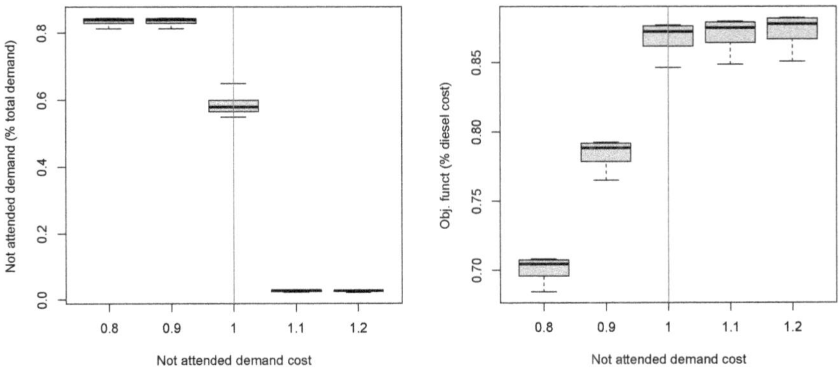

Figure 5. Impact of the penalty to unmet demand.

4. Discussion

According to the characteristics and results of the methodology presented, it can be concluded that it adapts to changes in the energy demand of the microgrid and also to changes in the availability of generation resources (air and solar radiation), making intensive use of renewable generation sources while possible and covering the gap from the use of non-renewable sources and energy storage in the battery system. This use conforms to the proposed strategies and, in this way, manages to satisfy the maximum possible demand from an installed fixed generation capacity.

For its part, the battery system and its management, which is presented as a contribution of this research, is especially relevant in seeking to achieve better coverage of demand. In this case, there are characteristics of the battery that are highly relevant to adequately meet the energy demand of the microgrid. Likewise, the use of a penalty cost for unattended demand points to a better microgrid management process, establishing a turning point for said penalty once it reaches the cost of the most expensive generation source.

That said, it is worth mentioning that, in many cases, the management model seeks to use the stored energy as soon as possible with respect to its production period; however, this generates periods in which the unserved demand reaches significant levels with respect to the overall demand. This situation can be addressed in future research, developing strategies that allow for the prioritization of demand at certain hours of the day and that also considers multiple types of demand, such that the energy use available in each period of time is prioritized for specific users or customers such as hospital infrastructure, food warehouses, etc.

On the other hand, the analysis of the sources of uncertainty that can affect the operation of the microgrid is proposed as future contributions, in such a way that the reliability of the management of generation sources is increased, taking into account different possibilities or scenarios.

5. Conclusions

This study presents an optimization-based methodology for managing the generation and use of energy from multiple sources within an isolated microgrid. This methodology implements specific management strategies according to the Colombian isolated communities context, components such as generators and battery systems that allow, despite trying to make the microgrid work at the lowest possible cost, other relevant aspects are taken into account, such as the minimization of unattended demand and proper functioning of these components. This methodology considers an initial situation in which there are isolated communities with an existing solution for energy access based on the use of diesel units. From this situation, the scenario is considered in which renewable generation sources and energy storage units are included to subsequently analyze their impact on the operating cost and sustainability of the microgrid.

This research offers a management model that contributes to improving energy access conditions in remote communities in Colombia, allowing the intelligent management of renewable generation units and energy storage units. This approach offers an alternative that allows isolated communities to rely less on fossil fuels, such as diesel, and thus make their microgrids more sustainable and robust.

Author Contributions: Conceptualization, M.E.-I., Á.J.-D., and P.M.-D.; methodology, M.E.-I. and P.M.-D.; software, M.E.-I.; validation, M.E.-I., Á.J.-D., and P.M.-D.; formal analysis, M.E.-I.; investigation, M.E.-I.; resources, M.E.-I. and Á.J.-D.; data curation, M.E.-I.; writing—original draft preparation, M.E.-I.; writing—review and editing, M.E.-I., Á.J.-D., and P.M.-D.; visualization, M.E.-I.; supervision, Á.J.-D. and P.M.-D.; project administration, M.E.-I., Á.J.-D., and P.M.-D.; funding acquisition, Á.J.-D. and P.M.-D. All authors have read and agreed to the published version of the manuscript.

Funding: This research was funded by the Colombia Scientific Program within the framework of the so-called Ecosistema Científico (Contract No. FP44842-218-2018).

Institutional Review Board Statement: Not applicable.

Informed Consent Statement: Not applicable.

Data Availability Statement: The models used in the computational experiment are available at GitHub in [32].

Acknowledgments: The authors gratefully acknowledge the support from the Colombia Scientific Program within the framework of the call Ecosistema Científico (Contract No. FP44842-218-2018). The authors also want to acknowledge Universidad de Antioquia for its support through the project "estrategia de sostenibilidad".

Conflicts of Interest: The authors declare no conflict of interest.

Abbreviations

The following abbreviations are used in this manuscript:

NCRES	Non-conventional renewable energy sources
UC	Unit Commitment
NIZ	Non-interconnected zones

Appendix A. Model Components

Table A1. Decision variables.

Variable	Description
x_{it}	Binary state (on/off) of the source i in the time interval t
g_{it}	Output power by the generation unit i in the time interval t [kW]
g_{Bt}	Output power by the battery unit in the time interval t [kW]
eb_t	Power intended to load the storage unit during the time interval t [kW]
b_t	Energy level in the storage unit in the time interval t [kWh]
el_t	Power generated at the time interval t destined to cover the demand [kW]
ew_t	Power discarded due to over generation in the time interval t [kW]
y_t	Binary state (discharging) of the battery in the time interval t
Sc	Counter of periods in which a discharging process starts
p_t^{min}	Binary state of the battery deep discharge level in period t
p_t^{max}	Binary state of the battery overcharge level in period t

Table A2. Technical parameters of the model.

Parameter	Description
d_t	Energy demand in the microgrid at time interval t [kW]
va_op_i	Variable generation cost per Kw contributed to the network by the generation unit i
$G_{S,test}$	Generation level for the solar panel under test conditions [kW]
$R_{S,test}$	Solar radiation test level for solar panel [kW/m^2]
R_t	Average solar radiation during the time interval t [kW/m^2]
ρ_W	Air density [kg/m^3]
A^S	Wind turbine swept area [m^2]
η_W	Wind turbine efficiency
W_{min}	Minimum wind speed for the turbine to start operating [m/s]
W_a	Wind speed for optimum turbine operation [m/s]
W_{max}	Maximum wind speed at which the turbine can operate [m/s]
W_t	Average wind speed during the time interval t [m/s]
g_{min}^D	Minimum power generated by the Diesel unit [kW]
g_{max}^D	Maximum power generated by the Diesel unit [kW]
η_{bat}	Battery system discharge efficiency
η_{inv}	Battery system charge efficiency
u_B	Installed battery capacity [kWh]
ϵ	Minimum power for the battery system to enter the discharge state [kW]
eb_{max}	Maximum amount of energy used to charge the battery per period of time [kW]
g_B^{max}	Maximum amount of energy to obtain from the battery per period of time [kW]
b_{min}	Battery charge level from which deep discharge is considered [kWh]
b_{max}	Battery charge level from which overcharge is considered [kWh]
L_{min}	Maximum number of periods in which deep discharge is allowed
L_{max}	Maximum number of periods in which overcharge is allowed

Appendix B. Installed Capacity for Experimental Locations

Table A3. Installed capacity for each location.

Localitation	Operation Capacity	Reserved Capacity
San Andrés (SA)	157,230 kW	18,700 kW
Providencia (P)	4482 kW	0 kW
Puerto Nariño (PN)	640 kW	130 kW

References

1. Guta, D.D. Determinants of household adoption of solar energy technology in rural Ethiopia. *J. Clean. Prod.* **2018**, *204*, 193–204. [CrossRef]
2. Arizkun, A.; García de la Cruz, J.M. La Energía. Retos y Problemas. *Dossieres EsF Econ. Sin Front.* **2017**, *24*, 20–24.
3. ONU. Energía—Desarrollo Sostenible. Available online: https://www.un.org/sustainabledevelopment/es/energy/ (accessed on 22 April 2021).
4. Valencia-López, D.; Carvajal Quintero, S.; Pineda-Agudelo, J. Design of demand management programs for the efficient use of electricity by industrial users. *Ing. Y Compet.* **2017**, *19*, 207–218.
5. Stirling, A. Diversity and sustainable energy transitions: Multicriteria diversity analysis of electricity portfolios. In *Analytical Methods for Energy Diversity and Security*; Elsevier Ltd.: Amsterdam, The Netherlands, 2008; pp. 1–29. [CrossRef]
6. Chowdhury, S.A.; Aziz, S.; Groh, S.; Kirchhoff, H.; Leal Filho, W. Off-grid rural area electrification through solar-diesel hybrid minigrids in Bangladesh: Resource-efficient design principles in practice. *J. Clean. Prod.* **2015**, *95*, 194–202. [CrossRef]
7. International Electrotechnical Commission. IEC 60050—International Electrotechnical Vocabulary—Details for IEV Number 617-04-22: "Microgrid". 2017. Available online: https://www.electropedia.org/iev/iev.nsf/display?openform&ievref=617-04-22 (accessed on 29 September 2021).
8. Fan, Z.; Kulkarni, P.; Gormus, S.; Efthymiou, C.; Kalogridis, G.; Sooriyabandara, M.; Zhu, Z.; Lambotharan, S.; Chin, W.H. Smart grid communications: Overview of research challenges, solutions, and standardization activities. *IEEE Commun. Surv. Tutor.* **2013**, *15*, 21–38. [CrossRef]
9. Armstrong, R.C.; Wolfram, C.; De Jong, K.P.; Gross, R.; Lewis, N.S.; Boardman, B.; Ragauskas, A.J.; Ehrhardt-Martinez, K.; Crabtree, G.; Ramana, M.V. The frontiers of energy. *Nat. Energy* **2016**, *1*, 1–8. [CrossRef]
10. Ellabban, O.; Abu-Rub, H.; Blaabjerg, F. Renewable energy resources: Current status, future prospects and their enabling technology. *Renew. Sustain. Energy Rev.* **2014**, *39*, 748–764. [CrossRef]
11. Arul, P.G.; Ramachandaramurthy, V.K.; Rajkumar, R.K. Control strategies for a hybrid renewable energy system: A review. *Renew. Sustain. Energy Rev.* **2015**, *42*, 597–608. [CrossRef]
12. Zahraee, S.M.; Khalaji Assadi, M.; Saidur, R. Application of Artificial Intelligence Methods for Hybrid Energy System Optimization. *Renew. Sustain. Energy Rev.* **2016**, *66*, 617–630. [CrossRef]
13. Liu, X.; Ding, M.; Han, J.; Han, P.; Peng, Y. Dynamic economic dispatch for microgrids including battery energy storage. In Proceedings of the 2nd International Symposium on Power Electronics for Distributed Generation Systems, PEDG 2010, Hefei, China, 16–18 June 2010; pp. 914–917. [CrossRef]
14. Ventosa, M.; Baíllo, Á.; Ramos, A.; Rivier, M. Electricity market modeling trends. *Energy Policy* **2005**, *33*, 897–913. [CrossRef]
15. Viteri, J.P.; Henao, F.; Cherni, J.; Dyner, I. Optimizing the insertion of renewable energy in the off-grid regions of Colombia. *J. Clean. Prod.* **2019**, *235*, 535–548. [CrossRef]
16. Wu, H.; Liu, X.; Ding, M. Dynamic economic dispatch of a microgrid: Mathematical models and solution algorithm. *Int. J. Electr. Power Energy Syst.* **2014**, *63*, 336–346. [CrossRef]
17. Wu, X.; Wang, X.; Bie, Z. Optimal generation scheduling of a microgrid. In Proceedings of the IEEE PES Innovative Smart Grid Technologies Conference Europe, Berlin, Germany, 14–17 October 2012. [CrossRef]
18. Liu, C.; Wang, X.; Liu, S.; Zhu, Z.; Wu, X.; Duan, J.; Hou, F.; Xie, L. Economic dispatch model considering battery lifetime for microgrid. *Dianli Zidonghua Shebei/Electric Power Autom. Equip.* **2015**, *35*, 29–36. [CrossRef]
19. Alomoush, M.I. Microgrid combined power-heat economic-emission dispatch considering stochastic renewable energy resources, power purchase and emission tax. *Energy Convers. Manag.* **2019**, *200*, 112090. [CrossRef]
20. Rigo-Mariani, R.; Sareni, B.; Roboam, X.; Turpin, C. Optimal power dispatching strategies in smart-microgrids with storage. *Renew. Sustain. Energy Rev.* **2014**, *40*, 649–658. [CrossRef]
21. Gil-González, W.; Montoya, O.D.; Holguín, E.; Garces, A.; Grisales-Noreña, L.F. Economic dispatch of energy storage systems in dc microgrids employing a semidefinite programming model. *J. Energy Storage* **2019**, *21*, 1–8. [CrossRef]
22. Nemati, M.; Braun, M.; Tenbohlen, S. Optimization of unit commitment and economic dispatch in microgrids based on genetic algorithm and mixed integer linear programming. *Appl. Energy* **2018**, *210*, 944–963. [CrossRef]
23. SerColombia—Asociacion de Energías Renovables. Potencial para Desarrollar Energías Renovables en Colombia Atrae a Inversionistas Extranjeros | Sala de Prensa | PROCOLOMBIA. 2020. Available online: https://ser-colombia.org/2020/noticias-del-sector/https-www-elinformador-com-co-index-php-general-185-economia-231603-potencial-para-desarrollar-energias-renovables-en-colombia-atrae-a-inversionistas-extranjeros/ (accessed on 21 January 2022).
24. IPSE. IPSE—Transformamos Territorios. Available online: https://ipse.gov.co/cnm/https://ipse.gov.co/ (accessed on 20 October 2021).
25. International Renewable Energy Agency. *Renewable Power Generation Costs in 2019*; IRENA: Masdar City, Abu Dhabi, 2018; p. 160.
26. UPME. Guia práctica para la aplicación de los incentivos tributarios de la Ley 1715 de 2014. *Minist. Minas Y Energ.* **2014**, *1*, 28.
27. Rodríguez-Urrego, D.; Rodríguez-Urrego, L. Photovoltaic energy in Colombia: Current status, inventory, policies and future prospects. *Renew. Sustain. Energy Rev.* **2018**, *92*, 160–170. [CrossRef]
28. ENAIR ENERGY. Aeronogenerador E200—La Última Tecnología. Available online: https://www.enair.es/es/aerogeneradores/e200 (accessed on 25 January 2022).
29. Pfenninger, S.; Staffell, I. Renewables Ninja. Available online: https://www.renewables.ninja/ (accessed on 19 February 2021).

30. NASA. MERRA. Available online: https://gmao.gsfc.nasa.gov/reanalysis/MERRA/ (accessed on 9 December 2020).
31. Azimian, M.; Amir, V.; Javadi, S.; Siano, P.; Alhelou, H.H. Enabling demand response for optimal deployment of multi-carrier microgrids incorporating incentives. *IET Renew. Power Gener.* **2022**, *16*, 547–564. [CrossRef]
32. Espitia-Ibarra, M. Microgrids Management Package. Available online: https://github.com/MateoEspitia/microgrids_management (accessed on 24 January 2022).

applied sciences

MDPI

Article

Mixed-Integer Linear Programming for Decentralized Multi-Carrier Optimal Energy Management of a Micro-Grid

Mohammad Faghiri [1], Shadi Samizadeh [1], Amirhossein Nikoofard [1,*], Mahdi Khosravy [2,*] and Tomonobu Senjyu [3,*]

[1] Faculty of Electrical Engineering, K. N. Toosi University of Technology, Tehran 1631714191, Iran; mohammad698476@email.kntu.ac.ir (M.F.); shadi.samizadeh@email.kntu.ac.ir (S.S.)
[2] Cross Labs, Cross-Compass Ltd., Tokyo 104-0045, Japan
[3] Department of Electrical and Electronics Engineering, Faculty of Engineering, University of the Ryukyus, Okinawa 903-0213, Japan
* Correspondence: a.nikoofard@kntu.ac.ir (A.N.); dr.mahdi.khosravy@ieee.org (M.K.); b985542@tec.u-ryukyu.ac.jp (T.S.)

Abstract: Increasing the load demand and penetration of renewable energy sources (RESs) poses real challenges for optimal energy management of distribution networks. Moreover, considering multi-carrier energy systems has increased the efficiency of systems, and provides an opportunity for using the advantages of RESs. In this regard, we adopted a new framework based on the new challenges in the multi-carrier energy micro-grid (MEMG). In the proposed method, a comprehensive MEMG was modeled that benefits from a large assortment of distributed energy resources (DERs), such as micro-turbines, fuel cells, wind turbines, and energy storage. Considering many DERs is necessary, because these resources could cover one another's disadvantages, which have a great impact on the total cost of the MEMG and decrease the emission impacts of fossil-fuel-based units. Furthermore, waste power plants, inverters, rectifiers, and emission constraints are considered in the proposed method for modeling a practical MEMG. Additionally, for modeling the uncertainty of stochastic parameters, a model based on a multilayer neural network was used in this paper. The results of this study indicate that using a decentralized model, along with stochastic methods for predicting uncertainty, can reduce operational costs in micro-grids and computational complexity compared with optimal centralized programming methods. Finally, the equations and results obtained from the proposed method were evaluated by experiments.

Keywords: optimal energy management; multi-agent system; multi-energy carrier; renewable energy sources; uncertainty

Citation: Faghiri, M.; Samizadeh, S.; Nikoofard, A.; Khosravy, M.; Senjyu, T. Mixed-Integer Linear Programming for Decentralized Multi-Carrier Optimal Energy Management of a Micro-Grid. *Appl. Sci.* **2022**, *12*, 3262. https://doi.org/10.3390/app12073262

Academic Editors: Davide Astiaso Garcia and Rodolfo Dufo López

Received: 9 January 2022
Accepted: 15 March 2022
Published: 23 March 2022

Publisher's Note: MDPI stays neutral with regard to jurisdictional claims in published maps and institutional affiliations.

Copyright: © 2022 by the authors. Licensee MDPI, Basel, Switzerland. This article is an open access article distributed under the terms and conditions of the Creative Commons Attribution (CC BY) license (https://creativecommons.org/licenses/by/4.0/).

1. Introduction

With the increasing need for electricity and the problems associated with centralized fossil fuel power plants—such as environmental pollution, exorbitant construction and maintenance costs, etc.—the use of MEMGs is a good solution; they allow the extensive utilization of renewable energies, distributed energy resources (DERs), and participation of consumers in the optimal management of power system operation, and using these systems can play a vital role in optimal energy consumption, system stability, and system reliability [1–4]. Furthermore, renewable energy sources have high uncertainty and an intermittent nature [5–8]. One solution to this problem is the micro-grid, which facilitates the response to load demand [9–12].

In terms of operation, micro-grid energy management systems (MEMSs) can be divided into centralized and decentralized (distributed) perspectives [13]. In centralized approaches, a central agent, which can process large amounts of data, is needed in order to gather information from other agents [14]. In multi-carrier micro-grids with distributed energy management, the privacy of agents is preserved. Moreover, each independent agent can optimize its costs in a parallel or sequential manner [15]. Providing an optimal approach to planning of system components is very important. The network used in this research consisted of micro-turbines, waste power plants, fuel cells, wind turbines, boilers, anaerobic reactors, inverters, rectifiers, and some energy storage units. It is also possible to exchange information between different levels of the system. This feature increases the reliability of the system and, compared with non-participating systems, achieves a better result. In this research, we sought to find the optimal performance strategy for system components, while meeting the electrical and thermal needs of customers. The micro-grid system is capable of exchanging electricity with high energy levels, as well as daily component performance scheduling.

The rest of this paper is organized as follows: In Section 2, a review of the related work, along with a description of the proposed methodology of this research, is provided. Section 3 presents the proposed MEMG structure. In Section 4, agents are modeled. In Section 5, the simulation of the proposed method to achieve optimal performance of MEMG agents is performed. Section 6 presents the simulation results, and we review the obtained results with different criteria and compare them with other methods in order to validate the proposed method.

2. Literature Review

In energy management systems with a single energy carrier, optimal performance and calculations are simpler, due to the lack of independence among energy carriers. Some research has been done to optimize the performance of such systems based on time-series analysis, factor-based optimization algorithms, etc. [16–19]. However, in a few cases, uncertainty in load demand is included in optimizing system performance. For example, in [20], the effect of the presence of DERs in optimizing the performance of energy management systems is investigated.

In multi-carrier energy management systems, computing and optimizing system performance is more complex. In [21], a model for optimizing the performance of the Poly Generation micro-grid of the University of Geneva is presented, showing that MEMGs can have economic and environmental benefits if they use the optimal strategy. Furthermore, in [22], an optimization model for a PG micro-grid in the presence of renewable energy sources is proposed. In [23], a real-time operational optimization method is presented. In [24–26], the problems of optimizing the performance of multi-carrier energy systems with centralized approaches, and from top to down, were investigated. On the other hand, in a few cases—such as [27,28]—the problems of optimal planning of the performance of the multi-carrier energy systems with decentralized and distributed approaches have been investigated.

Load demand is not considered in energy management systems with multiple energy carriers. The reasons for this include computational complexity, performance optimization of energy carriers, uncertainties in renewable energy production, and continuous fluctuation.

Hence, the authors of [27,28] could not consider the uncertainties. Considering uncertainties makes it difficult to provide an optimal approach but, on the other hand, it makes the optimal approach more efficient and reduces the operating costs of the system. In [29], optimization was achieved using PSO and GA to solve the problem of MEMG operational planning. Meanwhile, [30] used a stochastic model for electricity and natural gas pricing and load demand in real time. In [31], a micro-grid management approach is presented, considering random load and predicting the demand. In [26], a new method for a multi-agent system (MAS) is presented, which is a combination of ANFIS and GA.

In [15], deep learning is used to model uncertainties. In [32], optimization of the performance of the system is achieved using the gray wolf optimization method. In [33], the evolutionary vertical sequencing protocol is used to model coordination between high-level agents, and a two-layer MLIP is used for low-level uncertainty. In [34], micro-grid energy management with a decentralized approach is achieved using reinforcement learning. In [35], a game-theory-based optimization model is presented to configure the capacity of energy carrier agents.

In multi-carrier systems, providing an optimal approach is complex because, in such systems, uncertainties related to energy production in renewable sources, fluctuations in load demand, and uncertainty in market price exist. Reviewing the related work in energy management of multi-carrier energy networks indicates that such systems can reduce costs and pollution if optimally operated. Due to the importance of the optimal performance of MEMGs, studies in this field have been considered. However, most of these studies have not considered demand response programs and uncertainties related to the output of renewable energy. To overcome these limitations, this study presents a multi-carrier system (MCS) for planning the optimal performance of MEMG agents, considering uncertainties related to renewable energy production and energy demand fluctuations. The effect of using demand response programs is also presented, with the two objectives of minimizing operational and environmental costs. The efficiency of the proposed method is to simplify the complex MEMG model and reduce the calculations so as to apply uncertainty in the relationships of MG agents.

3. Introducing the Proposed System

As shown in Figure 1, the multi-agent combination, in which each agent performs its tasks to achieve the overall goal of the system, is called MCS. Generally, MCS is divided into three layers: upstream network, MG, and field, as shown in Figure 2. These three layers consist of eight agents. The agents are the upstream network, micro-grid, thermal, hydrogen, RB unit, renewable, storage, and load collector. The upstream grid agent is located in the first layer, which includes the natural gas grid and the electricity grid. This agent is used as an additional resource in case of a lack of energy production.

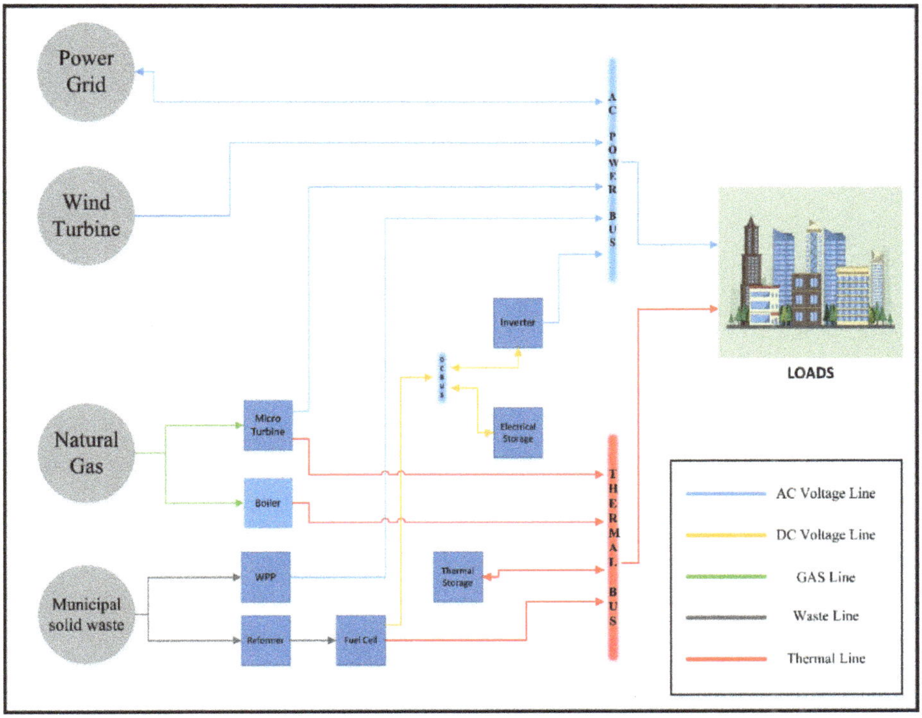

Figure 1. Structure of the proposed MEMG.

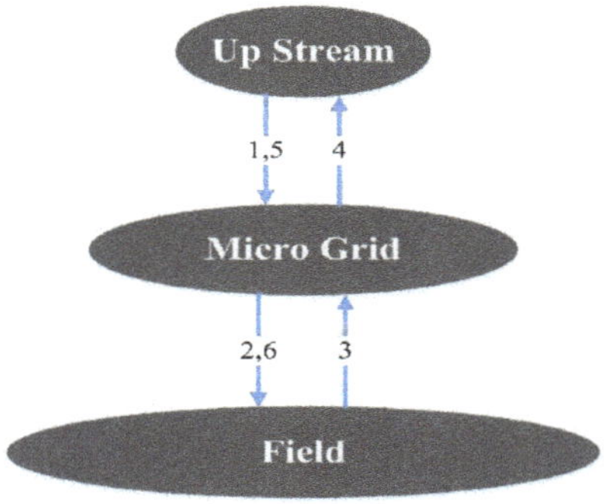

Figure 2. Architecture of the MCS and date exchange.

The micro-grid agent is located in the second layer, which is responsible for coordinating the production and consumption of electrical and thermal energy. This task is performed under the optimal performance of agents while observing the constraints.

The other six agents associated with the production or consumption of electrical and thermal energy and hydrogen are located in the field layer. The overall structure of the proposed method of this study is shown in Figure 3. As shown in Figure 3, in the first step, we select the data related to wind speed and energy demand, and apply them to the LSTM block as input data. In the second step, using the recursive neural network (LSTM) method, we predict the diagrams related to electrical energy data of the wind turbine output, energy demand, and energy price. In the third step, we use a mixed-integer linear programming method and optimize the total cost function, while meeting the existing constraints. This step is carried out according to the modeling of MEMG agents, which is discussed in the next section. Moreover, uncertainty data are modeled with the LSTM block. In the fourth step, the optimal performance of each agent is determined, while minimizing the objective function.

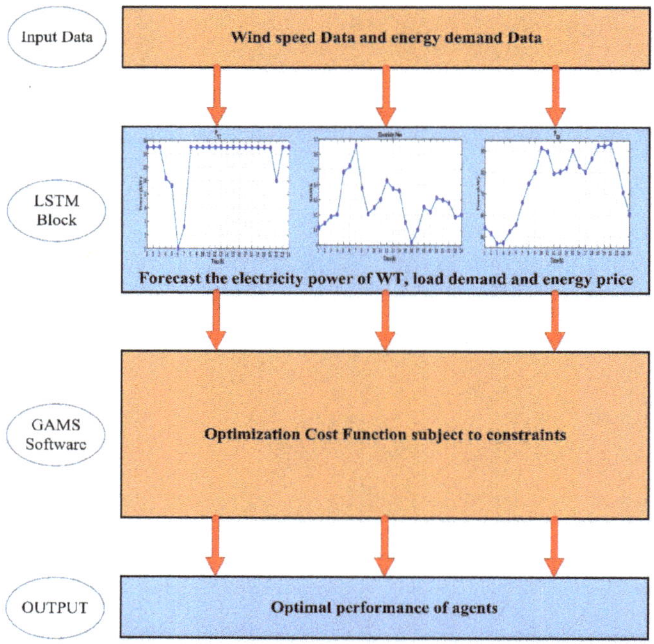

Figure 3. The overall structure of the proposed method.

4. Modeling Agents

4.1. Upstream Network

This agent must announce the hourly price of buying and selling electricity and natural gas, as well as the constraints on energy exchange in the network of the micro-grid operator.

$$Price_{NET}(t) = \pm P_{NET}(t)\ C_{NET}(t)\Delta t \tag{1}$$

$$P_{NET,\ min} \leq P_{NET}(t) \leq P_{NET,\ max} \tag{2}$$

4.2. Micro-Grid Agent

This agent must transmit information about energy costs to field layer agents; it is also responsible for monitoring the optimal performance of field layer agents while observing the constraints imposed by the upstream network and reducing the operating costs of the micro-grid.

The electrical equilibrium equation is defined as follows:

$$P_T(t) + P_{WPP}(t) + P_{WT}(t) + P_{INV}(t) - P_{REC}(t) \pm P_{NET}(t) = P_{ED}(t) \tag{3}$$

The AC power in the inverter is calculated using Equation (4):

$$P_{INV,\ AC}(t) = P_{Inv,\ DC}(t)\alpha_{Inv} \tag{4}$$

Additionally, the AC power in the rectifier is obtained from Equation (5).

$$P_{Rec,\ AC}(t) = \frac{P_{Rec,\ DC}(t)}{\alpha_{REC}} \tag{5}$$

The thermal equilibrium equation is also defined as follows:

$$P_{TT}(t) + P_{TFC}(t) + P_B(t) + P_{TS}(t) = P_{TD}(t) \tag{6}$$

The micro-grid agent checks the above equilibrium equations, and the system must operate in such a way that the above conditions are met.

The amounts of air pollutants emitted from the operation of micro-turbines, fuel cells, rubbish burning units, and the boiler in the micro-grid, in kg/kWh, are obtained from Equation (7):

$$Emission = \sum_{t=1}^{24}\{E_T(t) + E_{FC}(t) + E_{WPP}(t) + E_B(t)\} \tag{7}$$

Micro-grid performance must be optimized with the following constraints:

$$\frac{Emission}{\sum_{t=1}^{24} P_{ED}(t)} \leq Emission_{max} \tag{8}$$

where $Emission_{max}$ is the maximum value of the pollutants, and is equal to 0.66 kg/kWh.

The objective function of total costs of the micro-grid is defined by the following equation:

$$\begin{aligned} Obj.\ Function = & \sum_{t=1}^{24}\Big\{C_{f,T}(t) + C_{OM,T}(t) + C_{S,T}(t) + C_{f.\ FC} + C_{OM,FC}(t) + C_{S,FC}(t) \\ & + C_{f,WPP}(t) + C_{OM,WPP}(t) + C_{S,WPP}(t) + C_{OM,WT}(t) + C_{OM,TS}(t) \\ & + C_{OM,HT}(t) + C_{OM,ES}(t)\Big\} \end{aligned} \tag{9}$$

It should be noted that the electrical power of micro-turbine agents, fuel cells, rubbish burning units, electrical storage agents, and P_{Net} are considered to be decision variables.

4.3. Thermal Agent

This agent consists of two parts: micro-turbine, and boiler. The equation of the electrical output power of micro-turbines is as follows:

$$P_T(t) = \frac{\alpha_T HHV_{gas} Cons_T(t)}{\Delta t} \tag{10}$$

Additionally, the thermal output power of the micro-turbine is proportional to the electric power, which is as given in Equation (11):

$$P_{TT}(t) = K_{Th,T} P_T(t) \tag{11}$$

The thermal output power of the boiler is also given as Equation (12).

$$P_B(t) = \frac{\alpha_B HHV_{gas} Cons_B(t)}{\Delta t} \tag{12}$$

The costs of fuel, maintenance and repair, and switching on and off of the micro-turbine are also calculated by Equations (13)–(15), respectively.

$$C_{f,T}(t) = P_T(t) Price_{gas} \Delta t \tag{13}$$

$$C_{OM,T}(t) = u_T(t) P_T(t) Price_{OM,T} \Delta t \tag{14}$$

$$C_{S,T}(t) = S_T |u_T(t) - u_T(t-1)| \Delta t \tag{15}$$

The amounts of air pollutants produced by micro-turbines and boilers can be calculated through Equations (16) and (17), respectively.

$$E_T(t) = u_T(t) P_T(t) ER_T \Delta t \tag{16}$$

$$E_B(t) = u_B(t) P_B(t)\, ER_B \Delta t \tag{17}$$

4.4. Hydrogen Agent

This agent includes *FC* and *HT*; it must announce the characteristics of the above two parts to the micro-grid agent. Electric and thermal output power in *FC* is calculated by Equations (18) and (19), respectively.

$$P_{FC}(t) = \frac{\alpha_{FC} \alpha_{ref} HHV_{methane} Cons_{FC}(t)}{\Delta t} \tag{18}$$

$$P_{TFC}(t) = K_{Th,FC} P_{FC}(t) \tag{19}$$

Costs related to fuel consumption, maintenance, and turning on and off of the *FC* are formulated as in Equations (13)–(15). Moreover, the amount of air pollutants produced by *FC* is similar to that given in Equation (16), according to the specifications of the *FC*.

The amount of hydrogen stored in the hydrogen tank is formulated as follows (20):

$$V_{tank}(t) = V_{tank}(t-1) + \Delta V_{tank}(t) \tag{20}$$

$$\Delta V_{tank}(t) = \pm \frac{E_{H_2}(t) P_{H_2}}{HHV_{H_2}} \tag{21}$$

where P_{H_2} is the density of hydrogen, which is equal to 0.085 g/L. The constant HHV_{H_2} is considered to be 142 MJ/Kg.

4.5. Rubbish Burning Agent

The rubbish burning agent includes the RB power plant; it is also responsible for announcing the status and characteristics of the RB power plant to the micro-grid operator.

Moreover, the source of waste supply for this agent is municipal solid waste. The electrical output power of this unit is calculated using Equation (10). Furthermore, the costs of fuel consumption, maintenance and repair, and turning on and off of the RB unit are formulated according to Equations (13)–(15). The amount of pollutants produced by the RB power plant is similar to that given by Equation (16).

4.6. Renewable Agent

In recent years, artificial-intelligence-based methods have been known as a promising tool to model the different stochastic parameters, such as load demand, generation of renewable energy sources, and electric vehicle behavior [36,37]. In this paper, a method based on long short-term memory (LSTM) neural networks is used for modeling the stochastic parameters. The LSTM networks are very popular in time-series forecasting because they are robust against the vanishing gradient problem [38]. Interested readers are referred to [37,38] for more information about the LSTM network structure and its formulation.

4.7. Storage Agent

The storage agent must report the status and characteristics of the electrical and thermal storage units to the micro-grid agent. In this section, the amount of electric charge stored by the system is calculated using Equation (22):

$$V_{ES}(t) = V_{ES}(t-1) + V_{ES,Ch}(t) - V_{ES,dch}(t) \tag{22}$$

The amount of heat stored by the system is also calculated with the same equation (Equation (22)). An equation similar to Equation (14) satisfies the maintenance and repair costs of the storage system.

4.8. Load Collector Agent

As a renewable agent for modeling the uncertainty of the load controller agent, an LSTM neural network was used.

4.9. Agents' Connection

According to Figure 2, the communication between agents takes place in six steps. Figure 2 shows the sequence of information exchange in the proposed MCS. In Figure 2, the numbers indicate the sequence of messages. It should also be noted that messages are sent on an hourly basis. The connections between agents in the system are such that in the first step, the upstream network agent announces information about energy purchase costs and constraints to the micro-grid. In the second step, the micro-grid agent requests the status of the agents from the field layer agents; then, in the third step, the field layer agents respond to the micro-grid request. In the fourth step, the micro-grid agent sends the status of the energy shortage and the purchase request to the upstream network agent in order to return the confirmation of the purchase or sale of electricity to the micro-grid in the fifth step. Finally, in the sixth step, the micro-grid agent sends instructions related to the performance of the field agents to each agent.

4.10. LSTM

A recursive neural network (R-NN) is a modified version of conventional neural networks [39]. In deep R-NNs, the descriptive version of R-NNs, known as LSTM networks, can be used to solve the problem of gradient vanishing in hidden layers. In the mentioned LSTM, various operational gates are considered, as shown in Equations (23)–(27) [40].

$$i_t = \sigma\left(WiS_t^{(l-1)}\right) + WhiS_{(t-1)} + bi \tag{23}$$

$$f_t = \sigma\left(Wi\varphi S_t^{(l-1)}\right) + Wh\varphi S_{(t-1)} + bf \tag{24}$$

$$c_t = f_t c_{(t-1)} + i_t \tanh\left(Wi\gamma S_t^{(l-1)}\right) + Wh\gamma S_{(t-1)} + bc \tag{25}$$

$$o_t = \sigma\left(WioS_t^{(l-1)}\right) + WhoS_{(t-1)} + bo \tag{26}$$

$$S_t = o_t \tanh(c_t) \tag{27}$$

where
$Wi,\ Wi\varphi,\ Wi\gamma \in\ R^{r\times nh}$
$Whi,\ Wh\varphi,\ Wh\gamma \in\ R^{nh\times nh}$
and
$bi,\ bf,\ bc,\ bo \in\ R^{1\times nh}$ [41].

5. Linearization

In this step, in order to reduce the computational costs and problem-solving time, we linearize the equations for modeling MEMG agents via the following methods:

- Linearizing by multiplying two binary variables u_1, u_2 [42]:

$$z = u_1 \times u_2 \tag{28}$$

So (28) will be linearized by (29).

$$z \le u_1,\ z \le u_2\ ,\ z \ge u_1 + u_2 - 1 \tag{29}$$

- Linearizing by multiplying a binary variable u_1 and a continuous variable x_1 [43]:

$$z = u_1 \times x_1 \tag{30}$$

So (30) is linearized by the inequalities of (31).

$$z \le x_1,\ z \le M \times u_1\ ,\ z \ge x_1 - M(1 - u_1) \tag{31}$$

where M is a large constant;
- Linearizing quadratic cost function: To linearize quadratic cost function, we use the piecewise linear (P.W.L) method described in [44].

6. Simulation

To validate the proposed method, we used the proposed MCS method in the described MEMG. All simulations were conducted with an Intel® Core (TM) i7-10810u CPU with a frequency of 1.61 GHz and with GAMS software.

6.1. Input Data

In this research, information about uncertainties regarding wind turbine energy production as well as energy demand was predicted using the LSTM networks, as can be seen in the diagrams of Figures 4–7. The data from Ontario province in Canada were used as input data for the LSTM network based on [45,46]. Hourly data on wind speed, electricity prices, and energy demand over three years from 1 January 2018 to 30 December 2020 were investigated. It should be noted that the energy price data are for Ontario in Canada. Given that retail prices are commonly used for MGs, the Ontario market price data were scaled at an appropriate rate. The specifications of the micro-turbines, fuel cells, boilers, and the waste power plant are shown in Table 1 [47,48]. Moreover, the total cost in Equation (9) is minimized by considering the constraints in the system with GAMES software and a mixed-integer linear programming method. The nonlinear form of the total cost equation makes the calculations difficult. Therefore, once the nonlinear equation is minimized, the total cost equation is first linearized, and then the minimization is carried out. Furthermore, to compare the proposed method and the validation of this method, we used a conventional centralized approach to optimize the performance of agents in order to reduce the initial costs of the MG. In the centralized method, uncertainties related to wind speed and the total energy demand are not considered, and the actual amount is not predicted.

Table 1. Specifications of the micro-turbine, fuel cell, boiler, and waste power plant.

	Emission Factors (kg/MWh)			Start/Stop Cost (USD)	O&M Cost (USD/kWh)	Electrical Power Range (kW)		Thermal Power Range (kW)		Efficiency (%)	Fuel Cost	$K_{Thermal}$
	NO_x	CO_2	SO_2			Min	Max	Min	Max			
Micro-Turbine	0.2	724	0.0036	0.11	0.005	6	30	15.6	78	26	0.41 USD/m^3	2.6
Fuel Cell	0.013	489	0.0027	0.148	0.008	3	25	4.2	35	40	0.12 USD/kWh	1.4
Boiler	1.81	845	2.545	-	-	-	-	3	80	90	-	-
Waste Power Plant	0.2	300	0.1	0.12	0.006	6	30	-	-	30	0.02 USD/kWh	-

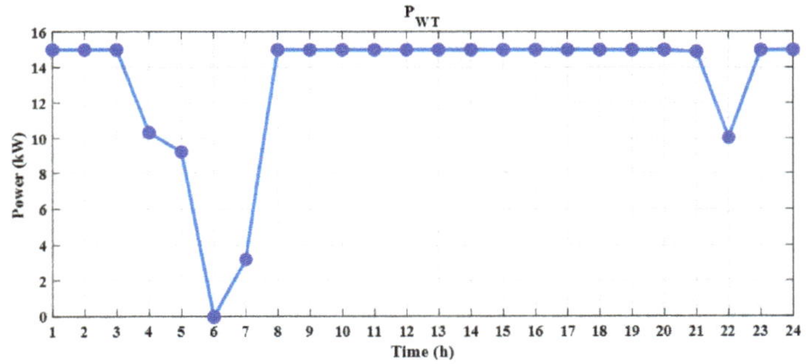

Figure 4. Daily electric power generated by wind turbines.

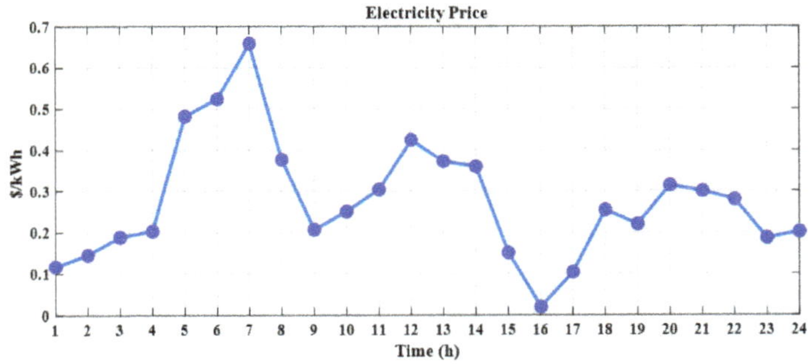

Figure 5. Daily electricity prices.

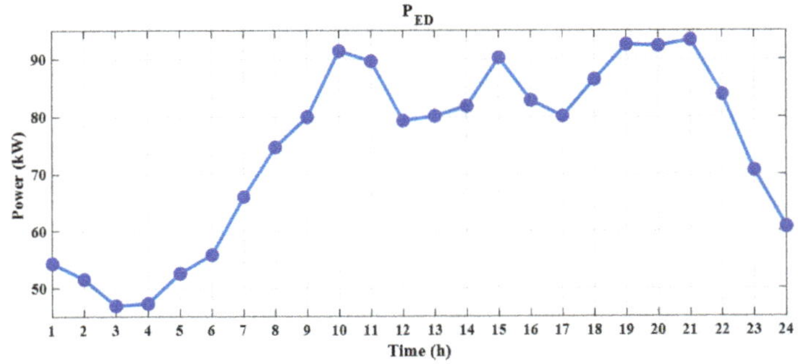

Figure 6. Daily electricity load demand.

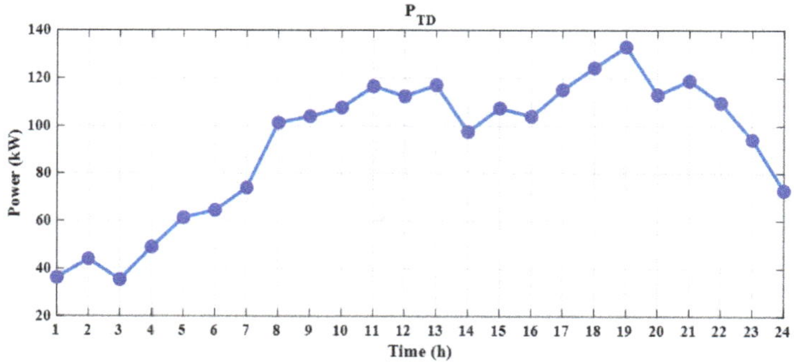

Figure 7. Daily thermal load demand.

6.2. Results

The results of the amounts of energy production or consumption in each of the network agents are shown in Figures 8 and 9. To validate the proposed method, the optimization results in linear and nonlinear methods, as well as the common optimization method, are shown in Table 2, regardless of the uncertainties. As shown in Table 2, using the proposed method reduces MEMG operating costs by 34% compared to conventional centralized models. The USD 26.6 decrease is due to a reduction in the charge and discharge cycles (ES). According to the diagram in Figure 8, it is clear that the electrical energy exchanged between the MG and the upstream grid in one day is equal to 354.5 KW. Since power generation with an MG is always assumed to be cheaper than purchasing power from the upstream grid, micro-grids have reduced operating costs. On the other hand, according to the data in Table 1, the electrical energy produced in WPP is cheaper than the micro-turbines and FC units. According to Figure 8, it can be seen that the amount of electrical energy produced by the WPP is higher than the FC and micro-turbine units, which is also one of the reasons for reducing the cost of the MEMG. As shown in Figure 9, from points 2 to 6, the thermal energy produced by the FC and micro-turbine is more than the heat load, and the thermal storage system is charging. On the other hand, from points 9 to 13 and 16 to 22, since more heat load is generated and stored than thermal energy, the boiler responds to the heat load demand. Moreover, the use of the proposed method reduces the emission of pollutants by the micro-turbine, RB, FC, and boiler compared to the conventional centralized method. In addition, Table 2 shows that the use of the proposed method leads to a reduction in CPU optimization time. This reduction in time indicates a reduction in the computation in the proposed method. It is clear that by linearizing the equations related to MG agents, the simulation time decreases significantly due to the linearization of equations and the reduction in the complexity of the optimization calculations.

Table 2. Results of the proposed MCS-based method and centralized method.

Case	Total Cost (USD/Day)	Total Emission (kg/Day)	CPU Optimization Time (s)
Nonlinear MCS	51.7	1080	32
Linear MCS	51.9	1081.25	2.6
Centralized	144.3	1330.81	69

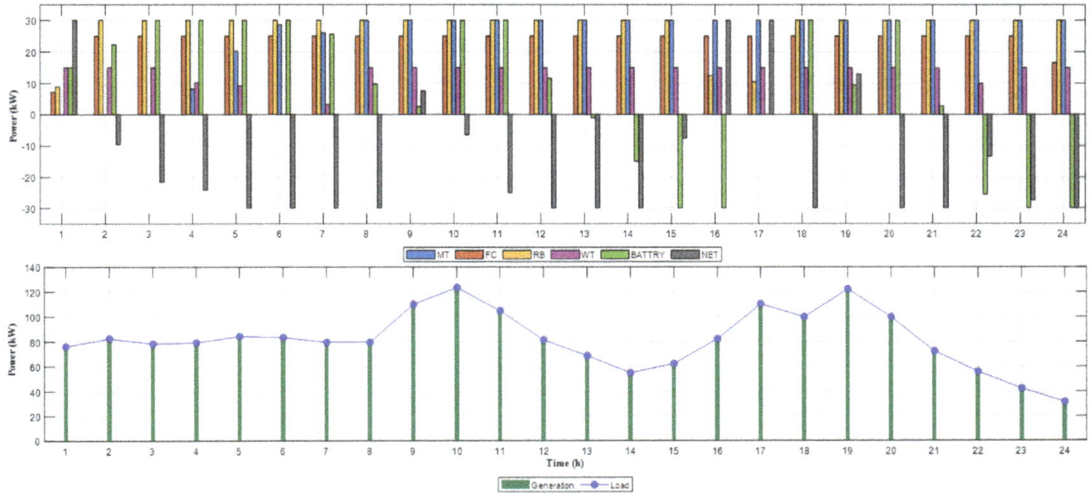

Figure 8. Optimal management of electrical elements of the micro-grid.

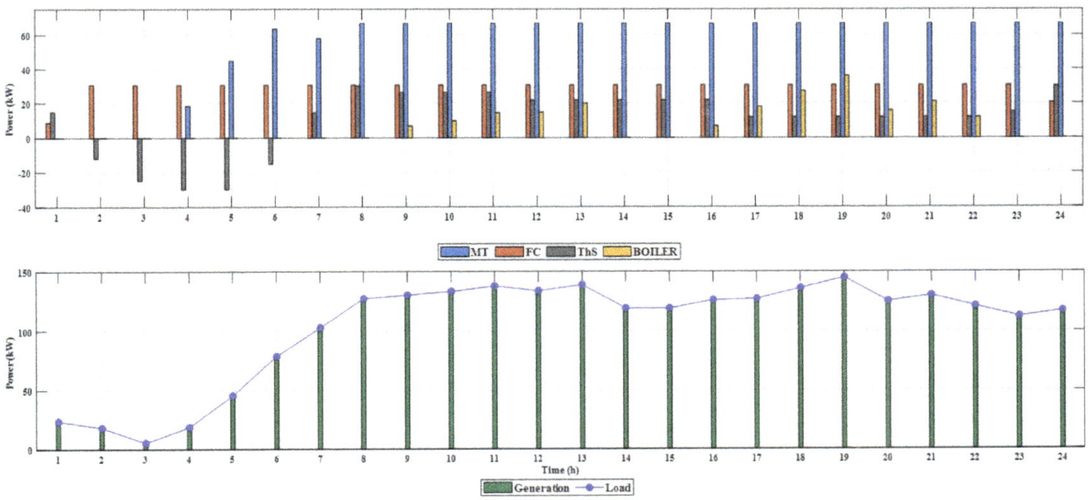

Figure 9. Optimal management of thermal elements of the micro-grid.

7. Conclusions

In this study, an applied method for optimal management of the performance of an MEMG is presented, considering the uncertainties associated with the prediction of daily demand. The agents of the energy system are at three decentralized levels, and are interrelated at these levels. The equations are formulated according to the relationships between agents at these three levels.

The proposed method was tested on an MEMG. The results indicate a reduction in operational network costs and the complexity of computations compared to centralized methods. On the other hand, linearization of equations was carried out. This linearization can reduce the computational complexity of the proposed method.

Therefore, energy management systems with an MCS-based modeling approach are a suitable solution for optimal energy management and reducing the demand of micro-consumers (urban buildings) from upstream networks, electricity, and natural gas networks, reducing greenhouse gas emissions. In future work, the design of an MG should be considered so that the MG can make operational decisions that affect the market price.

Author Contributions: Conceptualization, S.S.; methodology, M.F. and S.S.; software, M.F.; validation, M.K. and T.S.; formal analysis, S.S. and A.N.; investigation, M.K.; resources, A.N.; data curation, M.F. and S.S.; writing—original draft preparation, M.F.; writing—review and editing, A.N. and M.K.; visualization, S.S.; supervision, A.N. and T.S.; funding acquisition, A.N. All authors have read and agreed to the published version of the manuscript.

Funding: This research received no external funding.

Institutional Review Board Statement: Not applicable.

Informed Consent Statement: Not applicable.

Data Availability Statement: Publicly available datasets were analyzed in this study. This data can be found here: [http://www.ieso.ca/power-data] (8 January 2022), [http://climate.weather.gc.ca/historicaldata/searchhistoricdatae.html] (8 January 2022).

Conflicts of Interest: The authors declare no conflict of interest.

Abbreviations

$C_{NET}(t)$	Cost of power exchange	$Cons_T(t)$	Fuel consumption of micro-turbine
$P_{NET}(t)$	Exchanged power	$K_{Th,T}$	Constant that relates P_T and P_{TT}
$Price_{NET}(t)$	Price of power exchange	α_B	Boiler efficiency
$P_T(t)$	Electrical power of micro-turbine	$Cons_B(t)$	Fuel consumption of boiler
$P_{WPP}(t)$	Electrical power of waste power plant	$Price_{gas}$	Price of natural gas
$P_{WT}(t)$	Electrical power of wind turbine	$u_T(t)$	Status of micro-turbine (0 or 1)
$P_{INV,AC}(t)$	Electrical AC power of inverter	$Price_{OM,T}$	Price of micro-turbine O&M
$P_{REC,AC}(t)$	Electrical AC power of rectifier	S_T	Start/stop rate of micro-turbine
$P_{ED}(t)$	Electrical power demand	ER_T	Emission rate of micro-turbine
$P_{TT}(t)$	Thermal power of micro-turbine	ER_B	Emission rate of boiler
$P_{TFC}(t)$	Thermal power of fuel cell	α_{FC}	Fuel cell efficiency
$P_B(t)$	Thermal power of boiler	α_{ref}	Reformer efficiency
$P_{TS}(t)$	Charging/discharging of thermal storage	$HHV_{methane}$	Higher heating value of methane
$P_{TD}(t)$	Thermal power demand	$Cons_{FC}(t)$	Fuel consumption of fuel cell
$E_T(t)$	Emissions of micro-turbine	$K_{Th,FC}$	Constant that relates P_{FC} and P_{TFC}
$E_{FC}(t)$	Emissions of fuel cell	$V_{ES}(t)$	Quantity of electrical storage
$E_{WPP}(t)$	Emissions of waste power plant	$V_{ES,Ch}(t)$	Charging energy of electrical storage
$E_B(t)$	Emissions of boiler	$V_{ES,dch}(t)$	Discharging energy of electrical storage
$C_{f,T}(t)$	Fuel cost of micro-turbine	$i(t), f(t), c(t)$	Data vector of cell block, forget, and input gates at time t
$C_{OM,T}(t)$	O&M cost of micro-turbine	bi, bf, bc, bo	Bias vector for cell block, forget, input, and output gates
$C_{S,T}(t)$	Start/stop cost of micro-turbine	$o(t)$	Data vector of output gate at time t
$C_{f.\ FC}(t)$	Fuel cost of fuel cell	$s(t)$	State vector of current layer at state t
$C_{OM,FC}(t)$	O&M cost of fuel cell	$s(t)^l$	State vector of layer l at state t
$C_{S,FC}(t)$	Start/stop cost of fuel cell	$Whi, Wh\varphi, Wh\gamma, Who$	Weight vector for output of previous state input gate, forget gate, cell block, and output gate
$C_{f,WPP}(t)$	Fuel cost of waste power plant	$Wi, Wi\varphi, Wi\gamma, Wio$	Weight vector for input of current state input gate, forget gate, cell block, and output gate
$C_{OM,WPP}(t)$	O&M cost of waste power plant	Δt	Period of time
$C_{S,WPP}(t)$	Start/Stop cost of waste power plant	$u_b(t)$	Status of boiler (0 or 1)
$C_{OM,WT}(t)$	O&M cost of wind turbine	$V_{tank}(t)$	Amount of stored hydrogen
$C_{OM,TS}(t)$	O&M cost of thermal storage	$E_{H_2}(t)$	Charging/discharging output of the HT
$C_{OM,HT}(t)$	O&M cost of hydrogen tank	$P_{INV,DC}(t)$	Electrical DC power of inverter
$C_{OM,ES}(t)$	O&M cost of electrical storage	$P_{REC,DC}(t)$	Electrical DC power of rectifier
α_T	Micro-turbine efficiency	α_{INV}	Inverter efficiency
HHV_{gas}	Higher heating value of gas	α_{REC}	Rectifier efficiency

References

1. Li, Y.; Liang, W.; Tan, R. Optimal design of installation capacity and operation strategy for distributed energy system. *Appl. Therm. Eng.* **2017**, *125*, 756–766. [CrossRef]
2. Allison, J. Robust multi-objective control of hybrid renewable microgeneration systems with energy storage. *Appl. Therm. Eng.* **2017**, *114*, 1498–1506. [CrossRef]
3. Mohseni, S.; Moghaddas-Tafreshi, S.M. A multi-agent system for optimal sizing of a cooperative self-sustainable multi-carrier microgrid. *Sustain. Cities Soc.* **2018**, *38*, 452–465. [CrossRef]
4. Väisänen, S.; Mikkilä, M.; Havukainen, J.; Sokka, L.; Luoranen, M.; Horttanainen, M. Using a multi-method approach for decision-making about a sustainable local distributed energy system: A case study from Finland. *J. Clean. Prod.* **2016**, *137*, 1330–1338. [CrossRef]
5. Yang, G.; Zhai, X. Optimization and performance analysis of solar hybrid CCHP systems under different operation strategies. *Appl. Therm. Eng.* **2018**, *133*, 327–340. [CrossRef]
6. Luo, Z.; Wu, Z.; Li, Z.; Cai, H.; Li, B.; Gu, W. A two-stage optimization and control for CCHP microgrid energy management. *Appl. Therm. Eng.* **2017**, *125*, 513–522. [CrossRef]
7. Kumar, K.P.; Saravanan, B. Recent techniques to model uncertainties in power generation from renewable energy sources and loads in microgrids–A review. *Renew. Sustain. Energy Rev.* **2017**, *71*, 348–358. [CrossRef]
8. Melamed, M.; Ben-Tal, A.; Golany, B. A multi-period unit commitment problem under a new hybrid uncertainty set for a renewable energy source. *Renew. Energy* **2018**, *118*, 909–917. [CrossRef]
9. Bernal-Agustín, J.L.; Dufo-Lopez, R. Simulation and optimization of stand-alone hybrid renewable energy systems. *Renew. Sustain. Energy Rev.* **2009**, *13*, 2111–2118. [CrossRef]
10. Van Roy, J.; Leemput, N.; Geth, F.; Büscher, J.; Salenbien, R.; Driesen, J. Electric vehicle charging in an office building microgrid with distributed energy resources. *IEEE Trans. Sustain. Energy* **2014**, *5*, 1389–1396. [CrossRef]
11. Lasseter, R.; Akhil, A.; Marnay, C.; Stephens, J.; Dagle, J.; Guttromsom, R.; Meliopoulous, A.S.; Yinger, R.; Eto, J. Integration of distributed energy resources. In *The CERTS Microgrid Concept*; No. LBNL-50829; Lawrence Berkeley National Lab.(LBNL): Berkeley, CA, USA, 2002.
12. Nazari-Heris, M.; Abapour, S.; Mohammadi-Ivatloo, B. Optimal economic dispatch of FC-CHP based heat and power micro-grids. *Appl. Therm. Eng.* **2017**, *114*, 756–769. [CrossRef]
13. Prinsloo, G.; Dobson, R.; Mammoli, A. Synthesis of an intelligent rural village microgrid control strategy based on smartgrid multi-agent modelling and transactive energy management principles. *Energy* **2018**, *147*, 263–278. [CrossRef]
14. Malekpour, A.R.; Pahwa, A. Stochastic networked microgrid energy management with correlated wind generators. *IEEE Trans. Power Syst.* **2017**, *32*, 3681–3693. [CrossRef]
15. Afrasiabi, M.; Mohammadi, M.; Rastegar, M.; Kargarian, A. Multi-agent microgrid energy management based on deep learning forecaster. *Energy* **2019**, *186*, 115873. [CrossRef]
16. Elsied, M.; Oukaour, A.; Gualous, H.; Brutto, O.A.L. Optimal economic and environment operation of micro-grid power systems. *Energy Convers. Manag.* **2016**, *122*, 182–194. [CrossRef]
17. Niknam, T.; Golestaneh, F.; Malekpour, A. Probabilistic energy and operation management of a microgrid containing wind/photovoltaic/fuel cell generation and energy storage devices based on point estimate method and self-adaptive gravitational search algorithm. *Energy* **2012**, *43*, 427–437. [CrossRef]
18. Hussain, A.; Bui, V.H.; Kim, H.M. Optimal operation of hybrid microgrids for enhancing resiliency considering feasible islanding and survivability. *IET Renew. Power Gener.* **2017**, *11*, 846–857. [CrossRef]
19. Mohammadi, S.; Soleymani, S.; Mozafari, B. Scenario-based stochastic operation management of microgrid including wind, photovoltaic, micro-turbine, fuel cell and energy storage devices. *Int. J. Electr. Power Energy Syst.* **2014**, *54*, 525–535. [CrossRef]
20. Nikmehr, N.; Najafi-Ravadanegh, S.; Khodaei, A. Probabilistic optimal scheduling of networked microgrids considering time-based demand response programs under uncertainty. *Appl. Energy* **2017**, *198*, 267–279. [CrossRef]
21. Bracco, S.; Delfino, F.; Pampararo, F.; Robba, M.; Rossi, M. A mathematical model for the optimal operation of the University of Genoa Smart Polygeneration Microgrid: Evaluation of technical, economic and environmental performance indicators. *Energy* **2014**, *64*, 912–922. [CrossRef]
22. Bracco, S.; Brignone, M.; Delfino, F.; Pampararo, F.; Rossi, M.; Ferro, G.; Robba, M. An optimization model for polygeneration microgrids with renewables, electrical and thermal storage: Application to the savona campus. In *2018 IEEE International Conference on Environment and Electrical Engineering and 2018 IEEE Industrial and Commercial Power Systems Europe (EEEIC/I&CPS Europe)*; IEEE: Piscataway Township, NJ, USA, 2018; pp. 1–6.
23. Urbanucci, L.; Testi, D.; Bruno, J.C. An operational optimization method for a complex polygeneration plant based on real-time measurements. *Energy Convers. Manag.* **2018**, *170*, 50–61. [CrossRef]
24. Sigarchian, S.G.; Malmquist, A.; Martin, V. The choice of operating strategy for a complex polygeneration system: A case study for a residential building in Italy. *Energy Convers. Manag.* **2018**, *163*, 278–291. [CrossRef]
25. La Scala, M.; Vaccaro, A.; Zobaa, A.F. A goal programming methodology for multiobjective optimization of distributed energy hubs operation. *Appl. Therm. Eng.* **2014**, *71*, 658–666. [CrossRef]
26. Kampouropoulos, K.; Andrade, F.; Sala, E.; Espinosa, A.G.; Romeral, L. Multiobjective optimization of multi-carrier energy system using a combination of ANFIS and genetic algorithms. *IEEE Trans. Smart Grid* **2016**, *9*, 2276–2283. [CrossRef]

27. Karavas, C.S.; Kyriakarakos, G.; Arvanitis, K.G.; Papadakis, G. A multi-agent decentralized energy management system based on distributed intelligence for the design and control of autonomous polygeneration microgrids. *Energy Convers. Manag.* **2015**, *103*, 166–179. [CrossRef]
28. Karavas, C.S.; Arvanitis, K.; Papadakis, G. A game theory approach to multi-agent decentralized energy management of autonomous polygeneration microgrids. *Energies* **2017**, *10*, 1756. [CrossRef]
29. Wasilewski, J. Optimisation of multicarrier microgrid layout using selected metaheuristics. *Int. J. Electr. Power Energy Syst.* **2018**, *99*, 246–260. [CrossRef]
30. Roustai, M.; Rayati, M.; Sheikhi, A.; Ranjbar, A. A scenario-based optimization of Smart Energy Hub operation in a stochastic environment using conditional-value-at-risk. *Sustain. Cities Soc.* **2018**, *39*, 309–316. [CrossRef]
31. Giaouris, D.; Papadopoulos, A.I.; Patsios, C.; Walker, S.; Ziogou, C.; Taylor, P.; Voutetakis, S.; Papadopoulou, S.; Seferlis, P. A systems approach for management of microgrids considering multiple energy carriers, stochastic loads, forecasting and demand side response. *Appl. Energy* **2018**, *226*, 546–559. [CrossRef]
32. Zhang, J.; Wang, X.; Ma, L. An optimal power allocation scheme of microgrid using grey wolf optimizer. *IEEE Access* **2019**, *7*, 137608–137619. [CrossRef]
33. Mishra, S.; Bordin, C.; Tomasgard, A.; Palu, I. A multi-agent system approach for optimal microgrid expansion planning under uncertainty. *Int. J. Electr. Power Energy Syst.* **2019**, *109*, 696–709. [CrossRef]
34. Samadi, E.; Badri, A.; Ebrahimpour, R. Decentralized multi-agent based energy management of microgrid using reinforcement learning. *Int. J. Electr. Power Energy Syst.* **2020**, *122*, 106211. [CrossRef]
35. Jin, S.; Wang, S.; Fang, F. Game theoretical analysis on capacity configuration for microgrid based on multi-agent system. *Int. J. Electr. Power Energy Syst.* **2021**, *125*, 106485. [CrossRef]
36. Jahangir, H.; Gougheri, S.S.; Vatandoust, B.; Golkar, M.A.; Golkar, M.A.; Ahmadian, A.; Hajizadeh, A. A Novel Cross-Case Electric Vehicle Demand Modeling Based on 3D Convolutional Generative Adversarial Networks. *IEEE Trans. Power Syst.* **2021**, *37*, 1173–1183. [CrossRef]
37. Jahangir, H.; Tayarani, H.; Gougheri, S.S.; Golkar, M.A.; Ahmadian, A.; Elkamel, A. Deep Learning-Based Forecasting Approach in Smart Grids With Microclustering and Bidirectional LSTM Network. *IEEE Trans. Ind. Electron.* **2020**, *68*, 8298–8309. [CrossRef]
38. Sadeghi, S.; Jahangir, H.; Vatandoust, B.; Golkar, M.A.; Ahmadian, A.; Elkamel, A. Optimal bidding strategy of a virtual power plant in day-ahead energy and frequency regulation markets: A deep learning-based approach. *Int. J. Electr. Power Energy Syst.* **2021**, *127*, 106646. [CrossRef]
39. Jahangir, H.; Tayarani, H.; Ahmadian, A.; Golkar, M.A.; Miret, J.; Tayarani, M.; Gao, H.O. Charging demand of plug-in electric vehicles: Forecasting travel behavior based on a novel rough artificial neural network approach. *J. Clean. Prod.* **2019**, *229*, 1029–1044. [CrossRef]
40. Khodayar, M.; Wang, J. Spatio-temporal graph deep neural network for short-term wind speed forecasting. *IEEE Trans. Sustain. Energy* **2018**, *10*, 670–681. [CrossRef]
41. Gougheri, S.S.; Jahangir, H.; Golkar, M.A.; Ahmadian, A.; Golkar, M.A. Optimal participation of a virtual power plant in electricity market considering renewable energy: A deep learning-based approach. *Sustain. Energy Grids Netw.* **2021**, *26*, 100448. [CrossRef]
42. Norouzi, N.; Tavakkoli-Moghaddam, R.; Ghazanfari, M.; Alinaghian, M.; Salamatbakhsh, A. A new multi-objective competitive open vehicle routing problem solved by particle swarm optimization. *Netw. Spat. Econ.* **2012**, *12*, 609–633. [CrossRef]
43. Glover, F.; Woolsey, E. Converting the 0–1 polynomial programming problem to a 0–1 linear program. *Oper. Res.* **1974**, *22*, 180–182. [CrossRef]
44. Soroudi, A. *Power System Optimization Modeling in GAMS*; Springer: Berlin/Heidelberg, Germany, 2017.
45. Power Data. Available online: http://www.ieso.ca/power-data (accessed on 20 May 2021).
46. Renewable Energy Policy Network for the 21st Century. Available online: http://climate.weather.gc.ca/historicaldata/searchhistoricdatae.html (accessed on 20 May 2021).
47. Moghaddam, A.A.; Seifi, A.; Niknam, T.; Pahlavani, M.R.A. Multi-objective operation management of a renewable MG (micro-grid) with back-up micro-turbine/fuel cell/battery hybrid power source. *Energy* **2011**, *36*, 6490–6507. [CrossRef]
48. Zhang, L.; Xiang, J. The performance of a grid-tied microgrid with hydrogen storage and a hydrogen fuel cell stack. *Energy Convers. Manag.* **2014**, *87*, 421–427. [CrossRef]

applied sciences

MDPI

Article

Optimal Incorporation of Photovoltaic Energy and Battery Energy Storage Systems in Distribution Networks Considering Uncertainties of Demand and Generation

Hussein Abdel-Mawgoud [1], Salah Kamel [1], Marcos Tostado-Véliz [2,*], Ehab E. Elattar [3,*] and Mahmoud M. Hussein [1,4]

1 Department of Electrical Engineering, Faculty of Engineering, Aswan University, Aswan 81542, Egypt; hussein.abdelmawgoud@yahoo.com (H.A.-M.); skamel@aswu.edu.eg (S.K.); mahmoud_hussein@aswu.edu.eg (M.M.H.)
2 Electrical Engineering Department, University of Jaen, EPS, 23700 Linares, Spain
3 Department of Electrical Engineering, College of Engineering, Taif University, P.O. Box 11099, Taif 21944, Saudi Arabia
4 Department of Electrical Engineering, Faculty of Energy Engineering, Aswan University, Aswan 81528, Egypt
* Correspondence: mtostado@ujaen.es (M.T.-V.); e.elattar@tu.edu.sa (E.E.E.)

Abstract: In this paper, the Archimedes optimization algorithm (AOA) is applied as a recent meta-heuristic optimization algorithm to reduce energy losses and capture the size of incorporating a battery energy storage system (BESS) and photovoltaics (PV) within a distribution system. AOA is designed with revelation from Archimedes' principle, an impressive physics law. AOA mimics the attitude of buoyant force applied upward on an object, partially or entirely dipped in liquid, which is relative to the weight of the dislodged liquid. Furthermore, the developed algorithm is evolved for sizing several PVs and BESSs considering the changing demand over time and the probability generation. The studied IEEE 69-bus distribution network system has different types of the load, such as residential, industrial, and commercial loads. The simulation results indicate the robustness of the proposed algorithm for computing the best size of multiple PVs and BESSs with a significant reduction in the power system losses. Additionally, the AOA algorithm has an efficient balancing between the exploration and exploitation phases to avoid the local solutions and go to the best global solutions, compared with other studied algorithms.

Keywords: photovoltaics; BESS; optimization; AOA algorithm; uncertainty; distribution network

Citation: Abdel-Mawgoud, H.; Kamel, S.; Tostado-Véliz, M.; Elattar, E.E.; Hussein, M.M. Optimal Incorporation of Photovoltaic Energy and Battery Energy Storage Systems in Distribution Networks Considering Uncertainties of Demand and Generation. *Appl. Sci.* **2021**, *11*, 8231. https://doi.org/10.3390/app11178231

Academic Editor: Fabio La Foresta

Received: 1 July 2021
Accepted: 1 September 2021
Published: 5 September 2021

Publisher's Note: MDPI stays neutral with regard to jurisdictional claims in published maps and institutional affiliations.

Copyright: © 2021 by the authors. Licensee MDPI, Basel, Switzerland. This article is an open access article distributed under the terms and conditions of the Creative Commons Attribution (CC BY) license (https://creativecommons.org/licenses/by/4.0/).

1. Introduction

Recently, the penetration of PV systems into the electric grid has been increased in most countries to take advantage of the environment as well as the economic benefits. PV energy systems do not emit polluting gases such as traditional energy sources, and the owners of PV energy systems obtain incentives from utilities by selling the output energy from their PV units at a high price [1–3]. PV output is variable during the day as it depends on the variable natural source [4,5]. The designing, optimization, and planning of PV has been presented in [6–8]. The allocation of the PV energy system near the loads in the distribution system leads to improvement in voltage profile and to a decrease the emission, cost, and system losses as in [9,10]. The optimal planning of PV in a realistic case has been presented in [11]. In [12,13], an analytical method has been applied to decrease the system loss by incorporating PV in distribution networks. In [14], the optimal allocation of electric vehicles with a combination of PV and battery storage to reduce the total system cost is presented. Additionally, the optimal planning of PV with electric vehicles in distribution networks to decrease the system loss is presented in [15]. Incorporating PV in the distribution system to decrease the system loss and to improve the bus system voltage is introduced in [16,17]. Nevertheless, the high penetration of the PV energy system with the variation in demand exposes the distribution systems to various problems, such as fluctuations of power, voltage rise, extraordinary energy losses, and a low stability of voltage [1,18,19]. Therefore, load curtailment and integration of energy storage systems has been exploited to decrease power fluctuations and overcome any system constraint violations [1].

Nowadays, the integration of DG into distribution systems considering the energy losses has enticed attention. Research on the allocation of renewable DG (e.g., solar, wind and biomass) has been proposed to reduce energy loss, considering the variations of power generation and load demand [11,14,15]. According to the presence load type characteristics, optimum DG power factor dispatch for every load level would be an integral part for reducing energy losses. Nevertheless, most of the current research assumed that the DG systems work at predefined power factors.

In contrast to generation systems that depend on conventional generation techniques such as gas turbines and reciprocating engines, PV energy resources are non-dispatchable and intermittent, depending on temperature and radiation. The BESS technologies create a chance to transform the PV energy resources from non-dispatchable systems to dispatchable systems with similar conventional resources [20,21]. Throughout the past two decades, a hybrid system of BESS and PV has been considered for the applications of stand-alone systems [22–24]. Recently, the hybrid system of PV and BESS has been exploited as one of the furthermost worthwhile solutions in grid-connected applications to increase the penetration of the PV energy system in distribution systems. Such hybrid design helps alleviate the influences of intermittency in the PV energy systems, and offers many benefits for the owners of PV system, customers, and utilities. Many researchers have devoted their efforts to this interesting topic [18–20,25–31]. A hybrid PV-BESS has been evolved for applications of load demand-side to improve the efficiency of the electrical system [25,26]. In [27], authors have proposed an optimal BESS charging and discharging schedule in a PV grid-connected system for shaving of peak demand. Authors in [28] have presented a methodology to calculate the BESS size for shaving of peak load and power balance used in case of connecting PV energy system with the grid. In [29], authors have presented a methodology to compute the BESS size for raising the penetration of the PV energy system in case of residential system load with the objectives of voltage regulation and decreasing in maximum output power and yearly cost. A discharging and charging strategy for the BESS has been suggested to alleviate abrupt changes in the output power of the PV energy system and boost the evening peak load in the case of a residential load system [30]. The authors in [18] have proposed a conception of voltage regulation voltages in the distribution systems with high penetration of PV energy system by adjusting the output power of BESS at customer-side. BESS has been controlled and sized to diminish the fluctuation in the PV output power [20,31]. In [19], the authors have evolved the BESS's best discharging and charging schedule on an hourly basis to alleviate the discontinuity of PV output by reducing the energy loss.

Generally, the previous review indicates that significant research has been published on the size and discharging and discharging schedules of the BESS exploited in the case of connecting the PV energy system with the grid. Nevertheless, most of the research introduced has supposed that the optimal power factor dispatch for every hybrid PV-BESS throughout all of the time intervals is ignored as well and the size of PV units exploited in hybrid PV-BESS is prespecified. Based on the characteristics of the loads served, every PV-BESS hybrid that can provide reactive and active power with the optimum power factor may positively reduce energy losses in distribution systems.

This paper shows a conception of involving a hybrid PV and BESS in residential, industrial, and commercial distribution systems, taking into account the system energy loss. Where the PV energy system is deemed as a non-dispatchable energy source as its power output cannot be controlled, BESS is considered as a dispatchable energy source as its power output can be controlled. In this paper, AOA is applied to reduce the energy losses and capture the size of incorporating a PV energy system and BESS in a distribution network. However, the paper contributions can be summarized as follows:

- A new application of the Archimedes optimization algorithm for minimizing the energy losses and capture the size of incorporating battery energy storage system and photovoltaics in a distribution system.
- The developed algorithm is evolved for sizing several PVs and BESSs considering the changing demand over time and the probability generation.
- Validating the developed algorithm using IEEE 69-bus distribution network system which has different types of the load, such as residential, industrial, and commercial loads.
- The simulation results indicate the robustness of the proposed algorithm for computing the best size of multiple PVs and BESSs, with a significant reduction in the power system losses.

The remainder of this paper is constructed as following: the load, BESS, and PV modeling are introduced in Section 2. Also the problem formulation of BESS with PV is introduced in Section 2. The methodology of the proposed Archimedes optimization algorithm (AOA) has been presented in Section 3. The cases study on a 69-bus industrial, commercial, and residential distribution have been presented and discussed in Section 4. Section 5 offers the conclusions of the paper.

2. Problem Formulation

The two buses of the main feeder in the distribution network with a combination of PV and BES can be represented in Figure 1.

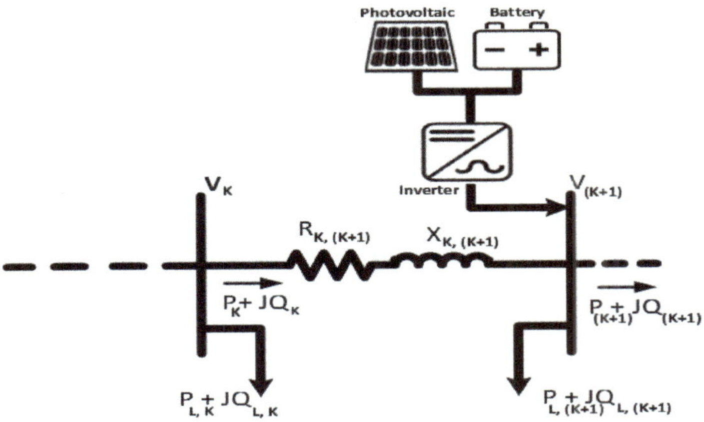

Figure 1. Two buses of radial distribution network.

Forward/backward sweep algorithm is utilized to obtain the system load flows. The reactive and real power flows are calculated by Equations (1) and (2), respectively [32].

$$P_K = P_{(K+1)} + P_{L,(K+1)} + R_{K,(K+1)}\left(\frac{(P_{(K+1)} + P_{L,(K+1)})^2 + (Q_{(K+1)} + Q_{L,(K+1)})^2}{|V_{(K+1)}|^2}\right) \quad (1)$$

$$Q_K = Q_{(K+1)} + Q_{L,(K+1)} + X_{K,(K+1)}\left(\frac{(P_{(K+1)} + P_{L,(K+1)})^2 + (Q_{(K+1)} + Q_{L,(K+1)})^2}{|V_{(K+1)}|^2}\right) \quad (2)$$

where, $P_{(K+1)}$ and $Q_{(K+1)}$ represent the real and reactive power flow from bus (K) to the next bus system, respectively. The reactive and real load flows between (K) and ($K + 1$) buses are Q_K and P_K, respectively. the reactance and resistance between (K) and ($K + 1$) buses are $X_{K,(K+1)}$ and $R_{K,(K+1)}$, respectively. The reactive and real load at bus ($K + 1$) are $Q_{L,(K+1)}$ and $P_{L,(K+1)}$, respectively.

The voltage magnitude of system buses is evaluated by Equation (3).

$$V_{(K+1)}{}^2 = V_K^2 - 2\left(P_K R_{K,(K+1)} + Q_K X_{K,(K+1)}\right) + \left(R_{K,(K+1)}{}^2 + X_{K,(K+1)}{}^2\right)\frac{(P_K{}^2 + Q_K{}^2)}{V_K^2} \quad (3)$$

where $V_{(K+1)}$ and $V_{(K)}$ are the system voltage at buses $(K+1)$ and $(\mathbf{K})$, respectively. Installation of PV and BESS in RDS changes the load flows through the system branches. Therefore, Equations (1) and (2) are modified to Equations (4) and (5), respectively.

$$P_K = P_{(K+1)} + P_{L,(K+1)} + R_{K,(K+1)}\left(\frac{\left(P_{(K+1)} + P_{L,(K+1)}\right)^2 + \left(Q_{(K+1)} + Q_{L,(K+1)}\right)^2}{\left|V_{(K+1)}\right|^2}\right) - P_{(PV+BESS),(K+1)} \quad (4)$$

$$Q_K = Q_{(K+1)} + Q_{L,(K+1)} + X_{K,(K+1)}\left(\frac{\left(P_{(K+1)} + P_{L,(K+1)}\right)^2 + \left(Q_{(K+1)} + Q_{L,(K+1)}\right)^2}{\left|V_{(K+1)}\right|^2}\right) - Q_{(PV+BESS),(K+1)} \quad (5)$$

where, $Q_{(PV+BES),(K+1)}$ and $P_{(PV+BES),(K+1)}$ are the injection reactive and real power from BESS and PV units at bus $(\boldsymbol{K}+\mathbf{1})$, respectively.

The ratio of system losses with incorporating BESS and PV to the system losses without incorporating BESS and PV in RDS is formulated as single objective function as shown in Equation (6).

$$F_o = \sum\left(\frac{\sum_{h=1}^{24} P_{loss}(h)^{after\ (PV+BESS)}\Delta(h)}{\sum_{h=1}^{24} P_{loss}(h)^{before\ (PV+BESS)}\Delta(h)}\right),\quad h = 1, 2, 3, 4, \ldots\ldots.. 24\,h \quad (6)$$

where, $P_{loss}(t)^{before\ (PV+BES)}$ and $P_{loss}(h)^{after\ (PV+BESS)}$ are the system losses before and after incorporating BESS and PV in distribution system at time (h).

The inequality and equality constraints are formulated as shown next [33–36]:

2.1. Equality Constraints

These constraints include power flow balance equations. Therefore, the power generation from substation and PV with BESS should be equal to the system loss and system load demand as shown next.

$$P_{rf} + \sum_{g=1}^{G} P_{PV+BES}(g) = \sum_{j=1}^{m} P_{L,j} + \sum_{nb=1}^{NB} P_{loss}(nb) \quad (7)$$

$$Q_{rf} + \sum_{g=1}^{G} Q_{PV+BES}(g) = \sum_{j=1}^{m} Q_{L,j} + \sum_{nb=1}^{NB} Q_{loss}(nb) \quad (8)$$

where $\boldsymbol{NB}$ and m are the overall number of branches and buses, respectively. $Q_{loss}(nb)$ and $P_{loss}(nb)$ are the reactive and real system loss at branch (j), respectively. G are the overall number of PV with BESS. Q_{rf} and P_{rf} represents the reactive and active power drawn from substation in RDS, respectively.

2.2. Inequality Constraints

These constraints include system operating constraints such as system voltage limits, PV generation with BESS limits and branch current limits as follows:

2.2.1. Voltage Limits

The operating bus voltage should be between high (V_{up}) and low (V_{lo}) voltage limits as shown in Equation (9).

$$V_{lo} \leq V_j \leq V_{up} \quad (9)$$

where, V_j represent the voltage at bus j.

2.2.2. Sizing Limits of (PV + BESS)

$$\sum_{g=1}^{G} P_{PV+BES}(g) \leq \left(\sum_{j=1}^{m} P_{L,j} + \sum_{nb=1}^{NB} P_{loss}(nb) \right) \tag{10}$$

$$\sum_{m=1}^{G} Q_{PV-BES}(m) \leq \left(\sum_{j=1}^{m} Q_{L,j} + \sum_{nb=1}^{NB} Q_{loss}(nb) \right) \tag{11}$$

$$P_{PV,low} \leq P_{PV} \leq P_{PV,high} \tag{12}$$

where, $P_{PV,high}$ and $P_{PV,low}$ are the maximum and minimum power generation limits of PV.

2.2.3. Sizing Limits of Battery

$$E_{BESS,L} \leq E_{BESS,j}(h) \leq E_{BESS,H} \tag{13}$$

where, $E_{BESS,L}$ and $E_{BESS,H}$ are the low and high magnitudes of battery energy stored.

2.2.4. Line Constraints

The current should be lower than the maximum current ($I_{max,b}$) through the branch (b) [37].

$$I_b \leq I_{max,b} \; b = 1, 2, 3 \ldots, Nb \tag{14}$$

2.3. Modeling and Sizing of PV and BES

2.3.1. Load Modelling

The distribution network system studied in this paper have has various daily load demand configurations, such as residential load as shown if in Figure 2, industrial load as shown in Figure 3, and commercial load as indicated in Figure 4 [38,39]. All previous load demand patterns are based on the voltage and time with reactive and actual load voltage indexes. Time-varying load demands are modelled from Equations (15) and (16), as shown below [40]:

$$P_w(t) = P_{ow}(t) \times V_w^{N_p} \tag{15}$$

$$Q_w(t) = Q_{ow}(t) \times V_w^{N_q} \tag{16}$$

where Q_k and P_k represent the reactive and real power at bus k; Q_{ok} and P_{ok} are the reactive and real load at bus k. V_k represents the voltage at bus k, and N_q and N_p represent the reactive and real load voltage indices that are demonstrated in Table 1 [40].

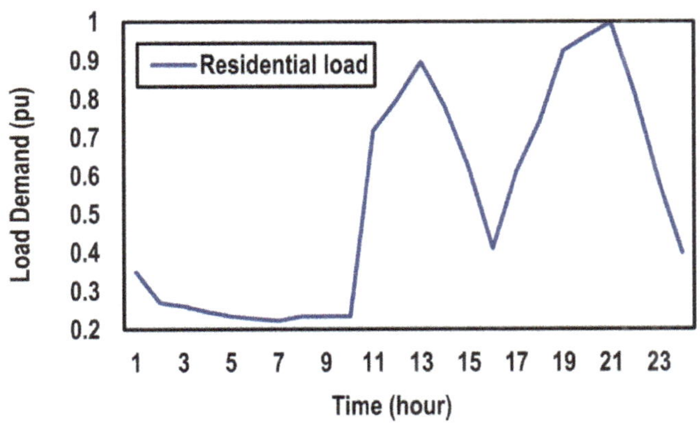

Figure 2. Normalized daily residential load demand curve.

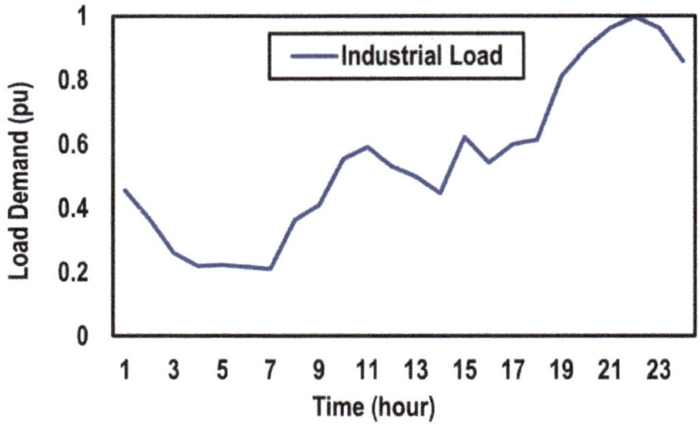

Figure 3. Normalized daily industrial load demand curve.

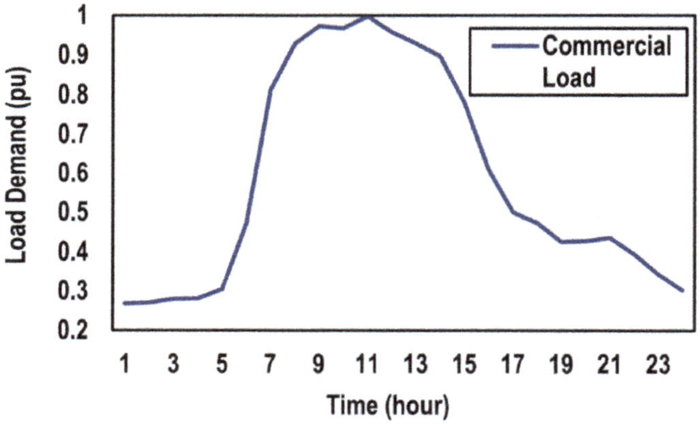

Figure 4. Normalized daily commercial load demand curve.

Table 1. The used parameters.

The Used Parameters	The Proposed Value
Number of search agents	20
Maximum iteration	2000
voltage limits	**0.9 pu** $\leq V_i \leq$ **1.05 pu**
Limits of active output generation from PV with BESS	**0.3 MW** $\leq P_{PV+BES,i} \leq$ **3 MW**
the real load voltage indices (N_p) for industrial, residential and commercial load models	0.18, 0.92 and 1.51, respectively
the reactive load voltage indices (N_q) for industrial, residential and commercial load models	6, 4.04 and 3.4, respectively

2.3.2. PV Modelling

The solar radiation probabilistic nature can be designated according to the probability density function (PDF) of Beta as follows [36,41]:

$$f_b(s) = \begin{cases} \frac{\Gamma(\alpha+\beta)}{\Gamma(\alpha)\,\Gamma(\beta)} s^{(\alpha-1)}(1-s)^{(\beta-1)} & 0 \le s \le 1,\ \alpha, \beta \ge 0 \\ 0 \ otherwise \end{cases} \tag{17}$$

where $f_b(s)$ refers to the s distribution function of Beta and s refers to the arbitrary variable of solar radiation in kilowatt per meter square; α and β refer to the parameters of $f_b(s)$ which are computed exploiting the standard deviation (σ) and mean deviation (μ) as shown in (18). The value of standard deviation and mean deviation are presented in [42].

$$\beta = (1-\mu)\left(\frac{\mu(1+\mu)}{\sigma^2} - 1\right);\ \alpha = \frac{\mu \times \beta}{1-\mu} \tag{18}$$

The PV module output power output depends on the solar radiation and surrounding temperature as well as the PV module characteristics itself. The maximum output power related to solar radiation s, $P_o(s)$, can be expressed as [19]:

$$P_o(s) = N \times FF \times V_y \times I_y \tag{19}$$

where,

$$FF = \frac{V_{MPP} \times I_{MPP}}{V_{oc} \times I_{sc}}; \tag{20}$$

$$V_y = V_{oc} - k_v \times T_{cy}; \tag{21}$$

$$I_y = s[I_{sc} + k_i \times (T_c - 25)]; \tag{22}$$

$$T_{cy} = T_A + s(\frac{N_{OT} - 20}{0.8}) \tag{23}$$

where, N refers to the module's number; T_{cy} and T_A refer to the cell temperature and ambient temperature (C^0), respectively; Ki and Kv refer to the coefficient of current temperature (A/C^0) and coefficient of voltage temperature (V/C^0), respectively; FF refers to fill factor; NOT refers to rated working temperature of cell per C^0; I_{sc} and refer to short circuit current (A) and open circuit voltage (V), respectively; ***VMPP*** and ***IMPP*** refer to voltage at maximum power point and current at maximum power point, respectively; $P_o(s)$ refers to the PV module maximum output power at solar radiation (s). The prospective output power at solar radiation (s) is computed according to Equation (10). Therefore, the overall prospective output during the identified interval time t, P_t (t = 1 h in this study) can be expressed as follows:

$$P_t = \int_0^1 P_o(s) f_b(s) ds \tag{24}$$

2.3.3. BESS Modelling

BESS is supposed to be linked to an alternating current (AC) system through bidirectional DC/AC converters [43]. In this paper, BESS works at unity power factor to discharge or charge active power. In another meaning, the BESS can work as a generator throughout the period of discharging and a load throughout the period of charging. The energy variation of BESS at bus k in time interval t is evaluated as the following [44]:

$$E_{BESSk}(t) = E_{BESSk}(t-1) - \frac{P^{disch}_{BESSk}(t)}{\eta_d}\Delta t,\ for\ P_{BESSk}(t) > 0 \tag{25}$$

$$E_{BESSk}(t) = E_{BESSk}(t-1) - \eta_c\, P^{ch}_{BESSk}(t)\Delta t, for\ P_{BESSk}(t) \le 0 \tag{26}$$

$$\eta_{BES} = \eta_{Ch} \times \eta_{Dch} \quad (27)$$

where E_{BESSk} refers to the overall energy stored inside the BESS; P^{disch}_{BESSk} and P^{ch}_{BESSk} refer to the BESS discharged and charged power, respectively; η_d and η_c refer to the BESS efficiency in case of discharging and charging, respectively; Δt indicates the duration of time interval t.

2.3.4. Sizing BES and PV

BESS is installed at the same location of PV in RDS. Therefore, the optimal sizing of BESS with PV are presented in [42]. Therefore, the charging and discharging energies of batteries at time (t) are calculated by Equations (28) and (29).

$$E^{DC}_{BESS,j} = \int_0^t P^{DC}_{BESS,j}(t)dt = \sum_{T=1}^{24} P^{DC}_{BESS,j}(t)\Delta t \quad (28)$$

$$E^{C}_{BESS,j} = \int_0^t P^{C}_{BESS,j}(t)dt = \sum_{t=1}^{24} P^{C}_{BESS,j}(t)\Delta t \quad (29)$$

$(E_{(PV+BES),j})$ is a combination energy of BESS and PV at bus (*j*) which is determined by Equation (30). PV energy is determined by Equation (31).

$$E_{(PV+BES),j} = E^{DS}_{PV,j} + E^{DC}_{BES,j} \quad (30)$$

$$E_{(PV),j} = E^{DS}_{PV,j} + E^{C}_{BES,j} \quad (31)$$

where $E^{DC}_{BESS,j}$ is the discharging energy of BESS to the distribution system (DS). $E^{DS}_{PV,j}$ is the injection power energy from PV to DS and $E^{C}_{BESS,j}$ the charging energy which is drawing from PV to BESS.

Round-trip efficiency can be determined by the ratio of discharging energy to the charging energy as shown below:

$$\eta_{BES} = \frac{E^{DC}_{BES,j}}{E^{C}_{BES,j}} \quad (32)$$

Consequently, PV energy is updated to Equation (33) as follows:

$$E_{PV,j} = \frac{E_{(PV+BESS),j} - (1-\eta_{BESS})E^{GR}_{PV,j}}{\eta_{BESS}} \quad (33)$$

$$P_{PV,j} = K^o_{PV}E_{PV,j} \quad (34)$$

$$K^o_{PV} = \frac{P^o_{PV}}{E^o_{PV}} \quad (35)$$

The high value of PV output during the day is evaluated by Equation (36).

$$P_{PV,j} = K^o_{PV}\left(\frac{E_{(PV+BESS),j} - (1-\eta_{BESS})E^{GR}_{PV,j}}{\eta_{BESS}}\right) \quad (36)$$

where, E^o_{PV} and P^o_{PV} are the energy and maximum output of PV during the day, respectively.

BESS sizing is determined by Equation (37).

$$E^{Ch}_{BES,j} = \frac{E_{(PV+BES),j} - E^{Gr}_{PV,j}}{\eta_{BES}} \quad (37)$$

3. Optimization Methodology

3.1. Frame Design

In general, the recommended Archimedes optimization algorithm describes what occurs when objects that have different volumes and weights are dipped into a liquid. The following subsections indicate how the AOA was based on the phenomena elucidated in Archimedes' principle. Then, we explain how this law of physics is applied along with an algorithm of optimization [45].

3.1.1. Principle of Archimedes

The Archimedes principle declares that when dipping an object partially or completely into a liquid, the liquid goes flat out at an upward force on this object equivalent to liquid's weight dislodged by this object. Figure 5 describes that when an object is dipped into a liquid, it will be exposed to an upward force, named buoyant force, equivalent to the weight of the liquid dislodged by this object [46].

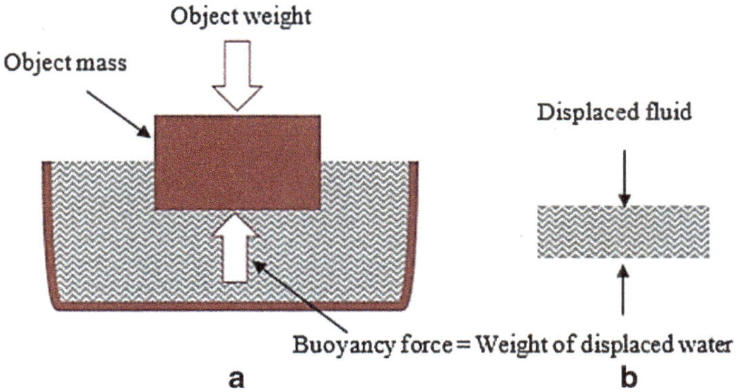

Figure 5. (**a**) An object is dipped into a liquid, and (**b**) the volume of liquid dislodged [45].

3.1.2. Theory

Figure 6 indicates when some objects dipped into the same liquid and every one attempts to achieve the state of equilibrium. The speed at which each immersed object reaches to the state of equilibrium varies due to its different density and volume. Any object will be in the state of equilibrium when the buoyant force F_b is equivalent to the weight (W_o) of this object:

$$F_b = W_o, \tag{38}$$

$$\rho_b v_b a_b = \rho_O v_O a_O \tag{39}$$

where ρ_b and ρ_o are the density of the liquid and the dipped object, respectively, v_b and v_o are the volume of the liquid and the dipped object, respectively, and a_o and a_b are the gravity or acceleration of the liquid and the dipped object, respectively. This previous equation can be reorganized as the following:

$$a_o = \frac{\rho_b v_b a_b}{\rho_o v_o} \tag{40}$$

Figure 6. Some objects dipped in the same liquid [45].

In the case of presence of another force acting on the object, such as colliding with another adjacent object (r), then the state of equilibrium will be:

$$F_b = W_o, \tag{41}$$

$$W_b - W_r = W_o, \tag{42}$$

$$\rho_b v_b a_b - \rho_r v_r a_r = \rho_0 v_0 a_0 \tag{43}$$

3.2. Archimedes Optimization Algorithm

AOA is an algorithm that depends on a population. In this algorithm, the individuals of the population are the dipped objects. Resembling other metaheuristic algorithms that depend on population, AOA likewise initiates search procedure with preliminary population of objects, called candidate solutions, with arbitrary densities, accelerations, and volumes. At this phase, every object is likewise started with its arbitrary situation in liquid. Afterward, assessing the fitness of preliminary population, AOA works in repetitions until the end limit is achieved. After each repetition, AOA modernizes the volume and density for each object. The object's acceleration is modernized based on the state of collision with any other nearby object. Modernizing, density, acceleration and volume define the object's new location. In the following sub-section, the mathematical expression steps for AOA are explained.

Steps of AOA Algorithm

The mathematical construction for the algorithm of AOA is presented in this sub-section. Theoretically, AOA is deemed as a universal algorithm where it involves both exploitation and exploration procedures. The pseudocode of the AOA is indicated in Algorithm 1; it includes the preliminary population of objects, population assessment, and modernizing parameters. Mathematically, the stages of the suggested AOA are indicated as follows:

1 **Preparation** Set the locations of overall objects using (44):

$$\boldsymbol{O_i = lb_i + rand\ (i, Dim) \times (ub_i - lb_i), i = 1{,}2, \ldots .., N} \tag{44}$$

where O_i refers to the ith object from the population that have N (search agents) objects. ub_i and lb_i are the higher and lower limits of the search scope, respectively. Dim represents the dimension variables.

Set the initial value of density (***den***) and volume (***vo***) for every ith object according to Equations (45) and (46).

$$\boldsymbol{den_i = rand(i, Dim)} \tag{45}$$

$$\boldsymbol{vol_i = rand(i, Dim)} \tag{46}$$

where rand refers to a random number within [0,1]. Finally, set the initial value of ith object acceleration (***acc***) using (47):

$$\boldsymbol{acc_i = lb_i + rand(i, Dim) \times (ub_i - lb_i)} \tag{47}$$

In this step, assess preliminary population and nominate the object that has the best fitness value. Specify, the best location (x_{best}), the best density (den_{best}), the best volume (v_{best})), and the best acceleration (acc_{best}).

2 **Modernize volumes and densities** The volume and the density for every object i at the repetition $(t + 1)$ is modernized according to (48) and (49):

$$vol_i^{t+1} = vol_i^t + rand \times (vol_{bset} - vol_i^t) \tag{48}$$

$$den_i^{t+1} = den_i^t + rand \times (den_{bset} - den_i^t) \tag{49}$$

where vol_{bset} is the volume correlated to the best object that has been obtained so far, and *rand* is a random number that is distributed uniformly.

Algorithm 1 AOA Pseudo code.

```
Procedure AOA (size of population N, maximum repetition t_max, C1, C2, C3, and C4)
    Preliminary objects population combined with random locations, volumes and densities
according to (44), (45), (46) and (47), respectively.
    Assess preliminary population and nominate one of them that has best fitness significance.
    Set repetition counter t = 1
    While t ≤ t_max do
        For every object I do
            Modernize volume and density for every object according to (49)
            Modernize the factors of transfer and decreasing of density TF and d according to
(50) and (51), respectively.
            If TF ≤ 0.5 then Exploration stage
                Modernize the object acceleration according to (52) and normalize this
acceleration according to (54)
                Modernize the object location according to (55)
            else Exploitation stage
                Modernize the object acceleration according to (53) and normalize this
acceleration according to (54)
                Modernize direction flag F according to (57)
                Modernize the object location according to (56)
            end if
        end for
        Assess every object and nominate one of them that has best fitness significance.
        Set t = t + 1
    End while
    return object that has best fitness significance
end Procedure
```

3 In the AOA algorithm, the population objects (search agents) are searching for the best promising area in all of the search space by the exploration phase and then searching for the best location (best object) in this promising area by the exploitation phase. ***TF*** is a factor that is changing with iteration to transfer the algorithm from the exploration phase to the exploitation phase through the simulation time, and can be evaluated as follows:

$$TF = exp(\frac{t - t_{max}}{t_{max}}) \tag{50}$$

where the ***TF*** factor rises progressively with increasing time till up to 1; t_{max} and t are the maximum repetitions number and repetition number, respectively. Likewise, density decreasing factor d also supports the proposed AOA on universal to local inspection. It reduces with increasing time according to (51):

$$d^{t+1} = exp(\frac{t_{max} - t}{t_{max}}) - (\frac{t}{t_{max}}) \tag{51}$$

where d^{t+1} reduces with increasing time that provides the capability to converge in the previously specified promising zone. To guarantee a balance between the exploration and the exploitation in the proposed AOA, this variable must be handled appropriately. The text continues here.

4 Exploration step (colliding among objects happens). If $TF \leq 0.5$, colliding among objects happens, an arbitrary material (mr) must be nominated and the acceleration of for repetition $t + 1$ according to (52) must be modernized:

$$acc_i^{t+1} = \frac{den_{mr} + vol_{mr} \times acc_{mr}}{den_i^{t+1} \times vol_i^{t+1}} \tag{52}$$

where acc_i, den_i, and vol_i are the acceleration, the density, and the volume of the object I, whereas acc_{mr}, vol_{mr} and den_{mr} are the acceleration, the volume, and the density of arbitrary material. It is significant to indicate that $TF \leq 0.5$ guarantees exploration through one third of repetitions. Using a value other than 0.5 will affect the behavior of changing from exploration to exploitation steps.

5 Exploitation step (no colliding among objects). If $TF > 0.5$, there is no colliding among objects, modernize the acceleration of the object for repetition $(t + 1)$ according to (53):

$$acc_i^{t+1} = \frac{den_{best} + vol_{best} \times acc_{best}}{den_i^{t+1} \times vol_i^{t+1}} \tag{53}$$

where acc_{best} refers to the best object acceleration.

6 Normalize the object acceleration. Normalize the object acceleration to compute the percentage of variation according to (54):

$$acc_{i-norm}^{t+1} = u \times \frac{acc_i^{t+1} - acc_{min}}{acc_{max} - acc_{min}} + l \tag{54}$$

where l and u represent the scope of normalization and put it at 0.1 and 0.9, respectively. The acc_{i-norm}^{t+1} calculates the percentage of the period that every agent will alteration. The value of acceleration will be great when the object is away from the global optimum, which means that the object will be in the exploration stage; other than that, it will be in the exploitation stage. This clarifies how the inspection modifies from the exploration stage to the exploitation stage. In an ordinary case, the factor of acceleration initiates with high value and reduces with increasing time. This aids search agents to move away from local solutions and at the same time transfer towards the global best solution. However, it is significant to state that there may still a small number of search agents that require extra time to stay in the exploration stage than in the normal case. Therefore, the proposed AOA attains the equilibrium between the exploration stage and the exploitation stage.

7 Modernize location If $TF \leq 0.5$ (exploration stage), the ith object's location for following repetition $t + 1$ according to (55)

$$x_i^{t+1} = x_i^t + C_1 \times rand \times acc_{i-norm}^{t+1} \times d \times (x_{rand} - x_i^{t+1}) \tag{55}$$

where C_1 referes to a constant that equals 2. Other than that, when $TF > 0.5$(exploitation stage), the objects modernize their locations according to (56).

$$x_i^{t+1} = x_{best}^t + F \times C_2 \times rand \times acc_{i-norm}^{t+1} \times d \times (T \times x_{best} - x_i^t) \tag{56}$$

where C_2 referes to a constant that equals 6. T rises with increasing time and it is proportional to transfer factor and it is determined according to $T = C_3 \times TF$. Additionally, it rises with increasing time through the scope $[C_3 \times 0.3, 1]$ and it possesses a particular percentage from the best location, at first. It begins with small percentage which causes a huge difference between the best location and the present location; consequently, the random walk step-size will be big. As the search continues, this percentage will rise progressively to reduce the difference between the best location and the present location. This results in an appropriate equilibrium between the exploration and the exploitation. F is the flag to vary the motion direction according to (57):

$$F = \begin{cases} +1 \; if \, P \leq 0.5 \\ -1 \; if \, P > 0.5 \end{cases} \tag{57}$$

where $\mathbf{P} = 2 \times rand - C_4$.

8 Assessment Assess every object exploiting function f and recollect the best solution found yet. Designate x_{best}, vol_{best}, den_{best}, and acc_{best}.

4. Simulation Results and Dissections

The IEEE 69-bus radial distribution system (RDS) includes 69 buses with a reactive load of 2694.6 kVAr and an active load of 3801.5 kW as shown in Figure 7 [47]. The results are obtained under base values of 12.66 kV and 10 MVA. The used parameters and the system constraints are given in Table 1. This paper studies the optimal allocation of PV alone or with BES in residential, industrial, and commercial system loads.

4.1. Residential Load

In this case, the overall reactive and real load demand during 24 h are 34.43 MVAr and 48.57 MW, respectively. Without integration PV and BES in RDS, the total reactive and real loss during 24 h are 0.85 MVAr and 1.87 MW, respectively. Installing one PV alone in RDS at bus 61 reduces the system loss to 1.39 MW. Additionally, installing two PV alone in RDS at buses 61 and 17 reduces the system loss to 1.35 MW. The total energies of one and two PV alone in RDS during the day are illustrated in Figures 8 and 9. Table 2 illustrates the locations and sizes of PV, the total energy of PV, and the injection energy from PV to the grid. Therefore, installing three PV alone reduces the system loss to 1.34 MW at buses 61, 18, and 11. From Figure 10, the total energy of three PV alone is 15.64 MWh.

From Table 3, simultaneous integration of PV and BES gives better results than integration of PV alone in RDS. Installing one BESS and PV in RDS decreases the system loss to 0.71 MW at bus 61. The energies of PV and BESS during the day by incorporating one PV with BESS in RDS are illustrated in Figures 11 and 12. Installing two and three PV with BES in RDS decrease the system loss to 0.61 MW and 0.59 MW, respectively. The energies of PV and BESS during the day by incorporating two PV with BESS in RDS are illustrated in Figures 13 and 14. Additionally, energies of PV and BESS during the day by incorporating three PV with BESS in RDS are illustrated in Figures 15 and 16. Table 3 illustrates the locations and sizes of PV and BESS, the total energy of PV, the injection energy from PV to the grid, the charging energy from PV to BESS, and the discharging energy from BESS to the grid.

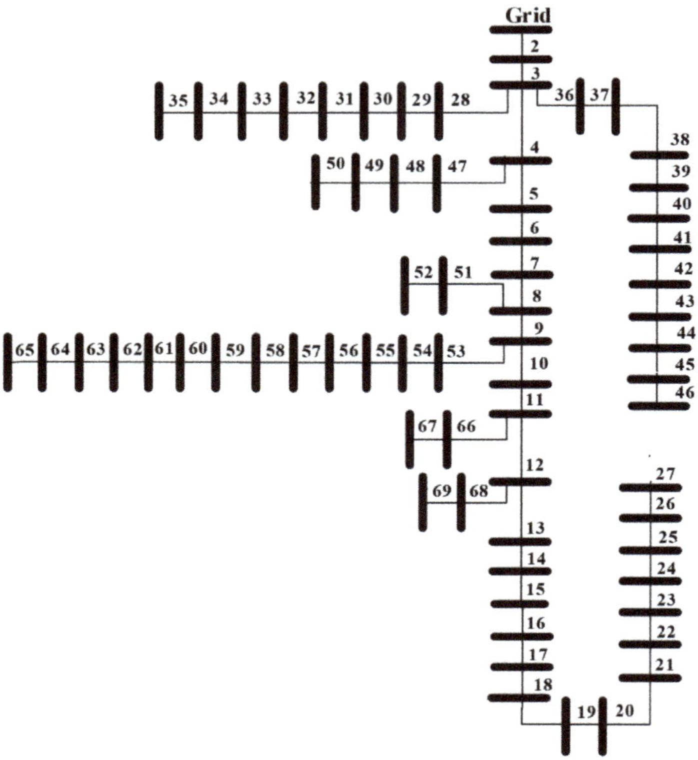

Figure 7. IEEE 69-bus RDS.

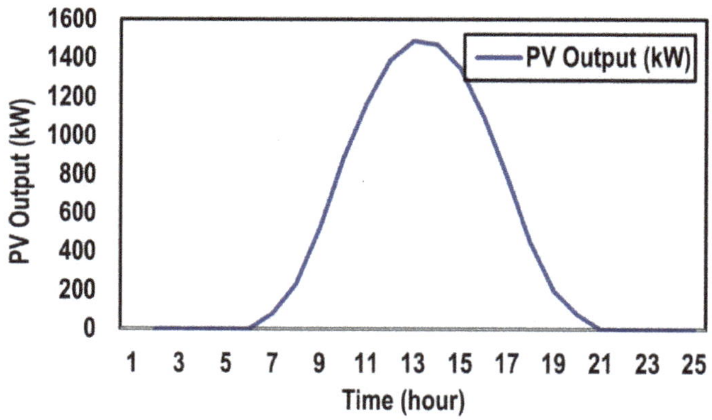

Figure 8. PV output during the day by installing one PV alone in residential system load.

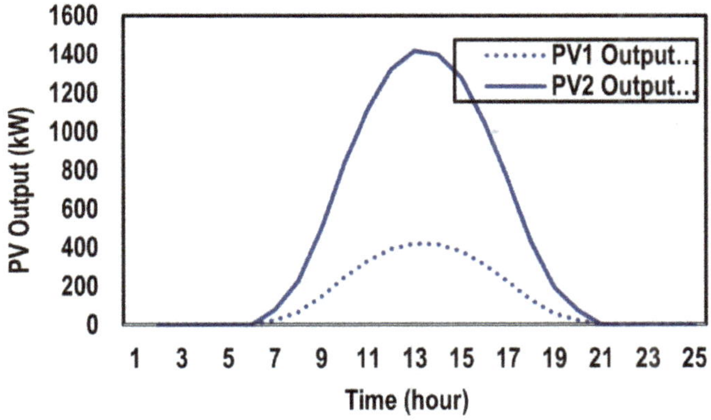

Figure 9. PV output during the day by installing two PV alone in residential system load.

Table 2. The obtained results with and without installing PV alone in residential system loads.

Item		Position (Size (kW))	PV Energy (kWh)	Total PV Energy (kWh)	P_{loss} (kW)
Residential Load	Without PV	-	-	-	1867.977
	1-PV	61 (1489)	61 (11,207)	11,207	1389.4
	2-PV	61 (1417.5) 17 (419.2)	61 (10,668) 17 (3155.3)	13,823.3	1349.2
	3-PV	61 (1369) 18 (302.8) 11 (406.1)	61 (10,304) 18 (2279) 11 (3056.6)	15,639.6	1341.6

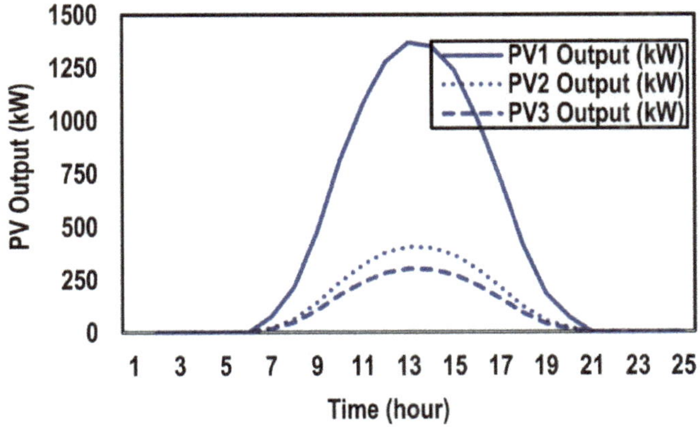

Figure 10. PV output during the day by installing three PV alone in residential system load.

Table 3. The obtained results with and without installing PV with BES in residential system loads.

Item			Position (Size (kW))	PV Energy (kWh)	$E_{PV\ to\ grid}$ (kWh)	Charging Energy (kWh)	Discharging Energy (kWh)	P_{loss} (kW)
Residential Load	Without PV and BES		-	-	-	-	-	1867.977
	1	PV	61 (3693.2)	27796	11666	-	-	711.9071
		BES	61 (2467.5)	-	-	16,130	12,358	
	2	PV	61 (3466.1) 17 (1163.1)	61 (26,088) 17 (7797.1)	61 (10,971) 17 (3279.6)	-	-	613.1804
		BES	61 (2323) 17 (693.992)	-	-	61 (15,116) 17 (4517.5)	61 (11,581) 17 (3460.9)	
	3	PV	61 (3345.6) 18 (745.66) 11 (1012.7)	61 (25,180) 18(5612.2) 11 (7622.3)	61 (10,589) 18 (2361.7) 11 (3202.5)	-	-	594.447
		BES	61 (2242.17) 18 (498.78) 11 (681.1)	-	-	61 (14,591) 18 (3250.5) 11 (4419.8)	61 (11,178) 18 (2490.2) 11 (3386)	

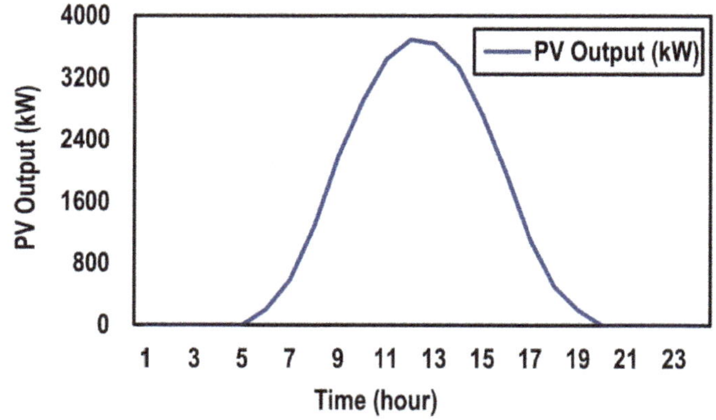

Figure 11. PV output during the day by installing one PV with BES in residential system load.

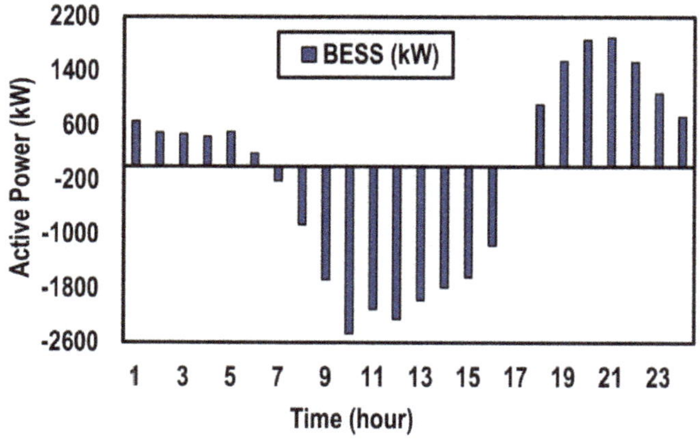

Figure 12. BES output during the day by installing one PV with BES in residential system load.

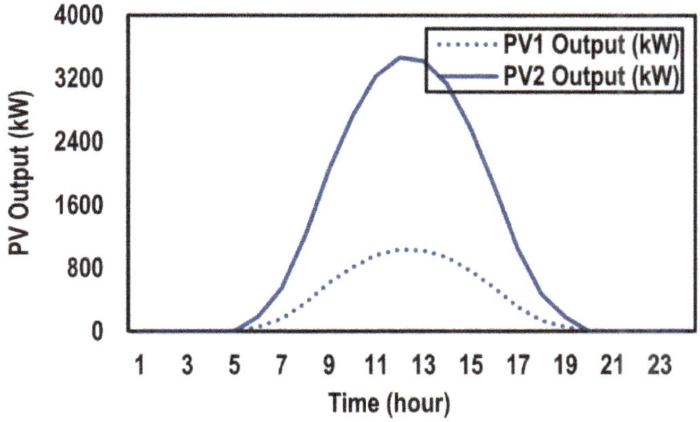

Figure 13. PV output during the day by installing two PV with BES in residential system load.

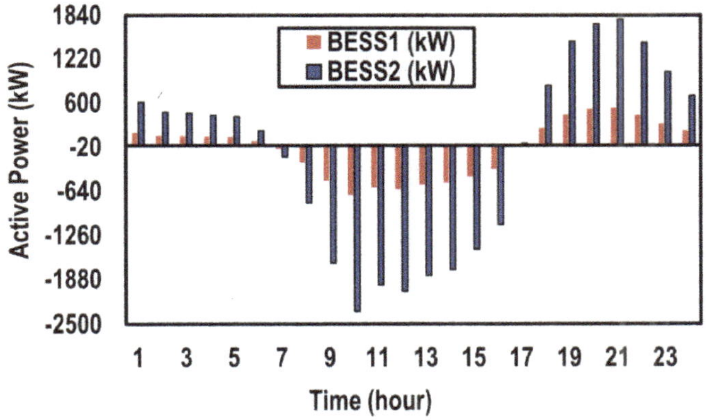

Figure 14. BES output during the day by installing two PV with BES in residential system load.

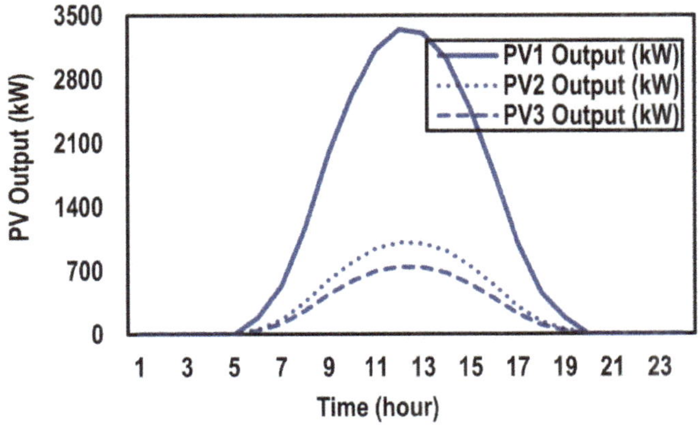

Figure 15. PV output during the day by installing three PV with BES in residential system load.

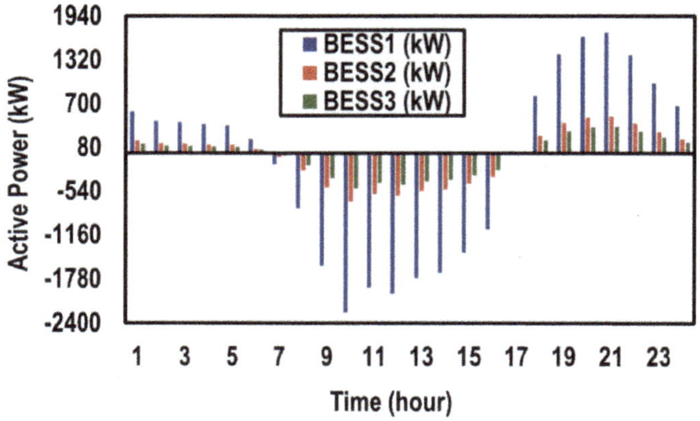

Figure 16. BES output during the day by installing three PV with BES in residential system load.

4.2. Industrial Load

In this case, the overall reactive and real load demand during 24 h are 35.64 MVAr and 50.29 MW, respectively. Without integrating PV and BES in RDS, the total active and reactive power loss during 24 h are 1.89 MW and 0.86 MVAr, respectively. The total system losses are decreased to 1.55 MW, 1.52 MW, and 1.52 MW by integrating one, two, and three PV alone in RDS, respectively. Table 4 presents the locations and sizes of PV, the total energy of PV, the injection energy from PV to the grid and the system power loss. From Figures 17 and 18, the total energies of one and two PV alone during the day are 9.56 MWh and 11.80 MWh, respectively. Additionally, the total energy of three PV alone during the day is 13.35 MWh as shown in Figure 19.

Table 4. The obtained results with and without installing PV alone in industrial system loads.

Item		Position (Size (kW))	PV Energy (kWh)	Total PV Energy (kWh)	P_{loss} (kW)
Industrial Load	Without PV	-	-	-	1890.1117
	1-PV	61 (1270.7)	61 (9563.6)	9563.6	1553.5
	2-PV	61 (1209.5) 17 (358.8)	61 (9102.9) 17 (2700.9)	11,803.8	1524.4
	3-PV	61 (1168) 18 (259.6) 11 (345.9)	61 (8790.6) 18 (1954.1) 11 (2603.7)	13,348.4	1518.93

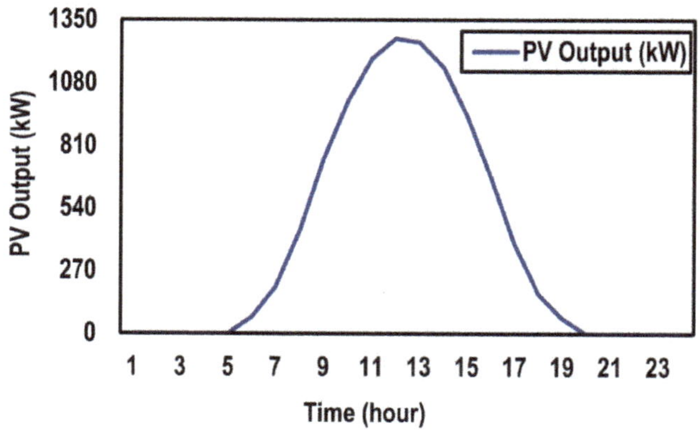

Figure 17. PV output during the day by installing one PV alone in industrial system load.

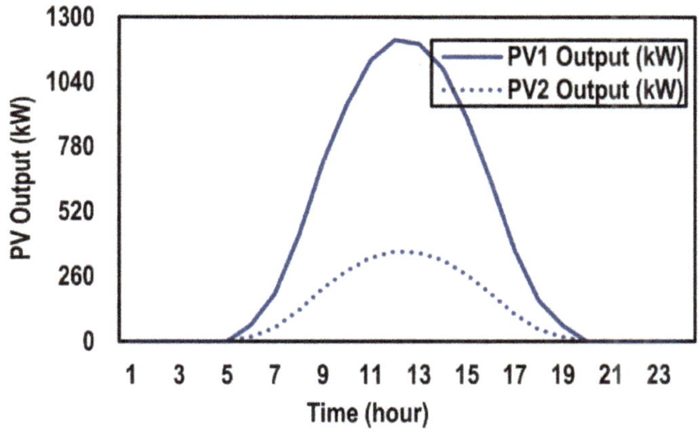

Figure 18. PV output during the day by installing two PV alone in industrial system load.

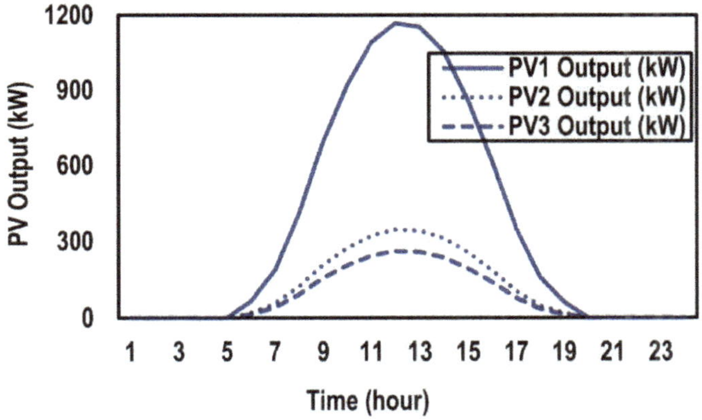

Figure 19. PV output during the day by installing three PV alone in industrial system load.

The optimal allocation of one PV with BES in RDS at bus 61 decreases the system loss to 0.72 MW. The total energies of PV and BESS during the day by incorporating one PV with BESS are presented in Figures 20 and 21. The system losses are decreased to 0.62 MW and 0.60 MW by integrating two and three PV with BES in RDS, respectively. Figures 22 and 23 illustrate the energies of PV and BESS during the day by installing two PV with BESS in RDS. Additionally, Figures 24 and 25 illustrate the energies of PV and BESS during the day by installing three PV with BESS in RDS. The total injection energies from PV to BESS and to the grid are presented in Table 5. Additionally, the charging and discharging energies of BESS are presented in Table 5 and Figure 25.

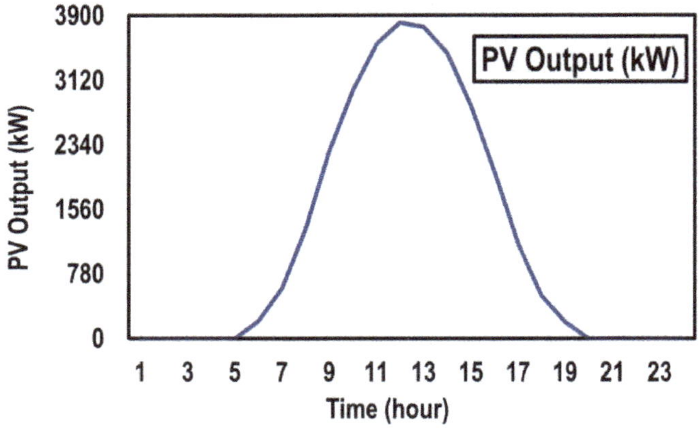

Figure 20. PV output during the day by installing one PV with BES in industrial system load.

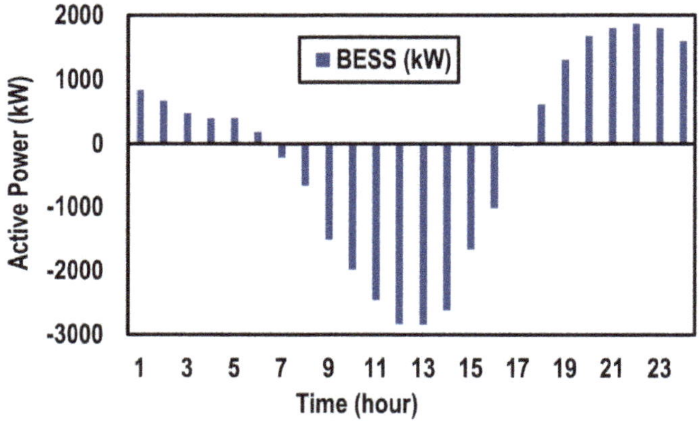

Figure 21. BES output during the day by installing one PV with BES in industrial system load.

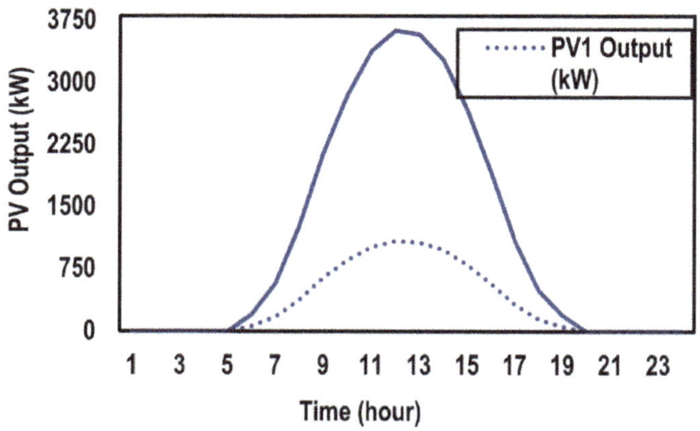

Figure 22. PV output during the day by installing two PV with BES in industrial system load.

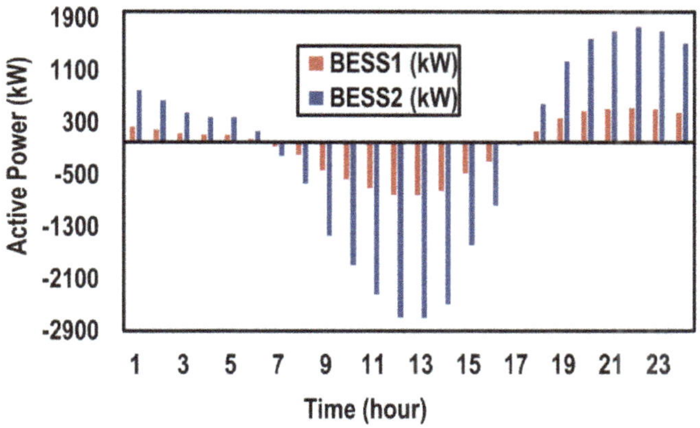

Figure 23. BES output during the day by installing two PV with BES in industrial system load.

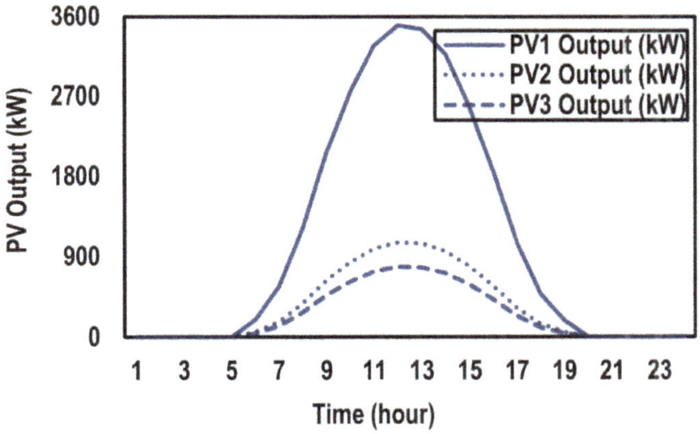

Figure 24. PV output during the day by installing two PV with BES in industrial system load.

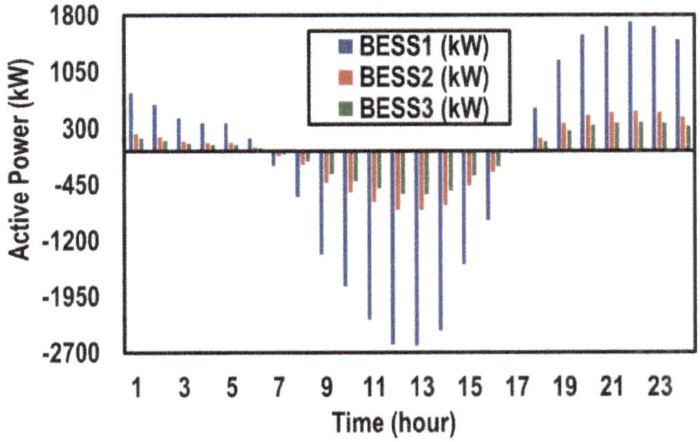

Figure 25. BES output during the day by installing three PV with BES in industrial system load.

Table 5. The obtained results with and without installing PV with BES in industrial system loads.

Item			Position (Size (kW))	PV Energy (kWh)	$E_{PV\ to\ grid}$ (kWh)	Charging Energy (kWh)	Discharging Energy (kWh)	P_{loss} (kW)
Industrial Load	Without PV and BES		-	-	-	-	-	1890.112
	1	PV	61 (3812.4)	28,694	10,722	-	-	720.7217
		BES	61 (2841.7)	-	-	17,972	13,807	
	2	PV	61 (3627.4) 17 (1084.2)	61 (27,302) 17 (8160)	61 (10,203) 17 (3051.9)	-	-	622.0804
		BES	61 (2703.8) 17 (807.842)	-	-	61 (17,099) 17 (5108.2)	61 (13,137) 17 (3924.5)	
	3	PV	61 (3501.1) 18 (780.37) 11 (1060)	61 (26,351) 18 (5873.4) 11 (7978)	61 (9848.4) 18 (2203.4) 11 (2964.7)	-	-	603.1228
		BES	61 (2609.5) 18 (580.93) 11 (790.95)	-	-	61 (16,502) 17 (3670.1) 11 (5013.3)	61 (12,678) 17 (2819.6) 11 (3851.6)	

4.3. Commercial Load

In this case, the overall reactive and real load demand during 24 h are 37.82 MVAr and 53.35 MW, respectively. Without integrating BESS and PV in RDS, the overall reactive and real loss during 24 h are 0.99 MVAr and 2.17 MW, respectively. The system power loss is reduced to 1.12 MW with installing one PV alone at bus 61. The optimal placement and sizing of two PV alone at buses 61 and 17 decreases the system loss to 1.04 MW as shown in Table 6. Additionally, the optimal sizing of three PV alone at buses 61, 18, and 11 with total energy of 22.81 MWh decreases the system loss to 1.02 MW. From Figures 26 and 27, the total energies of one and two PV and three PV alone during the day are presented in Figure 26, Figure 27, and Figure 28, respectively.

Table 6. The obtained results with and without installing PV alone in commercial system loads.

Item		Position (Size (kW))	PV Energy (kWh)	Total PV Energy (kWh)	P_{loss} (kW)
Commercial Load	Without PV	-	-	-	2173.851
	1-PV	61 (2168.2)	61 (16,319)	16,319	1124.5
	2-PV	61 (2063.1) 17 (611.6)	61 (15,527) 17 (4603)	20,130	1038.2
	3-PV	61 (1991.5) 18 (439.4) 11 (599.7)	61 (14,989) 18 (3306.8) 11 (4513.5)	22,809.30	1021.7

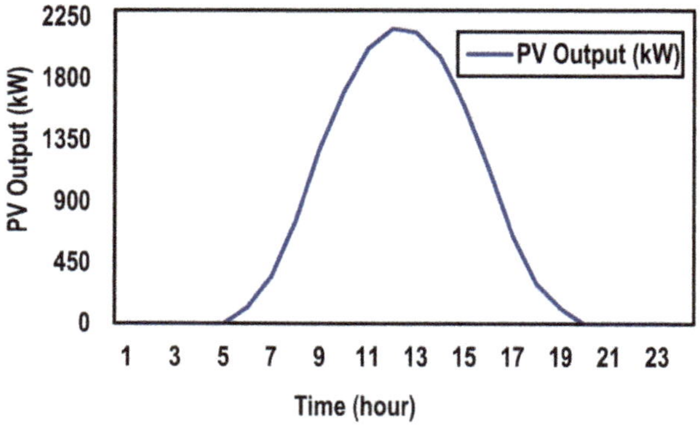

Figure 26. PV output during the day by installing one PV alone in commercial system load.

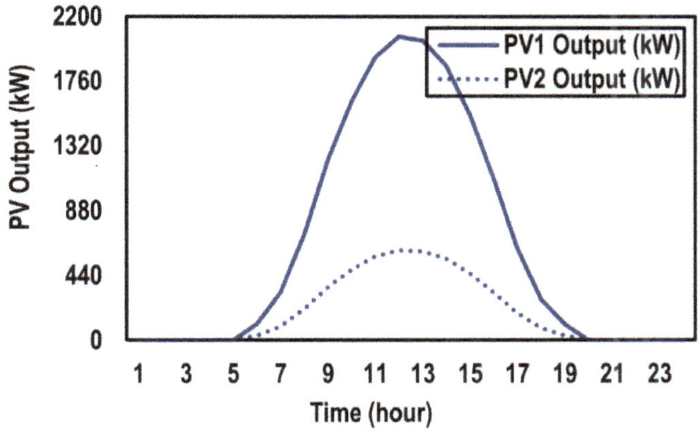

Figure 27. PV output during the day by installing two PV alone in commercial system load.

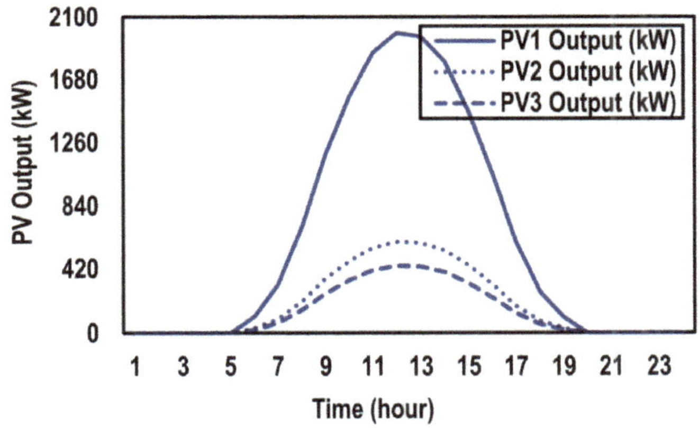

Figure 28. PV output during the day by installing three PV alone in commercial system load.

Installing one, two, and three PV with BES decreases the system losses to 0.83 MW, 0.71 MW, and 0.69 MW, respectively, as shown in Table 7. The total energy of PV and the charging and discharging energies of BES by integrating one PV with BES in RDS are illustrated in Figures 29 and 30. Figures 31 and 32 show the energies of two PV and the charging and discharging energies of BESS during the day, respectively. By incorporating three PV with BESS, the injection energies from PV to BESS and the grid during the day are shown in Figure 33, and the charging and discharging energies of BESS are shown in Figure 34. The results proved which the presented algorithm is an efficient to obtain the best global results when compared with modified HGSO algorithm and HGSO algorithm. This comparative study is illustrated in Table 8.

Table 7. The obtained results with and without installing PV with BES in commercial system loads.

Item			Position (Size (kW))	PV Energy (kWh)	$E_{PV\ to\ grid}$ (kWh)	Charging Energy (kWh)	Discharging Energy (kWh)	P_{loss} (kW)
Commercial Load	Without PV and BES		-	-	-	-	-	2173.851
	1	PV	61 (3832.2)	28,843	17,155	-	-	825.1585
		BES	61 (2064.9)	-	-	11,688	8936.1	
	2	PV	61 (3644.3) 17 (1089.2)	61 (27,429) 17 (8197.6)	61 (16,253) 17 (4853.2)	-	-	709.9147
		BES	61 (1945.8) 17 (582.223)	-	-	61 (11,175) 17 (3344.4)	61 (8544.5) 17 (2557.1)	
	3	PV	61 (3517.3) 18 (783.5) 11 (1065.6)	61 (26473) 18 (5897.1) 11 (8020.3)	61 (15685) 18 (3483.5) 11 (4777.6)	-	-	688.1289
		BES	61 (1878.5) 18 (420.22) 11 (564.685)	-	-	61 (10788) 18 (2413.6) 11 (3242.7)	61 (8248.5) 18 (1845.4) 11 (2479.4)	

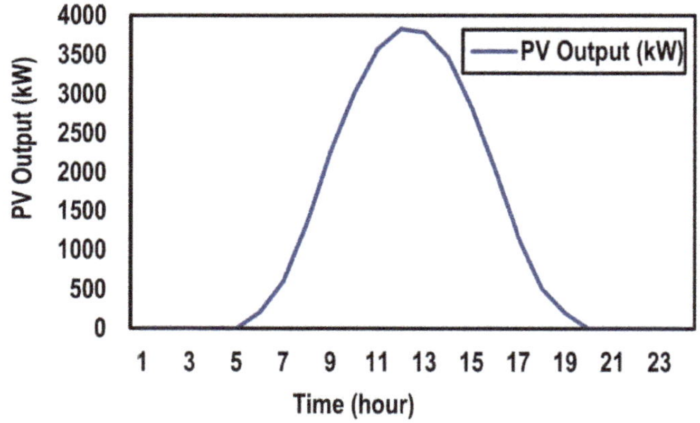

Figure 29. PV output during the day by installing one PV with BES in commercial system load.

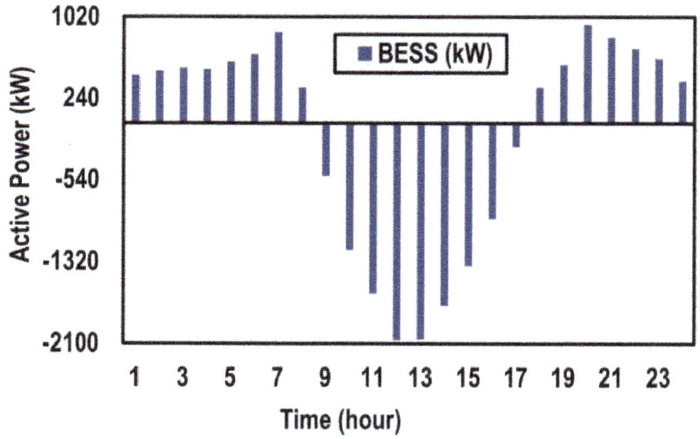

Figure 30. BES output during the day by installing one PV with BES in commercial system load.

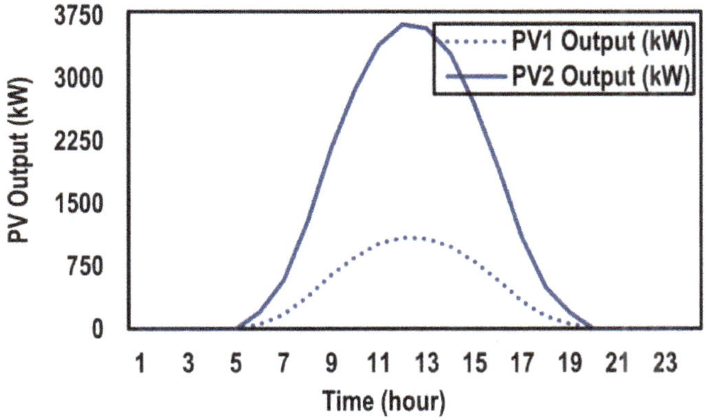

Figure 31. PV output during the day by installing two PV with BES in commercial system load.

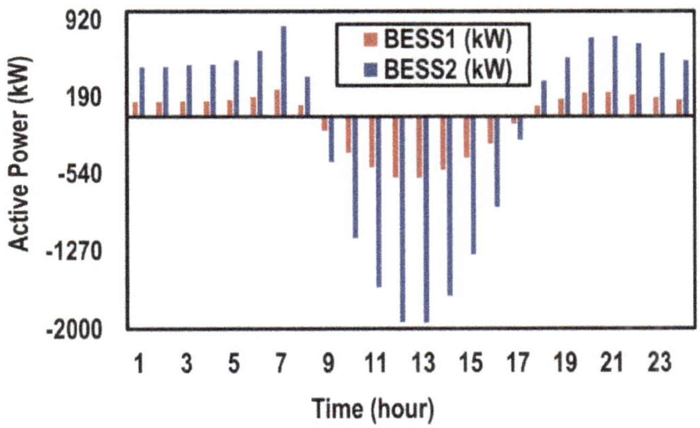

Figure 32. BES output during the day by installing two PV with BES in commercial system load.

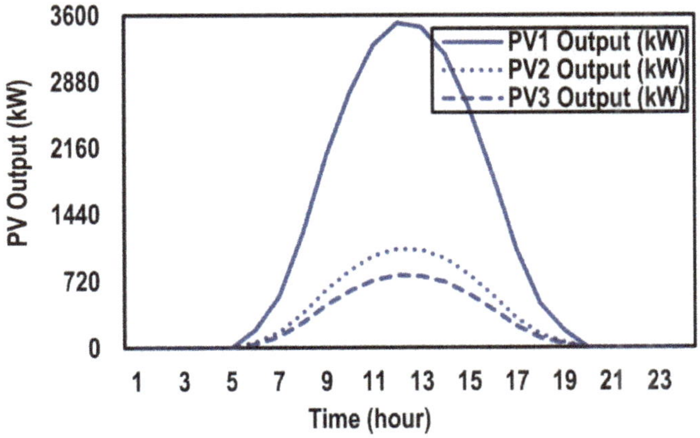

Figure 33. PV output during the day by installing three PV with BES in commercial system load.

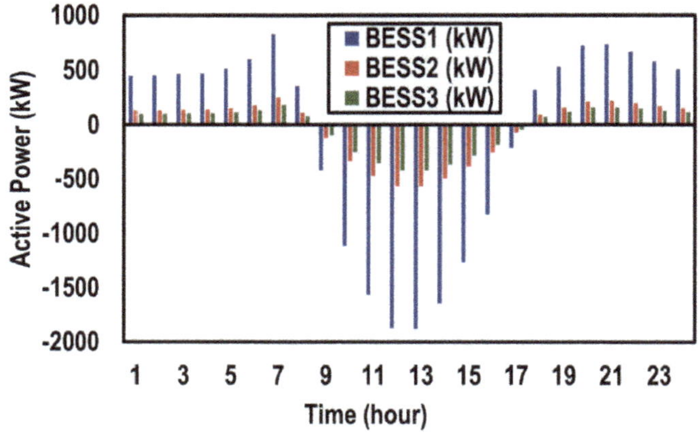

Figure 34. BES output during the day by installing three PV with BES in commercial system load.

Table 8. Comparison results between AOA, Modified HGSO, and HGSO algorithms in commercial system load.

Item	AOA	Modified HGSO [36]	HGSO [36]
P_{loss} (kW) Without PV and BES	2173.851	2173.851	2173.851
Location (PV size (kW))	61 (3517.3) 18 (783.5) 11 (1065.6)	61 (3517.488) 18 (784.1074) 11 (1064.323)	61 (3187.526) 18 (860.6001) 11 (934.0223)
Location (BES size (kW))	61 (1878.5) 18 (420.22) 11 (564.685)	61 (1878.5) 18 (420.843) 11 (563.386)	61 (1911.138) 18 (486.4727) 11 (595.803)
P_{loss} (kW)	688.1289	688.129	716.809

5. Conclusions

In this paper, an application for a recent optimization algorithm called the Archimedes optimization algorithm (AOA) has been proposed for reducing energy losses and to capture the size of incorporating battery energy storage system (BESS) and photovoltaic (PV) energy system in RDS. In this paper, all non-dispatchable PV energy systems have been transformed into a dispatchable energy resource with BESS integration with PV. AOA has been evolved for sizing several PV and BESS considering the changing demand over time and the probability generation. The proposed algorithm has been applied on the IEEE 69-bus distribution network with various daily demand configurations such as residential, industrial, and commercial loads demand. The obtained results demonstrate that the model can boost high penetration of the PV energy system accompanied with effective usage of BESS energy resources, which shows the strength of the presented algorithm for evaluating the best sizing of numerous PV and BESS with a significant reduction in energy losses. In addition, the AOA gives better results compared with other well-known optimization algorithms.

Author Contributions: Conceptualization, H.A.-M. and S.K.; data curation, M.T.-V. and E.E.E.; formal analysis, M.M.H.; methodology, M.T.-V. and E.E.E.; software, H.A.-M. and S.K.; supervision, M.M.H. and E.E.E.; validation, H.A.-M. and S.K.; visualization, M.T.-V.; writing—original draft, H.A.-M. and S.K.; writing—review and editing, M.T.-V., M.M.H. and E.E.E. All authors together organized and refined the manuscript in the present form. All authors have read and agreed to the published version of the manuscript.

Funding: This work was supported by Taif University Researchers Supporting Project number (TURSP-2020/86): Taif University, Taif, Saudi Arabia.

Institutional Review Board Statement: Not applicable.

Informed Consent Statement: Not applicable.

Data Availability Statement: Not applicable.

Conflicts of Interest: The authors declare no conflict of interest.

References

1. Omran, W.A.; Kazerani, M.; Salama, M.M.A. Investigation of methods for reduction of power fluctuations generated from large grid-connected photovoltaic systems. *IEEE Trans. Energy Convers.* **2011**, *26*, 318–327. [CrossRef]
2. Dincer, F. The analysis on photovoltaic electricity generation status, potential and policies of the leading countries in solar energy. *Renew. Sustain. Energy Rev.* **2011**, *15*, 713–720. [CrossRef]
3. Liu, F.; Li, R.; Li, Y.; Yan, R.; Saha, T. Takagi–Sugeno fuzzy model-based approach considering multiple weather factors for the photovoltaic power short-term forecasting. *IET Renew. Power Gener.* **2017**, *11*, 1281–1287. [CrossRef]
4. Rizwan, M.; Mujtaba, G.; Memon, S.A.; Lee, K.; Rashid, N. Exploring the potential of microalgae for new biotechnology applications and beyond: A review. *Renew. Sustain. Energy Rev.* **2018**, *92*, 394–404. [CrossRef]
5. Yang, M.; Huang, X. Ultra-short-term prediction of photovoltaic power based on periodic extraction of PV energy and LSH algorithm. *IEEE Access* **2018**, *6*, 51200–51205. [CrossRef]
6. Dawoud, S.M.; Lin, X.; Okba, M.I. Hybrid renewable microgrid optimization techniques: A review. *Renew. Sustain. Energy Rev.* **2018**, *82*, 2039–2052. [CrossRef]
7. Zia, M.F.; Elbouchikhi, E.; Benbouzid, M. Microgrids energy management systems: A critical review on methods, solutions, and prospects. *Appl. Energy* **2018**, *222*, 1033–1055. [CrossRef]
8. Siddaiah, R.; Saini, R. A review on planning, configurations, modeling and optimization techniques of hybrid renewable energy systems for off grid applications. *Renew. Sustain. Energy Rev.* **2016**, *58*, 376–396. [CrossRef]
9. Niknam, T.; Taheri, S.I.; Aghaei, J.; Tabatabaei, S.; Nayeripour, M. A modified honey bee mating optimization algorithm for multiobjective placement of renewable energy resources. *Appl. Energy* **2011**, *88*, 4817–4830. [CrossRef]
10. Taher, N. A new HBMO algorithm for multiobjective daily Volt/Var control in distribution systems considering distributed generators. *Appl. Energy* **2011**, *88*, 778–788.
11. Bakos, G.C. Distributed power generation: A case study of small scale PV power plant in Greece. *Appl. Energy* **2009**, *86*, 1757–1766. [CrossRef]
12. Wang, C.; Nehrir, M.H. Analytical approaches for optimal placement of distributed generation sources in power systems. *IEEE Trans. Power Syst.* **2004**, *19*, 2068–2076. [CrossRef]
13. Acharya, N.; Mahat, P.; Mithulananthan, N. An analytical approach for DG allocation in primary distribution network. *Int. J. Electr. Power Energy Syst.* **2006**, *28*, 669–678. [CrossRef]
14. Kasturi, K.; Nayak, M.R. Optimal Planning of Charging Station for EVs with PV-BES Unit in Distribution System Using WOA. In Proceedings of the 2017 IEEE 2nd International Conference on Man and Machine Interfacing (MAMI), Bhubaneswar, India, 21–23 December 2017; pp. 1–6.
15. Dixit, M.; Kundu, P.; Jariwala, H.R. Optimal placement of photo-voltaic array and electric vehicles in distribution system under load uncertainty. In Proceedings of the 2017 IEEE Power & Energy Society General Meeting, Chicago, IL, USA, 16–20 July 2017; pp. 1–5.
16. Yeh, H.G.; Gayme, D.F.; Low, S.H. Adaptive VAR control for distribution circuits with photovoltaic generators. *IEEE Trans. Power Syst.* **2012**, *27*, 1656–1663. [CrossRef]
17. Turitsyn, K.; Sulc, P.; Backhaus, S.; Chertkov, M. Options for control of reactive power by distributed photovoltaic generators. *Proc. IEEE* **2011**, *99*, 1063–1073. [CrossRef]
18. Sugihara, H.; Yokoyama, K.; Saeki, O.; Tsuji, K.; Funaki, T. Economic and efficient voltage management using customer-owned energy storage systems in a distribution network with high penetration of photovoltaic systems. *IEEE Trans. Power Syst.* **2013**, *28*, 102–111. [CrossRef]
19. Teng, J.H.; Luan, S.W.; Lee, D.J.; Huang, Y.Q. Optimal charging/discharging scheduling of battery storage systems for distribution systems interconnected with sizeable PV generation systems. *IEEE Trans. Power Syst.* **2013**, *28*, 1425–1433. [CrossRef]
20. Teleke, S.; Baran, M.E.; Bhattacharya, S.; Huang, A.Q. Rule-based control of battery energy storage for dispatching intermittent renewable sources. *IEEE Trans. Sustain. Energy* **2010**, *1*, 117–124. [CrossRef]
21. Hill, C.A.; Such, M.C.; Dongmei, C.; Gonzalez, J.; Grady, W.M. Battery energy storage for enabling integration of distributed solar power generation. *IEEE Trans. Smart Grid* **2012**, *3*, 850–857. [CrossRef]
22. Borowy, B.S.; Salameh, Z.M. Methodology for optimally sizing the combination of a battery bank and PV array in a wind/PV hybrid system. *IEEE Trans. Energy Convers.* **1996**, *11*, 367–375. [CrossRef]
23. Ekren, O.; Ekren, B.Y. Size optimization of a PV/wind hybrid energy conversion system with battery storage using response surface methodology. *Appl. Energy* **2008**, *85*, 1086–1101. [CrossRef]
24. Avril, S.; Arnaud, G.; Florentin, A.; Vinard, M. Multi-objective optimization of batteries and hydrogen storage technologies for remote photovoltaic systems. *Energy* **2010**, *35*, 5300–5308. [CrossRef]

25. Castillo-Cagigal, M.; Gutiérrez, A.; Monasterio-Huelin, F.; Caamaño-Martín, E.; Masa, D.; Jiménez-Leube, J. A semi-distributed electric demand-side management system with PV generation for self-consumption enhancement. *Energy Convers. Manag.* **2011**, *52*, 2659–2666. [CrossRef]
26. Wei-Fu, S.; Shyh-Jier, H.; Chin, E.L. Economic analysis for demand-side hybrid photovoltaic and battery energy storage system. *IEEE Trans. Ind. Appl.* **2001**, *37*, 171–177. [CrossRef]
27. Nottrott, A.; Kleissl, J.; Washom, B. Energy dispatch schedule optimization and cost benefit analysis for grid-connected, photovoltaic-battery storage systems. *Renew. Energy* **2013**, *55*, 230–240. [CrossRef]
28. Yu, R.; Kleissl, J.; Martinez, S. Storage size determination for grid-connected photovoltaic systems. *IEEE Trans. Sustain. Energy* **2013**, *4*, 68–81.
29. Tant, J.; Geth, F.; Six, D.; Tant, P.; Driesen, J. Multiobjective battery storage to improve PV integration in residential distribution grids. *IEEE Sustain. Energy* **2013**, *4*, 182–191. [CrossRef]
30. Alam, M.J.E.; Muttaqi, K.M.; Sutanto, D. Mitigation of rooftop solar PV impacts and evening peak support by managing available capacity of distributed energy storage systems. *IEEE Trans. Power Syst.* **2013**, *28*, 3874–3884. [CrossRef]
31. Daud, M.Z.; Mohamed, A.; Hannan, M.A. An improved control method of battery energy storage system for hourly dispatch of photovoltaic power sources. *Energy Convers. Manag.* **2013**, *73*, 256–270. [CrossRef]
32. Eminoglu, U.; Hocaoglu, M.H. Distribution systems forward/backward sweep-based power flow algorithms: A review and comparison study. *Electr. Power Compon. Syst.* **2008**, *37*, 91–110. [CrossRef]
33. Abdel-mawgoud, H.; Kamel, S.; Ebeed, M.; Youssef, A.-R. Optimal allocation of renewable dg sources in distribution networks considering load growth. In Proceedings of the 2017 Nineteenth International Middle East Power Systems Conference (MEPCON), Cairo, Egypt, 19–21 December 2017; pp. 1236–1241.
34. El-Fergany, A. Optimal allocation of multi-type distributed generators using backtracking search optimization algorithm. *Int. J. Electr. Power Energy Syst.* **2015**, *64*, 1197–1205. [CrossRef]
35. Ali, E.; Elazim, S.A.; Abdelaziz, A. Ant lion optimization algorithm for renewable distributed generations. *Energy* **2016**, *116*, 445–458. [CrossRef]
36. Abdel-Mawgoud, H.; Kamel, S.; Khasanov, M.; Khurshaid, T. A strategy for PV and BESS allocation considering uncertainty based on a modified Henry gas solubility optimizer. *Electr. Power Syst. Res.* **2021**, *191*, 106886. [CrossRef]
37. Aman, M.; Jasmon, G.; Bakar, A.; Mokhlis, H. A new approach for optimum simultaneous multi-DG distributed generation Units placement and sizing based on maximization of system loadability using HPSO (hybrid particle swarm optimization) algorithm. *Energy* **2014**, *66*, 202–215. [CrossRef]
38. Lopez, E.; Opazo, H.; Garcia, L.; Bastard, P. Online reconfiguration considering variability demand: Applications to real networks. *IEEE Trans. Power Syst.* **2004**, *19*, 549–553. [CrossRef]
39. Hung, D.Q.; Mithulananthan, N.; Lee, K.Y. Determining PV penetration for distribution systems with time-varying load models. *IEEE Trans. Power Syst.* **2014**, *29*, 3048–3057. [CrossRef]
40. Price, W.; Casper, S.G.; Nwankpa, C.O.; Bradish, R.W.; Chiang, H.-D.; Concordia, C.; Staron, J.V.; Taylor, C.W.; Vaahedi, E. Bibliography on load models for power flow and dynamic performance simulation. *IEEE Power Eng. Rev.* **1995**, *15*, 70.
41. Salameh, Z.M.; Borowy, B.S.; Amin, A.R. Photovoltaic module-site matching based on the capacity factors. *IEEE Trans. Energy Convers.* **1995**, *10*, 326–332. [CrossRef]
42. Hung, D.Q.; Mithulananthan, N.; Bansal, R. Integration of PV and BES units in commercial distribution systems considering energy loss and voltage stability. *Appl. Energy* **2014**, *113*, 1162–1170. [CrossRef]
43. Gabash, A.; Pu, L. Active-reactive optimal power flow in distribution networks with embedded generation and battery storage. *IEEE Trans. Power Syst.* **2012**, *27*, 2026–2035. [CrossRef]
44. Chen, S.; Gooi, H.B.; Wang, M. Sizing of energy storage for microgrids. *IEEE Trans. Smart Grid* **2011**, *3*, 142–151. [CrossRef]
45. Hashim, F.A.; Hussain, K.; Houssein, E.H.; Mabrouk, M.S.; Al-Atabany, W. Archimedes optimization algorithm: A new metaheuristic algorithm for solving optimization problems. *Appl. Intell.* **2020**, *51*, 1531–1551. [CrossRef]
46. Rorres, C. Completing book ii of archimedes's on floating bodies. *Math. Intell.* **2004**, *26*, 32–42. [CrossRef]
47. Savier, J.; Das, D. Impact of network reconfiguration on loss allocation of radial distribution systems. *IEEE Trans. Power Deliv.* **2007**, *22*, 2473–2480. [CrossRef]

applied sciences

MDPI

Article

A Novel Model for Enhancing the Resilience of Smart MicroGrids' Critical Infrastructures with Multi-Criteria Decision Techniques

Abdulaziz Almaleh [1,†], David Tipper [2,†], Saad F. Al-Gahtani [3,†] and Ragab El-Sehiemy [4,*,†]

1 Information Systems Department, King Khalid University, Abha 62529, Saudi Arabia
2 Department of Informatics and Networked Systems, School of Computing and Information, University of Pittsburgh, Pittsburgh, PA 15213, USA
3 Electrical Engineering Department, King Khalid University, Abha 61411, Saudi Arabia
4 Electrical Engineering Department, Faculty of Engineering, Kafrelsheikh University, Kafrelsheikh 33516, Egypt
* Correspondence: elsehiemy@eng.kfs.edu.eg; Tel.: +20-10-0128-1051
† These authors contributed equally to this work.

Featured Application: This paper is concerned with a multi-criteria technical and economical allocation procedure of microgrids in smart cities.

Abstract: Microgrids have the potential to provide reliable electricity to key components of a smart city's critical infrastructure after a disaster, hence boosting the microgrid power system's resilience. Policymakers and electrical grid operators are increasingly concerned about the appropriate configuration and location of microgrids to sustain post-disaster critical infrastructure operations in smart cities. In this context, this paper presents a novel method for the microgrid allocation problem that considers several technical and economic infrastructure factors such as critical infrastructure components, geospatial positioning of infrastructures, power requirements, and microgrid cost. In particular, the geographic allocation of a microgrid is presented as an optimization problem to optimize a weighted combination of the relative importance of nodes across all key infrastructures and the associated costs. Furthermore, the simulation results of the formulated optimization problem are compared with a modified version of the heuristic method based on the critical node identification of an interdependent infrastructure for positioning microgrids in terms of the resilience of multiple smart critical infrastructures. Numerical results using infrastructure in the city of Pittsburgh in the USA are given as a practical case study to illustrate the methodology and trade-offs. The proposed method provides an effective method for localizing renewable energy resources based on infrastructural requirements.

Keywords: microgrid architecture; smart infrastructure; resilience; optimization model; critical node identification

Citation: Almaleh, A.; Tipper, D.; Al-Gahtani, S.; El-Sehiemy, R. A Novel Model for Enhancing the Resilience of Smart MicroGrids' Critical Infrastructures with Multi-Criteria Decision Techniques. *Appl. Sci.* **2022**, *12*, 9756. https://doi.org/10.3390/app12199756

Academic Editor: Luis Hernández-Callejo

Received: 21 August 2022
Accepted: 26 September 2022
Published: 28 September 2022

Publisher's Note: MDPI stays neutral with regard to jurisdictional claims in published maps and institutional affiliations.

Copyright: © 2022 by the authors. Licensee MDPI, Basel, Switzerland. This article is an open access article distributed under the terms and conditions of the Creative Commons Attribution (CC BY) license (https://creativecommons.org/licenses/by/4.0/).

1. Introduction

The portion of the world population inhabiting an urban environment has grown in the last decade from about 33% to 55% [1]. That growth has produced enormous demand and stress on the infrastructures and systems that deliver essential city services, resulting in significant interest in developing smart cities. The main purpose of smart city schemes is to create intelligent infrastructures for cities by harnessing innovations in cyber-physical systems, data science, and information and communication technology (ICT). Moreover, smart infrastructures are more dependent on both ICT and electrical power for proper operation. This increased dependence can introduce new vulnerabilities and lower infrastructure resilience [2]. In particular, severe weather (e.g., snow/ice storms, typhoons, tornadoes, drought-induced wildfires, etc.) is a growing vulnerability concern as the frequency, intensity, and geographic scope of severe weather events are predicted to increase with climate change [3]. Currently, severe weather events [4] are the number one reason for power outages in the United States, which are in turn, the top reason for outages in ICT services. For that, intelligent infrastructures provide more consistent and reliable system performance, new features/functions, and increased sustainability.

Industrial scale microgrids are basically self-supporting power systems typically in the 1.5–5 MW range. They have been advocated as a mechanism to improve the availability of power to significant societal and business facilities such as hospitals, military bases, and factories. Additionally, microgrids are promoted as an approach to incrementally incorporate shared renewable power production, such as wind and solar, into the bulk power grid in the case of disaster [5]. Mircrogrids are also proposed in the literature as a solution to achieve climate adaptation and mitigation goals [6].

Figure 1 gives a typical microgrid architecture. As shown in Figure 1, the building blocks of the microgrid are the controller, electrical power switches, local energy supplies (e.g., solar cells, wind turbines, and diesel generators), energy storage, and various loads. Microgrids are designed to operate in standalone mode and joined mode to the primary grid. In the joined mode scenario, the microgrid serves as additional energy back up to the bulk power system, decreasing peak loads, enhancing power stability, and reducing harmful emissions [7]. In island mode, the microgrid disconnects from the bulk grid and functions as a standalone power supply. The microgrid controller manages the transitions between modes seeking to maintain voltage and synchronization while minimizing load dropping and disruption.

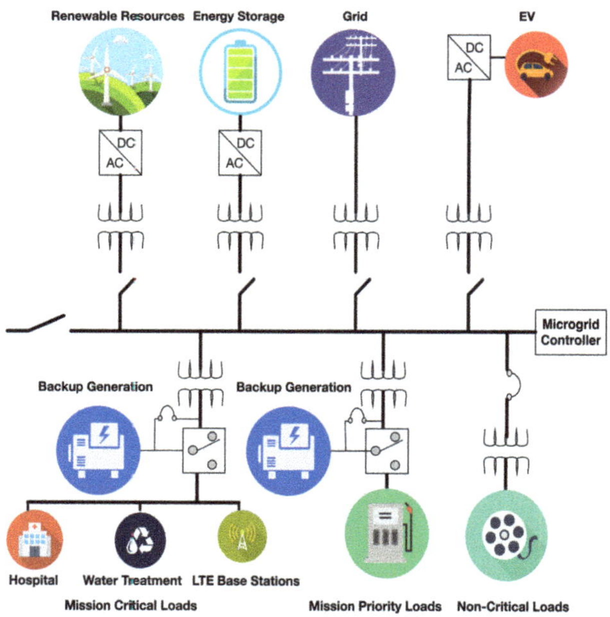

Figure 1. Sample microgrid architecture.

Since the available power is limited in island mode, the power loads are grouped into their significance categories: mission critical, mission priority, and non-critical. Mission critical loads are given the highest priority and consists of the essential components of the critical infrastructures of interest (e.g., hospitals, water treatment plants, and cellular network base stations). Mission priority loads are given second priority and would include loads that are important to society but not essential to the functioning of smart critical infrastructures (e.g., drug store, gas station, etc.). Lastly, non-critical loads would include residential and non-essential businesses (e.g., movie theaters).

Generally, microgrids have fixed geographical limits and are planned in island mode to produce power sufficient to sustain mission-critical loads inside the geographic boundaries. Thus, based on the power accessible in island mode, the microgrid controller may implement load shedding, dropping non-critical loads and a portion of the mission priority loads [5]. Furthermore, microgrids are required to possess the capability to shift efficiently from island mode to grid-connected operation, providing re-synchronization with minimal consequence to significant loads through the transition phases.

Unlike the nanogrids used in residential settings, a significant obstacle to the implementation of industrial size microgrids is the high cost in constructing, operating, and maintaining a microgrid. Currently, industrial capacity microgrids are mostly owned by an individual private organization. Recognizing the non-linear economic costs of implementing microgrids [8], the authors in [9] advocated for shared mid-size microgrids, with the expense borne by both the vital infrastructure proprietors with critical loads (e.g., cellular network operators and hospitals) and the government organizations that will utilize the infrastructures during disaster recovery. This shared community use approach could be facilitated by government-sponsored financing and tax credits. Furthermore, it may justify either re-insurance or bonding mechanisms to help in reducing the cost. Here, the goal of this work is to design and place microgrids based on minimizing the overall expenses and to ensure power flow connected over the most vital critical infrastructure parts. Therefore, the proposed design is able to achieve significant techno-economical merits for microgrids.

Related work on microgrids includes using microgrids to improve radial distribution power grid restoration after a natural disaster [5,10] and dynamically forming local microgrids around distributed generation sources after a disaster [11]. Kelly-Pitou [6] introduced the notion of employing a microgrid for the purpose of enhancing both power resilience and for alleviating climate change impacts. Nevertheless, this early work did not propose a method for determining the location of microgrids to improve the resilience of different critical infrastructures viewed as a group. In [12], the authors suggested using an algorithm to achieve multi-agent resource allocation in distributed scenarios through a shared microgrid, including residential and commercial buildings, with the least amount of information exchange between the users. However, their study lacks the commercial and residential segments by not assigning the priorities to critical nodes in the network.

The bulk of the research literature on critical infrastructures within smart-city schemes has focused on optimizing performance and providing new functionalities. Previous work covering smart-city resilience has focused specifically on developing frameworks [13] or "solidifying" critical infrastructures. Traditionally, policy-makers mandated or supported hardening techniques such as constructing flood barriers and rebuilding levees according to the probability of 1 in 100-year situations. However, when considering smart critical infrastructures and the increasing weather variability, planners need to move beyond physical hardening techniques, adopting new preparedness methods and policies that acknowledge the dependence on both power and ICT.

In [9], a structure for providing power and ICT is developed to enhance smart city critical infrastructure services in post-disaster conditions. The proposed method in this work employs multi-user microgrids to generate electricity concurrently with cellular-based communications, which are dynamically re-adjusted into a mesh network along with local edge computing to control/operate smart critical infrastructures. The main aim of the framework is to construct districts within a mid size zone that act as secure area, including essential critical infrastructure functions operating at limited but acceptable levels. Guaranteeing that the combination of microgrids, cellular-based catastrophe recovery mesh network, and edge computing are geographically located in this "socially planned" fashion will assist in reducing at-risk districts and bolster the economic argument for microgrids.

Related work on enhancing critical infrastructure resilience has focused on hardening [14,15] the essential elements in every infrastructure. Various techniques have been introduced in the literature for discovering the most vital components in a critical infrastructure, such as graph theoretic analysis, simulation-based analysis, stochastic modeling, and expert judgment [14]. Graph centrality measures have been utilized in developing critical infrastructure protection strategies, including vulnerability studies of power systems. A heuristic approach introduced in [16] uses five graph centrality measures, namely degree centrality, between centrality, centered centrality, eccentricity centrality, and radially. The authors evaluated the nodes with each of the five centrality measures. If a node is highly ranked by at least two measures, it is considered a critical node. The model was applied to assess the impact of potential attacks on the Swiss power network. However, the author utilized the proposed model considering the impact on the power network's infrastructure, neglecting the dependency on other infrastructures that could potentially have a higher effect. In [17], the authors introduced a scheme to determine the critical nodes of a smart power network according to the highest power flow in the system. However, their investigation concentrated on the betweenness centrality as the primary graph measure, which could lead to a bias in the outcome by neglecting other essential centrality measures in networks, such as node degrees and closeness. The authors of [18] considered the centrality metrics of a dependency risk graph, exploring the connection between dependency risk paths and graph centrality. They mapped different critical infrastructures into one graph and applied the centrality metrics on each node, assuming that the links are represented by the escalating failure values between nodes from different infrastructures. The primary motivation for that was to identify the critical infrastructure nodes between interdependent critical infrastructures that noticeably impact the essential routes of risk in the network and then to analyze cascading failures to different nodes or links in the network.

In [19], the authors proposed a model to identify the most critical nodes in interdependent critical infrastructures. They developed an integer linear programming optimization formulation that models the approach of an attacker who targets a collection of nodes with the intention of compromising or damaging them. They assumed that the attacker is motivated by three objectives: (i) minimizing the size of the largest connected component, (ii) maximizing the number of disconnected components, and (iii) minimizing the cost of an attack. All three objectives are based on graph theory metrics and can be used to determine where to hardened the infrastructure. Here, the problem of where to harden multiple infrastructures as a group using a microgrid is considered.

Noting that microgrids are the most costly element in our framework, our focus is on where to place a microgrid in order to promote smart critical infrastructure operations post-disaster. An holistic approach is adopted for the microgrid location problem, considering multiple critical infrastructures at once, and focuses on factors such as component importance within a critical infrastructure, the geospatial placement of infrastructures, power requirements, and microgrid cost. Optimization problems are formulated to determine the location of a microgrid in a geographic space that optimizes a weighted combination of the relative importance of nodes across all critical infrastructures and the cost. Furthermore, a simple heuristic method for positioning microgrids is presented and demonstrated. This method is compared with the optimization problem. Numerical results using Pittsburgh as a case study are given to illustrate the effectiveness of the methodology and its trade-offs.

This paper is structured as follows. Section 2 presents the proposed methodology, which determines the placement of microgrids. In Section 3, the numerical results and a discussion of implementing the proposed method are presented. Section 4 provides the findings of the study and future work.

2. Materials and Methods

2.1. Microgrid Location Methodology

Consider a neighborhood or section of a city where multiple infrastructures geographically overlaid co-exist, as illustrated in Figure 2. For instance, the three infrastructures shown could include an intelligent water network, a natural gas pipeline system, and a healthcare system consisting of a hospital and community health clinics. The proposed approach to determining the location of a microgrid essentially has four steps as follows:

1. First, each smart critical infrastructure is analyzed individually to determine the relative importance of each node/component of the infrastructure.
2. Second, a geographic grid of the city is defined, and the infrastructures are aligned spatially.
3. An optimization problem is formulated to determine the place in a geographic grid that optimizes a combination of the importance of nodes/components across infrastructures and the cost of the microgrid.
4. Following the three-step procedure, the analysis of critical infrastructures is considered to define the comparable significance of elements.

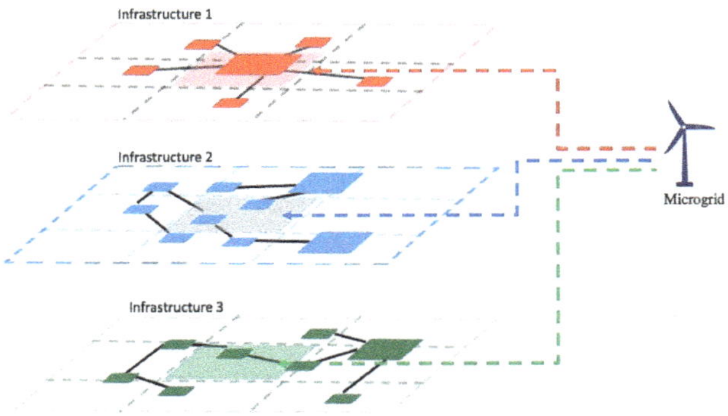

Figure 2. Geographic overlaid infrastructures connected to a microgrid.

2.2. Critical Infrastructure Analysis

Critical infrastructures are grouped into two classes according to the applicability of modeling the infrastructure with a network graph: (1) interconnected infrastructures and (2) standalone infrastructures.

2.2.1. Interconnected Infrastructure

Network science methods based on graph theory have been applied to analyze critical infrastructures that include the interconnected elements or operations, such as power grids [20], transportation networks [21], water systems [22], and optical backbone communication networks [23]. Interconnected infrastructures are modeled with a graph $G = (V, E)$, where V is the set of vertices or network nodes (e.g., power plants and substations, pumps and pipe junctions, and optical switches), and E is the set of edges or links or connections (e.g., power lines, water pipes, and optical fibers) joining the nodes. Given a graph model of an interconnected infrastructure, one can use network science methods in part to determine the relative importance of nodes in the infrastructure based on graph metrics. This paper adopted centrality metrics similar to [16], as explained in the following.

Degree Centrality

Node degree denotes the total number of neighbor nodes to which a node is immediately attached. The degree of centrality is a primary graph analysis measure that can be calculated as the total of edges connected to node v, which is called degree $deg(v)$, and identified as centrality degree $C_d(v)$, which is provided as follows:

$$C_d(v) = deg(v) \tag{1}$$

Betweenness Centrality

Betweenness centrality is a benchmark of centrality measure used in graph networks that reflects the shortest paths within couples of vertices s and t. It can be calculated as the percentage of shortest paths that cross through a vertex $v \in V(G)$. Hence, $C_b(v)$ can be written as follows:

$$C_b(v) = \sum_{s \neq v \neq t} \frac{\sigma_{st(v)}}{\sigma_{st}}, \tag{2}$$

where σ_{st} signifies the number of shortest routes from node s to node t, and $\sigma_{st(v)}$ equals the total number of these routes that cross within v.

Closeness Centrality

In any connected graph, the normalized closeness centrality $C_c(v)$ of a particular node $v \in V(G)$ equals the average distance of the shortest routes connecting node v to all additional nodes in the graph. The closeness can be defined as follows:

$$C_c(v) = \frac{1}{\sum_s dist(s,v)}, \tag{3}$$

In the above equation $dist(s,v)$ denotes the length connecting vertices s and v.

Power Requirements

The power $C_p(v)$ required for vertex $v \in V(G)$ is one of the factors considered and is assumed to be given.

Total Weighted Value (TWV_{II})

Combining the three centrality metrics forward beside the power requirements, we define a total weight value for interconnected infrastructure $TWV_{II}(v)$ for each node $v \in V(G)$. In calculating TWV, feature scaling [24] is used to bring all values into a common range by applying the *z-score* normalization approach based on the mean and standard deviation of each node metric across all nodes v [25], as shown for metric $C_d(v)$:

$$C_{d_{norm}} = \frac{C_d(v) - mean(C_d(v))}{standard\ deviation(C_d(v)} \tag{4}$$

We apply the normalization method to each metric, resulting in the following:

$$TWV_{II}(v) = C_{d_{norm}}(v) + C_{b_{norm}}(v) + C_{c_{norm}}(v) + C_{p_{norm}}(v) \quad . \tag{5}$$

2.2.2. Standalone Infrastructure

Many critical infrastructures are not typically presented as a graph (e.g., factories and healthcare). Instead, as standalone infrastructures, we utilize a weighted mixture of m context-dependent factors to discover the related importance of infrastructure elements. For instance, the characteristic factors of one healthcare element, such as a hospital, are the capacity of that hospital in terms of power requirements, bed number, and the population inhabited around the hospital.

Four critical infrastructures have been measured in this paper, as follows:

Healthcare

A healthcare infrastructure analysis has shown that hospitals are the most critical components. Therefore, ensuring that hospitals are running is crucial to disaster response and resilience. Accordingly, three parameters were employed to describe hospitals: (1) power consumption, (2) capacity, and (3) population surrounding a hospital area. Hospital size or capacity is typically determined through a specific quantity of beds that can be obtained from publicly available information. Employing the same hospital capacity further points to a helpful parameter called energy consumption over applying the formula conducted by Schneider Electric [26], as

$$P_H = BD * \overline{U} \tag{6}$$

This equation represents the hospital energy use as P_H, BD denotes the number of operated beds in the same hospital, and $\overline{U}$ signifies the bed's power use in kWh (assumed as 30,000 kWh/year in [26]). The population surrounding the hospital is another parameter assigned to measure the importance of healthcare nodes. It can be assigned using the density data from [27] and a one km radius circular space throughout that node (hospital).

Water System

Through natural disasters, ensuring a reserve of drinking water is crucial. Therefore, water treatment plants have been selected to be critical in this model to ensure the supply of such infrastructures. Their size can also be characterized by millions of gallons per day (MGD). Thus, the power consumption P_W was also estimated utilizing the following formula:

$$P_W = G * J \tag{7}$$

G in this equation reflects the portion of water processed in the MGD unit, and J represents the power required to process a million gallons [28], where the average water planet use around 1470 kWh/MG.

Cellular Network

The communication infrastructure is considered one of the most vital infrastructures to a society remaining functional. Due to the ubiquity of cell phones, cellular base stations are considered significant elements in the communication infrastructure. Furthermore, the authors of [9] discussed how the cellular network could be reconfigured to be used to provide disaster recovery communications. This study identified the most critical cellular network base stations by employing three factors: (1) geographic coverage, (2) population covered, and (3) power requirements. The geographic coverage for each base station was classified into short, medium, and long in terms of distance in miles.

Emergency Shelter

Through natural disasters, it is critical to provide emergency shelters with power. Governments commonly utilize event centers as a shelter when emergencies occur. For instance, the George R. Brown Convention Center in Houston, TX, covered thousands of people throughout hurricane Harvey [29]. Three inputs were selected to classify emergency shelters: energy usage, size, and capacity. Both data regarding the capacity and size of emergency shelters are openly accessible online. Nevertheless, power consumption was determined by calculating the size in ft^2 and then by multiplying it by the power average required for universal non-residential property space, which is 14 kWh per ft^2 [30].

Total Weighted Value (TWV_{SI})

The total weighted value TWV_{SI} related to every node v in a standalone infrastructure can be defined as follows:

$$TWV_{SI}(v) = \sum_{n=1}^{m} Parm_n(v) \tag{8}$$

where $Parm_n$ is the context-dependent factor for that infrastructure. As in the interconnected infrastructure case, feature scaling is used to put the parameters on the same scale.

2.3. Optimization Model Formulation

Consider a set of L smart infrastructures in an area such as a city. Each infrastructure include N^l nodes with $\{l = 1, 2, \ldots L\}$. The geographic space is divided into K zones denoted by Geo_i with $\{i = 1, 2, \ldots K\}$. The size and configuration of the geographic zones can be based on various properties of the region investigated (e.g., political boundaries, squares, etc.). Note that the geographical range of microgrid elements are normally limited to within 10 km^2 [31]. Here, for simplicity, the geographic space is spilt into K equal size squares, as shown in Figure 2.

F_i denotes the price of building and running a microgrid in Geo_i. Determining the total project cost F_i will be based on many determinants, such as the volume of power produced, the combination of power sources (i.e., fuel cells, wind turbines, solar cells, and batteries), the value of property, financing, and the construction and maintenance cost. Here, we assume that F_i is pre-computed by employing a relevant model [8], including the scaling toward the range of TWV. Let the decision variable x_i denote a binary variable equivalent to one if area Geo_i is selected as the location for a microgrid. We define S_i as the power production of a microgrid at position i. Let PC_{iv^l} denote the price regarding energy provided through a microgrid at position i passed to node v^l of the lth infrastructure. The decision variable y_{iv^l} implies the portion of power node v^l of infrastructure l obtained from a microgrid at location i. Given the notation above, the microgrid location problem is to be expressed as an optimization problem $P1$ as follows:

$$P1: \ Min \ \alpha \sum_{i=1}^{K} \sum_{l=1}^{L} \sum_{v^l \in Geo_i} -C_p(v^l) y_{iv^l} PC_{iv^l} TWV(v^l) + \beta \sum_{i=1}^{K} F_i * x_i \tag{9}$$

$$\sum_{l=1}^{L} \sum_{v^l \in Geo_i} C_p(v^l) y_{iv^l} \leq S_i * x_i \qquad \forall\, i \tag{10}$$

$$\sum_{v^l \in Geo_i} C_p(v^l) y_{iv^l} \leq \eta^l S_i * x_i \qquad \forall\, l \tag{11}$$

$$\sum_{i=1}^{K} x_i = 1 \tag{12}$$

$$0 \leq y_{iv^l} \leq 1 \quad \forall i,\, v^l; \qquad x_i \in \{0,1\}, \tag{13}$$

The objective (9) is to find the minimum cost location for the microgrid while powering the most critical infrastructure nodes. The objective function represents, in the first term, the expense of transferring power to a node v^l from a microgrid placed in Geo_i weighted by the importance $TWV(v^l)$ of the node. The $TWV(v^l)$ values are conducted using Equations (5) or (8) depending on the type of infrastructure. The second term in the objective function is the total microgrid project cost if installed in Geo_i. Finally, α and β in the objective function are weights that can be customized to trade-off infrastructure importance versus cost of building and operating. The first constraint ensures that the power transferred to the infrastructure loads is less than or equal to the production of the microgrid if it is located in Geo_i. The second constraint seeks to enforce the community/shared nature of the microgrid by ensuring that a single infrastructure can receive a maximum of η^l percent of the power produced by the microgrid in Geo_i. The η^l values are assumed and could reflect the infrastructure's financial contribution of infrastructure l to the cost of constructing and operating the microgrid. The constraint in (12) guarantees that only a single microgrid is built, and the constraints in (19) ensure the boundaries of the decision variables.

Note that several alternate formulations and extensions to the optimization model are possible. For example, one can relax the constraint that a microgrid in Geo_i can only power infrastructure nodes v^l in Geo_i. Instead, it assumes that the potential location of a microgrid is the center of each Geo_i and definesd the distance from a microgrid placed in Geo_i to node v^l as d_{iv^l}. Furthermore, d_{max} is defined as the maximum distance that a node can be located from a microgrid. In this case, the microgrid placement problem can be formulated as problem *P*2.

$$P2: \; Min \; \alpha \sum_{i=1}^{K} \sum_{l=1}^{L} \sum_{v^l=1}^{N^l} -C_p(v^l) y_{iv^l} PC_{iv^l} TWV(v^l) + \beta \sum_{i=1}^{K} F_i * x_i \tag{14}$$

$$\sum_{l=1}^{L} \sum_{v^l=1}^{N^l} C_p(v^l) y_{iv^l} \leq S_i * x_i \qquad \forall\, i \tag{15}$$

$$\sum_{v^l=1}^{N^l} C_p(v^l) y_{iv^l} \leq \eta^l S_i * x_i \qquad \forall\, l \tag{16}$$

$$\sum_{i=1}^{K} x_i = 1 \tag{17}$$

$$y_{iv^l} d_{iv^l} \leq d_{max} \qquad \forall\, i,\, v^l \tag{18}$$

$$0 \leq y_{iv^l} \leq 1 \quad \forall i,\, v^l; \qquad x_i \in \{0,1\}, \tag{19}$$

The cost of the microgrid is minimized while connecting nodes with greater importance in the infrastructures considered to the microgrid. The first three sets of constraints serve the same function as in model *P*1, that is, (12) ensures that the capacity of the microgrid is not exceeded, (13) limits the fraction of power that a single infrastructure can use, and (14) requires that only a single location for the microgrid is selected. In addition, the constraints defined by (15) limit the maximum distance that a node can be from the microgrid, thereby ensuring a practical geographic span for the system. Lastly, constraints (16) define the restrictions on the decision variables.

The optimization models *P*1 and *P*2 are mixed integer linear programming (MILP) problems that the bound and branch model can solve for undersized problem instances using standard optimization software (e.g., CPLEX, Gurobi). The outcomes of the models show the optimal location for a microgrid and have the benefit of selecting which infrastructure nodes attach to the pre-selected microgrid. In general, given the regulatory constraints on the size and ownership of microgrids, we expect that if multiple microgrids are deployed, they are built sequentially and support different consortia of infrastructure owners and community groups. In this scenario, the optimization models can be applied iteratively by re-running the model while modifying the power requirement and value of TWV, excluding nodes linked to a previously deployed microgrid.

2.4. Critical Node Identification

The branch and bound algorithm used to solve *P*1 and *P*2 is known to have scalability issues as the fundamental problem is NP-hard. Furthermore, the number of nodes/components in critical infrastructures can be quite large in a city. For example, consider the core (9 km × 9 km) area of the Pittsburgh, Pennsylvania metro area, which has a population of 2.3 million. According to the US Department of Homeland Security, the core area contains over 2700 critical infrastructure nodes, which include 80 water infrastructure nodes and 530 communication infrastructure nodes. Hence, optimizing all nodes/components in several infrastructures will be computationally complex. Here, the microgrid location optimization models are scaled by reducing the number of nodes in each infrastructure to only the most critical nodes determined by the TWV values. In effect, this reduces the search space over which the optimization models are solved, significantly speeding up the computation but at the expense of loss of global optimality guarantees. Various approaches targeting the selection of the most critical or essential nodes in each infrastructure have been introduced in the literature. For that, two methods are considered, as follows:

2.4.1. Combined Metric

In this method, the nodes v^l are arranged in descending order based on the $TWV(v^l)$ values. The nodes with the largest $TWV(v^l)$ values are considered the highest critical nodes. For simplicity, a size of 20 nodes has been selected to show the model's top nodes.

2.4.2. List of Lists

An alternate method is to rank the nodes according to each parameter/metric and then to combine the lists for an overall ranking. Hence, a prioritized list following descending order has been generated toward interdependent infrastructures using the values of C_d, C_b, C_c, and C_p. In the standalone infrastructures, the lists have been generated using the outcomes value of $Parm_n$ for every infrastructure. The positional ranking value in each list is taken as the score for that list. Next, all positional rank values are summed into a total score and sorted in ascending order from below to most crucial to discover the critical nodes (a low score implies a more important node).

2.5. Microgrid Location Heuristics

With subsequent determination of the critical nodes concerning every infrastructure, the geographic space for the smart city is aligned before the optimization implementation. Figure 3 illustrates an example where Geo_i is taken as a square of 3 km × 3 km. The optimization models $P1$ or $P2$ can then be solved over this reduced set of infrastructure nodes. Note that the set size selection can control the optimization model solution's computational run time. The larger the set size, the larger the search space, resulting in longer solution times and being closer to a global optimal.

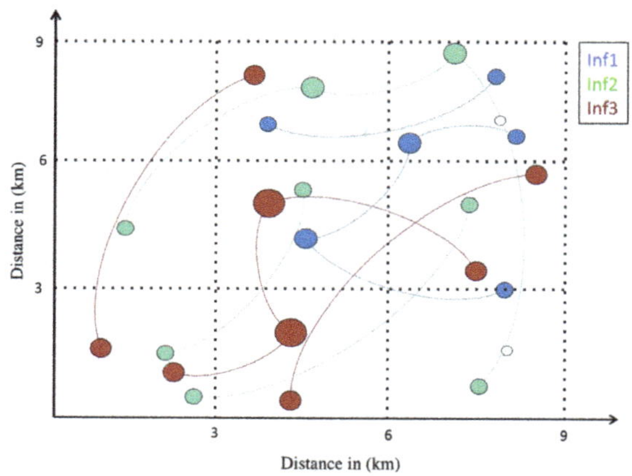

Figure 3. Critical node geographic representation.

Alternatively, we propose a simple heuristic based entirely on the TWV values (i.e., ignoring the cost). Let TI_i denote the total importance value of area Geo_i. Then, TI_i is determined by adding the TWV regarding every node positioned inside Geo_i as revealed below.

$$TI_i = \sum_{l=1}^{L} \sum_{v^l \in Geo_i} (TWV_{II}(v^l) + TWV_{SI}(v^l)) \quad . \tag{20}$$

Since node v^l is a part of a single infrastructure, just one of $TWV_{II}(v)$ or $TWV_{SI}(v)$ remains non-zero. Geo_i with the most significant TI_i value is chosen as the most optimal microgrid position. If many microgrids happen to be discovered, one reproduces the heuristic sequentially concerning every microgrid and coordinates the power and TWV rates in every repetition. Note that the heuristic is computationally simple and can be solved over the entire set of critical infrastructures nodes.

3. Results and Discussions

First, the two critical node selection methods of Section 2.4 are compared by developing a random graph of N nodes and E edges. The edges in this scenario can link to a node using a probability p. For simplicity in the calculation, parameter values $N = 200$, $E = 238$, and $p = 0.01$ were selected by drawing on previous work that used random graphs to model transportation networks and smart power grid networks [32]. The power requirements $C_P(v)$ of each node v were created by sampling a uniform [0, 1] random variable. Table 1 lists the twenty most critical nodes using the combined metric ranking. Table 2 shows the twenty most important nodes of the same infrastructure employing the list of lists ranking. As discussed earlier, the *z-score* normalization method was used for scaling the terms throughout our numerical results.

Table 3 contrasts the results of the two critical node identification methods for an individual network. Observe the variations within each rank concerning the most critical nodes; the collection of nodes in the highest twenty possess $\geq 80\%$ equivalent. Since both ranking methods produce comparable outcomes, the combined metric has been utilized for the rest of our study.

3.1. Case Study

As a case study illustrating the location problem, critical infrastructures in the city of Pittsburgh, Pennsylvania, were analyzed, which has a metropolitan area population of 2.3 million. We concentrate on the center section regarding the metro area, studying a 9 km × 9 km section centered on downtown. In addition, the four infrastructures of water, cellular communications, healthcare, and emergency shelter were studied and discussed.

Table 1. Top 20 most critical nodes utilizing the Combined Metric approach.

Node ID	$C_{b_{norm}}$	$C_{c_{norm}}$	$C_{d_{norm}}$	$C_{p_{norm}}$	Sum	Rank (Sum)
172	2.737	0.525	3.553	−0.023	6.792	1
197	3.874	0.537	2.075	0.130	6.616	2
133	2.881	0.529	2.814	−0.227	5.998	3
49	4.075	0.513	2.075	−0.872	5.791	4
144	2.461	0.493	1.337	1.450	5.740	5
160	1.390	0.465	2.075	1.416	5.346	6
1	2.481	0.548	1.337	0.680	5.046	7
182	1.548	0.496	2.075	0.750	4.870	8
32	2.739	0.502	1.337	0.209	4.786	9
5	1.878	0.533	2.075	0.276	4.762	10
10	1.024	0.455	1.337	1.468	4.284	11
188	1.177	0.450	2.075	0.556	4.258	12
195	1.135	0.442	1.337	1.289	4.203	13
162	1.952	0.500	2.075	−0.532	3.995	14
142	1.287	0.498	1.337	0.735	3.857	15
137	1.249	0.479	1.337	0.699	3.765	16
130	1.412	0.493	0.598	1.150	3.653	17
86	1.350	0.496	1.337	0.358	3.541	18
116	0.284	0.488	1.337	1.384	3.493	19
30	1.084	0.512	0.598	1.244	3.438	20

3.1.1. Healthcare

The healthcare infrastructure data are presented in Table 4, which were normalized and applied to evaluate the TWV_{SI} from (8). Three parameters were selected to represent each hospital's importance: capacity, power consumption, and population using Equation (6). The table shows that the UPMC Presbyterian, UPMC Shadyside, and UPMC Mercy Hospitals have the highest total weighted values.

Table 2. Top 20 most critical nodes utilizing the List of Lists approach.

Node ID	$C_{b_{norm}}$	$R(C_b)$	$C_{c_{norm}}$	$R(C_c)$	$C_{d_{norm}}$	$R(C_d)$	$C_{p_{norm}}$	$R(C_p)$	Sum	Rank (Sum)
144	2.461	7	0.493	28	1.337	11	1.450	7	53	1
1	2.481	6	0.548	1	1.337	11	0.680	64	82	2
160	1.390	22	0.465	48	2.075	3	1.416	12	85	3
182	1.548	20	0.496	23	2.075	3	0.750	55	101	4
197	3.874	2	0.537	2	2.075	3	0.130	97	104	5
10	1.024	34	0.455	53	1.337	11	1.468	6	104	5
30	1.084	31	0.512	13	0.598	34	1.244	26	104	5
5	1.878	13	0.533	5	2.075	3	0.276	85	106	8
116	0.284	54	0.488	31	1.337	11	1.384	14	110	9
142	1.287	25	0.498	21	1.337	11	0.735	57	114	10
130	1.412	21	0.493	29	0.598	34	1.150	34	118	11
32	2.739	4	0.502	18	1.337	11	0.209	89	122	12
176	0.758	40	0.500	20	0.598	34	1.202	29	123	13
172	2.737	5	0.525	8	3.553	1	−0.023	110	124	14
50	0.883	37	0.519	9	1.337	11	0.628	67	124	14
195	1.135	29	0.442	63	1.337	11	1.289	23	126	16
4	0.996	35	0.482	35	0.598	34	1.251	25	129	17
133	2.881	3	0.529	6	2.814	2	−0.227	120	131	18
132	0.303	52	0.508	15	0.598	34	1.156	32	133	19
137	1.249	26	0.479	38	1.337	11	0.699	62	137	20

Table 3. Comparison between node ranking in both approaches.

Node Ranking	Combined	List
1	172	144
2	197	1
3	133	160
4	49	182
5	144	197
6	160	10
7	1	30
8	182	5
9	32	116
10	5	142
11	10	130
12	188	32
13	195	176
14	162	172
15	142	50
16	137	195
17	130	4
18	86	133
19	116	132
20	30	137

3.1.2. Cellular Network

Table 6 shows the three parameters selected for measuring the importance of LTE base stations: coverage index, population, and power consumption. The coverage index has been assigned to a specific value from the high (3) to low (1). For simplicity purposes, the base stations were considered from a single operator (i.e., AT&T). The geographic coverage range of a base station and the population density around that station determine the population included. Finally, applying the energy model from [33], we measure the power requirement of each base station. The table displays the data obtained in the specified area of central Pittsburgh, USA.

Table 4. Healthcare.

Hospital	Power Consumption (kWh)	Population	Size (Beds)	TWV_{SI}
West Penn Hospital	9,510,000	12,915	317	−0.034
UPMC Montefiore	7,500,000	24,080	250	0.976
UPMC Mercy	14,460,000	15,714	482	2.361
Allegheny Hospital	11,430,000	5949	381	−0.402
UPMC St. Margaret	7,470,000	5200	249	−2.078
Children's Hospital	8,880,000	15,500	296	0.136
LifeCare Hospitals	7,080,000	9555.00	236	−1.529
UPMC Presbyterian	23,850,000	24,080.00	795	7.396
VA Healthcare	4,380,000	14,117	146	−1.854
UPMC Shadyside	15,600,000	15,916	520	2.841
Magee's Hospital	10,800,000	10,850	360	0.140
St. Clair Hospital	9,840,000	4569	328	−1.249
Children's Genetics	7,500,000	10,850	250	−1.156

3.1.3. Water System

Table 5 delivers Pittsburgh's City water treatment plants, using public data, including the power consumption for each water treatment facility, as calculated using the Equation (7). Then, Equation (8) is applied for the total weighted value for standalone infrastructure. The table shows that two of the five tested water treatment planets are considered critical: Pittsburgh WTP and Westview WTP.

Table 5. Water treatment plants (WTP).

Facility	Capacity (Million G/Day)	Power Consumption (kWh)	TWV_{SI}
Pittsburgh WTP	70,000,000	327,600,000	3.029
Brush WTP	1,500,000	7,020,000	−1.387
Plum Creek WTP	2,234,669	10,458,250.9	−1.340
Harmar Twp WTP	1,500,000	7,020,000	−1.387
Westview WTP	39,850,000	186,498,000	1.085

Table 6. LTE base stations.

eNB ID	Coverage Index	Population Covered	Power Consumption (kWh)	TWV_{SI}
780206	3	4475	25	2.156
780007	2	16,227	20	1.848
780165	2	5338	20	−0.141
780059	3	7119	25	2.639
780213	1	15,500	15	−0.739
780108	2	15,500	20	1.715
780399	1	12,915	15	−1.211
780184	1	5096	15	−2.639
780017	3	843	25	1.492
780527	1	8424	15	−2.032
780037	1	8424	15	−2.032
780364	1	24,080	15	0.828
780154	2	6446	20	0.061
780178	2	6446	20	0.061
780560	3	6446	25	2.516
780225	2	2710	20	−0.621
780163	3	7622	25	2.731

Table 6. *Cont.*

eNB ID	Coverage Index	Population Covered	Power Consumption (kWh)	TWV_{SI}
780032	1	2688	15	−3.079
780540	3	5948	25	2.425
780167	2	4593	20	−0.277
780477	2	4593	20	−0.277
780169	1	3316	15	−2.965
780218	1	6078	15	−2.460

3.1.4. Emergency Shelter

Table 7 lists the TWV regarding every emergency shelter in Pittsburgh city. Power usage and shelter capacities are the selected parameters, as explained in Section 2, to measure the shelter's total weighted value. The table also shows that the Convention Center has the highest values applying such parameters.

Table 7. Emergency shelter.

Facility	Power Consumption (kWh)	Size ft^2	Occupancy	TWV_{SI}
Petersen Center	224,000	16,000	12,508	−0.198
PPG Paints Arena	10,080,000	720,000	19,758	−0.017
Convention Center	21,000,000	1,500,000	109,445	2.225
Irish Centre	168,000	12,000	315	−0.503
Sigmas Center	140,000	10,000	150	−0.507
Sherwood Center	252,000	18,000	500	−0.499

Given the infrastructure data above in Tables 4–7, we made an initial analysis of the city, considering a grid of 3 km × 3 km squares as Geo_i. Figure 4 displays heat maps of the top significant nodes in each infrastructures. Figure 5 exhibits the heat map when all four infrastructures are overlain and considered a group. The results present the outputs of applying the proposed heuristic and optimization models to the data.

3.2. Heuristic

Table 8 shows the overall importance TI_i value for each Geo_i. The table also shows the fifth, sixth, and third squares are in the top of the list with the highest values at 16.3046, 7.6456, and 4.3614. In Table 9, the power demand of each square is broken down by individual infrastructure. Note that the power is included in scaled form in the TI values of Table 8. Interestingly, when the squares are ranked by power requirements only, they do not match the ranking based on total importance score except for the first and last position (squares 5 and 4, respectively). Considering both tables, the fifth, sixth, and third squares are the most important compared to the rest and thus are the most desirable for locating a microgrid.

3.3. Optimization Model

The optimization models $P1$ and $P2$ require estimating the total cost F_i for microgrid to construct and operate in a specfic location Geo_i. HOMER design software [34] was used to determine the cost of a 4 MW microgrid consisting of a mix of diesel generators, DC/AC converters, flat panel photovoltaic cells, 1.5 kW wind turbines, and 1 kW lithium acid batteries for storage. Furthermore, the cost of real estate assuming a greenfield deployment of the microgrid was estimated for each Geo_i Considering a lifetime of 23 years and a net present cost value of F_i covering the capital cost, replacement, salvage, operating and fuel, and repair, a discount value of 6% was found as $F_i = \{6, 6.2, 7.4, 7.2, 11, 11.5, 5.4, 5.6, 4.4\}$. The difference in F_i values is the real estate cost. The additional $P1$ optimization model variables were $S_i = 4MW \; \forall i$ and $\alpha = \beta = 0.5$. Table 10 shows the outcomes of solving optimization model $P1$ and the heuristic based on TI_i while varying the number of sequentially built microgrids. Table 10 clearly shows that Geo_5 is the preferred location for a microgrid due to the high number of critical nodes in that geographic space. Furthermore, Table 10 shows that both the optimization model and heuristic implemented sequentially favor the critical squares with high total weight value, even in the case of multiple microgrids.

Table 8. Total importance value for every zone.

Geo_i	Healthcare	Water	Cellular	Shelter	TI	Rank
1	0.5793	0.5598	0	1.1171	2.2562	5
2	0	0	0	0.0527	0.0527	8
3	0.3497	1	3.0117	0	4.3614	3
4	0	0	0	0	0	9
5	8.7257	0	3.5511	4.0278	16.3046	1
6	3.2216	0	4.3177	0.1063	7.6456	2
7	0.5609	0	0	0	0.5609	7
8	0	0	1.2411	0	1.2411	6
9	0	0.0107	2.9535	0	2.9642	4

Table 9. Total power demand for every zone (kWh).

Geo_i	Healthcare	Water	Cellular	Shelter	Total
9	0	3285	15	0	3300
8	0	0	45	0	45
7	27,333.3	0	0	0	27,333
6	94,416.7	0	95	44,467	138,979
5	189,000	0	90	88,336	277,426
4	0	0	0	0	0
3	20,750	98,280	75	0	119,105
2	0	0	0	389	389
1	33,000	55,949	0	22,400	111,349
Total	364,500	157,514	320	155,592	677,926

Table 10. Optimization vs. Heuristic represented by square numbers.

Microgrid Quantity	1	2	3
Heuristic Geo_i	5	5, 5	5, 5, 6
$P1$ Optimization Model Geo_i	5	5, 5	5, 5, 6

The trade-offs were also investigated through the infrastructure and cost weight by modifying the value of α and the value of β considering one microgrid, with the results presented in Table 11. The table shows that geographical zone number five is again selected, excluding the case where $\alpha = 0$ shows that the cost is only minimized.

Table 11. Optimization outcomes varying weights.

Geo_i	5	5	5	5	5	5	5	5	5	5	9
α	1	0.9	0.8	0.7	0.6	0.5	0.4	0.3	0.2	0.1	0
β	0	0.1	0.2	0.3	0.4	0.5	0.6	0.7	0.8	0.9	1

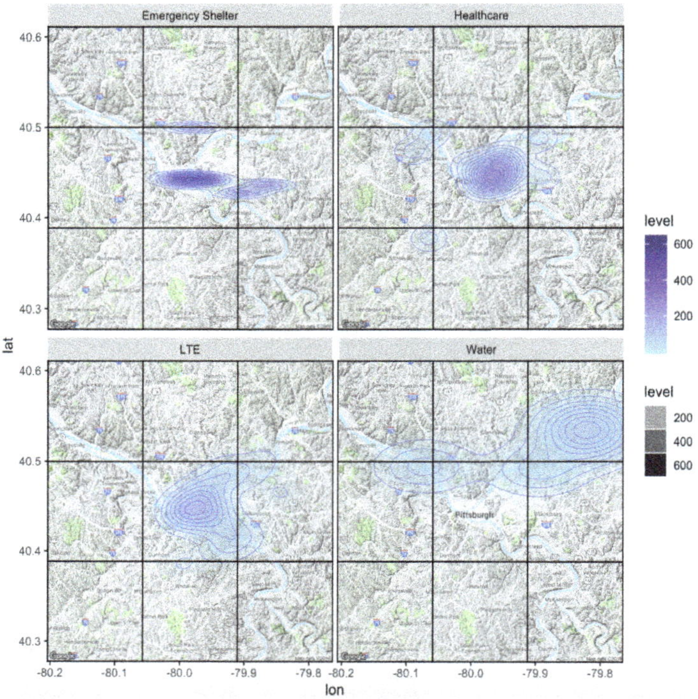

Figure 4. Critical node heat map for each individual infrastructure.

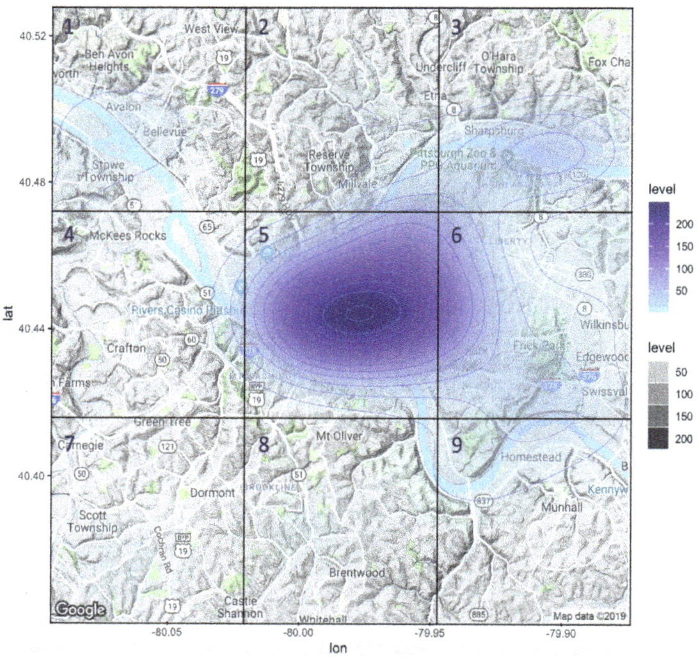

Figure 5. Heat map of critical nodes across all infrastructures.

The solution of optimization model *P*2 was studied next. As noted earlier, *P*2 allows a microgrid to supply important power nodes outside of the Geo_i it is located in, as long as the distance to the node is less than a given maximum (e.g., 6 km). The results of solving *P*2, using the same cost values F_i as discussed above, to place three microgrids sequentially are shown in Figure 6. In Figure 6, the squares identify the microgrid locations, and the circles denote the nodes connected to the microgrid. Notice that the microgrid in Geo_5 powers two nodes (cellular base stations) in Geo_1. Similarly, the microgrid in Geo_6 connects to nodes in Geo_5 and Geo_3 as well as nodes in Geo_6. Table 12 shows how the power created by each microgrid is shared among the four infrastructures for each microgrid. Notice that it varies with location. Comparing the solution of *P*2 with the results of problem *P*1, the total cost of three microgrids will be less than *P*2 (29.9 vs. 33.5 million).

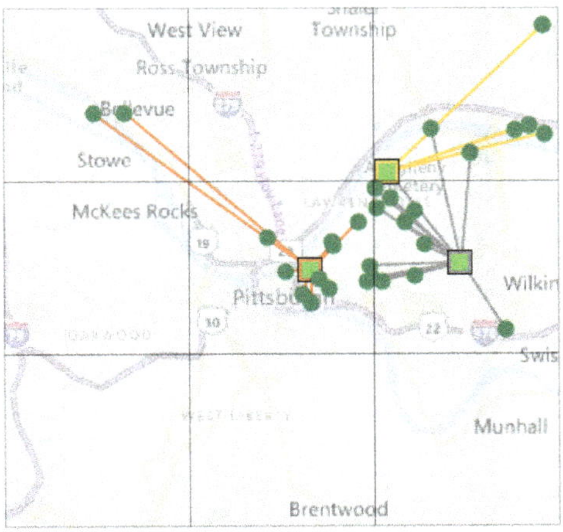

Figure 6. Microgrid locations from solution of *P*2.

Table 12. Percentage of power generated.

Geo_i	Hospital	Water	Cellular	Shelter	Total
5	68.13%	0.00%	0.03%	31.84%	100.00%
6	99.41%	0.00%	0.10%	0.49%	100.00%
3	17.42%	82.52%	0.06%	0.00%	100.00%

Furthermore, the effects of varying the capacity of the microgrids S_i in the optimization models were considered. Specifically, we varied the microgrid capacity over 3, 4, 5, and 10 MW and determined each case's corresponding cost F_i. The Pittsburgh case study's solution to both optimization algorithms does not change. For the solution to problem *P*2, the location of three microgrids is Geo_5, Geo_6, and Geo_3, regardless of the microgrid capacity. Similarly, the solution to *P*1 is Geo_5, Geo_5, and Geo_6 for the location of three microgrids for all microgrid capacities considered.

For this paper, which focuses on showing the applicability of shared Microgrid among interdependent infrastructure, we did not consider the existence of emergency generation assets required by regulation at critical infrastructure. However, such work can be extended in future work by addressing all the possible scenarios and applying all required policies.

4. Conclusions

In this work, it has been proven that emerging smart critical infrastructures will need disaster resilience that includes continuity of power and ICT support in addition to traditional infrastructure-specific methods. Furthermore, we advocated for using community-based, multi-ownership microgrids and studied where to locate microgrids to enhance the resilience of smart critical infrastructures. The suggested method takes a holistic view of considering multiple critical infrastructures and incorporates several factors, such as the component importance within critical infrastructure, the geospatial placement of infrastructures, power requirements, and microgrid cost. Furthermore, optimization models were proposed to determine the microgrid location to optimize a weight combination of cost and infrastructure node criticality. Additionally, a heuristic for determining the microgrid location based on infrastructure node importance was proposed. A case study demonstrating our method was presented for the city of Pittsburgh. From a resilience viewpoint, quantifying and perceiving which geographic zones in a city would most benefit from a microgrid will help provide a community-wide justification for microgrids. Future avenues of work include studying economic and regulatory models to improve the microgrid cost.

Author Contributions: Formal analysis, A.A. and D.T.; Methodology, A.A. and D.T.; Validation, S.F.A.-G.; Writing—original draft, A.A. and D.T.; Writing—review & editing, S.F.A.-G. and R.E.-S. All authors have read and agreed to the published version of the manuscript.

Funding: This research received no external funding

Institutional Review Board Statement: Not applicable

Informed Consent Statement: Not applicable

Data Availability Statement: Not applicable

Conflicts of Interest: The authors declare no conflict of interest.

References

1. World Bank Open Data. Available online: https://data.worldbank.org (accessed on 30 August 2021).
2. Almaleh, A.; Tipper, D. Risk-Based Criticality Assessment for Smart Critical Infrastructures. *Infrastructures* **2022**, *7*, 3. [CrossRef]
3. National Centers for Environmental Information (NCEI). Billion-Dollar Weather and Climate Disaster: Overview. July 2016. Available online: http://www.ncdc.noaa.gov/billions (accessed on 30 July 2022).
4. US Department of Energy. Economic Benefits of Increasing Electric Grid Resilience to Weather Outages. August 2013. Available online: http://energy.gov/sites (accessed on 10 June 2022).
5. Wang, Y.; Chen, C.; Wang, J.; Baldick, B. Research on Resilience of Power Systems Under Natural Disasters. *IEEE Trans. Power Syst.* **2016**, *31*, 1604–1613. [CrossRef]
6. Kelly-Pitou, K.; Ostroski, A.; Contino, B.; Grainger, B.; Kwasinski, A.; Reed, G. Microgrids and resilience: Using a systems approach to achieve climate adaptation and mitigation goals. *Electr. J.* **2017**, *30*, 23–31. [CrossRef]
7. Kamboj, A.; Chanana, S. Planning and operational strategy of a renewable energy microgrid considering reliability. In Proceedings of the IEEE International Conference on Power Electronics, Intelligent Control and Energy Systems, Delhi, India, 4–6 July 2016; IEEE: Piscataway, NJ, USA, 2016; pp. 1–6.
8. Hanna, R.; Ghonima, M.; Kleissl, J.; Tynan, G.; Victor, D. Evaluating Business Models for Microgrids: Interaction of Technology and Policy. *Energy Policy* **2017**, *103*, 47–61. [CrossRef]
9. Alqahtani, A.; Abhishek, R.; Tipper, D.; Medhi, D. Disaster Recovery Power and Communications for Smart Critical Infrastructures. In Proceedings of the IEEE International Conference on Communications, Kansas City, MO, USA, 20–24 May 2018; IEEE: Piscataway, NJ, USA, 2018.
10. Kabalcı, E.; Kabalcı, Y.; Siano, P. Design and implementation of a smart metering infrastructure for low voltage microgrids. *Int. J. Electr. Power Energy Syst.* **2022**, *134*, 107375. [CrossRef]
11. Chen, C.; Wang, J.; Qiu, F.; Zhao, D. Resilient Distribution System by Microgrids Formation After Natural Disasters. *IEEE Trans. Smart Grid* **2016**, *7*, 958–966. [CrossRef]
12. Dimitrov, P.; Piroddi, L.; Prandini, M. Distributed allocation of a shared energy storage system in a microgrid. In Proceedings of the 2016 American Control Conference (ACC), Boston, MA, USA, 6–8 July 2016; IEEE: Piscataway, NJ, USA, 2016; pp. 3551–3556.
13. Habibi Rad, M.; Mojtahedi, M.; Ostwald, M.J. Industry 4.0, disaster risk management and infrastructure resilience: A systematic review and bibliometric analysis. *Buildings* **2021**, *11*, 411. [CrossRef]
14. Lewis, T.G. *Critical Infrastructure Protection in Homeland Security: Defending a Networked Nation*; John Wiley & Sons: Hoboken, NJ, USA, 2014.
15. Banerjee, J.; Basu, K.; Sen, A. On Hardening Problems in Critical Infrastructure Systems. *Int. J. Crit. Infrastruct. Prot.* **2018**, *23*, 49–67. [CrossRef]
16. Bilis, E.I.; Kroger, W.; Nan, C. Performance of electric power systems under physical malicious attacks. *IEEE Syst. J.* **2013**, *7*, 854–865. [CrossRef]

17. Nasiruzzaman, A.; Pota, H. Critical node identification of smart power system using complex network framework based centrality approach. In Proceedings of the North American Power Symposium (NAPS), Boston, MA, USA, 4–6 August 2011; IEEE: Piscataway, NJ, USA, 2011; pp. 1–6.
18. Stergiopoulos, G.; Kotzanikolaou, P.; Theocharidou, M.; Gritzalis, D. Risk mitigation strategies for critical infrastructures based on graph centrality analysis. *Int. J. Crit. Infrastruct. Prot.* **2015**, *10*, 34–44. [CrossRef]
19. Faramondi, L.; Setola, R.; Panzieri, S.; Pascucci, F.; Oliva, G. Finding critical nodes in infrastructure networks. *Int. J. Crit. Infrastruct. Prot.* **2018**, *20*, 3–15. [CrossRef]
20. Nandanoori, S.P.; Guan, S.; Kundu, S.; Pal, S.; Agarwal, K.; Wu, Y.; Choudhury, S. Graph neural network and koopman models for learning networked dynamics: A comparative study on power grid transients prediction. *IEEE Access* **2022**, *10*, 32337–32349. [CrossRef]
21. Gupta, B.B.; Gaurav, A.; Marín, E.C.; Alhalabi, W. Novel Graph-Based Machine Learning Technique to Secure Smart Vehicles in Intelligent Transportation Systems. *IEEE Trans. Intell. Transp. Syst.* **2022**. [CrossRef]
22. Doyal, A. *Prioritizing Water Distribution Network Asset Maintenance Using Graph Theory Methods*; Technical Report; Air Force Institute Of Technology: Wright-Patterson, OH, USA, 2022.
23. Long, X.; Tipper, D.; Gomes, T. Measuring the Survivability of Networks to Geographic Correlated Failures. *Opt. Switch. Netw.* **2014**, *14*, 117–133. [CrossRef]
24. Juszczak, P.; Tax, D.; Duin, R.P. Feature scaling in support vector data description. In Proceedings of the 8th Annual Conference of the Advanced School for Computing and Imaging 2002, Lochem, The Netherlands, 19–21 June 2002; pp. 95–102.
25. Bruce, P.; Bruce, A. *Practical Statistics for Data Scientists: 50 Essential Concepts*; O'Reilly Media, Inc.: Sebastopol, CA, USA, 2017.
26. Global Specialist in Energy Management And Automation Schneider Electric, 2017. Available online: https://www.schneider-electric.us/en/ (accessed on 30 July 2022).
27. McCoy, J.; Johnston, K.; Environmental Systems Research Institute. *Using ArcGIS Spatial Analyst: GIS by ESRI*; Environmental Systems Research Institute: Redlands, CA, USA, 2001.
28. Pacific Gas Electric and California Public Utilities Commission. *Municipal Water Treatment Plant Energy Baseline Study Characteristics*; California Public Utilities Commission: San Francisco, CA, USA, 2006.
29. Houston Convention Center Shelters One More Harvey Survivor: An Opera Company. The New York Times, 2017. Available online: https://www.nytimes.com/2017/09/26/arts/music/houston-grand-opera-hurricane-harvey.html (accessed on 22 June 2022).
30. Energy Information Administration. *Annual Energy Outlook 2010*; Department of Energy, Energy Information Administration: Washington, DC, USA, 2010; Volume 92010, pp. 1–15.
31. Badger, J.; Volker, P.J.H. Efficient large-scale wind turbine deployment can meet global electricity generation needs. *Proc. Natl. Acad. Sci. USA* **2017**, *114*, E8945. [CrossRef] [PubMed]
32. Schoonenberg, W.C.; Farid, A.M. Modeling smart cities with hetero-functional graph theory. In Proceedings of the 2017 IEEE International Conference on Systems, Man, and Cybernetics (SMC), Banff, AB, Canada, 5–8 October 2017; IEEE: Piscataway, NJ, USA, 2017; pp. 1627–1632.
33. Arnold, O.; Richter, F.; Fettweis, G.; Blume, O. Power consumption modeling of different base station types in heterogeneous cellular networks. In Proceedings of the Future Network and Mobile Summit, Florence, Italy, 16–18 June 2010; IEEE: Piscataway, NJ, USA, 2010; pp. 1–8.
34. Homer Energy. Available online: http://www.homerenergy.com (accessed on 30 July 2022).

MDPI
St. Alban-Anlage 66
4052 Basel
Switzerland
www.mdpi.com

Applied Sciences Editorial Office
E-mail: applsci@mdpi.com
www.mdpi.com/journal/applsci

Disclaimer/Publisher's Note: The statements, opinions and data contained in all publications are solely those of the individual author(s) and contributor(s) and not of MDPI and/or the editor(s). MDPI and/or the editor(s) disclaim responsibility for any injury to people or property resulting from any ideas, methods, instructions or products referred to in the content.

www.ingramcontent.com/pod-product-compliance
Lightning Source LLC
Chambersburg PA
CBHW040819250926
51850CB00007BA/1213